PRODUCT DESIGN.

Graphics with Materials Technology

Lesley Cresswell
Alan Goodier
Douglas Fielding-Smith
Derek Sheehan
Barry Lambert
Colin Chapman

Edexcel
Success through qualifications

Heinemann Educational Publishers
Halley Court, Jordan Hill, Oxford OX2 8EJ
a division of Reed Educational & Professional
Publishing Ltd

OXFORD   MELBOURNE   AUCKLAND
JOHANNESBURG   BLANTYRE   GABORONE
IBADAN   PORTSMOUTH (NH) USA   CHICAGO

Heinemann is a registered trademark of Reed
Educational & Professional Publishing Ltd

First published in 2002

06 05 04 03 02
10 9 8 7 6 5 4 3 2 1

British Library Cataloguing in Publication Data
A catalogue record for this book is available from the
British Library

ISBN 0 435 75771 7

Designed by Wendi Watson
Typeset by ⤳ Tek-Art, Croydon, Surrey
Illustrated by ⤳ Tek-Art, Croydon, Surrey
Printed and bound in Great Britain by
The Bath Press Ltd, Bath

Acknowledgements
The authors and publishers would like to thank the
following:

Gillian Whitehouse at Edexcel for her kind
support and guidance as well as Jon Attwood
and all the students who have contributed
artwork and examples of student work. For the
use of logos, thank you to British Farm Standard
p.146; British Telecom p.128 (BT plus PIPER
device and British Telecom logo are registered
trademarks of British Telecommunications public
limited company); Envirowise p.223; IBM p.217
(IBM and the IBM logo are registered trademarks
of International Business Machines Corporation
in the United States and other countries);
Microsoft p.217.

The publishers would like to thank the following for
permission to use photographs:

Aardman Animations p.129; Advertising Archive
pp. 112 (bottom), 113 (bottom); AKG p.274
(left); Alfa-Robot p.248; Art Archive p.176; Art
Directors-Tripp p27 (right); Benetton p.218 (top);
Bettmann/Corbis pp.109, 110, 111 (left);
Blazepoint/Kiosk Information Systems pp.260,
261; Robert Bircher p.142; Gareth Boden pp.37,
98; John Callan/Shout pp.180, 217, 241; Giles
Chapman p.142; Christies pp.110, 116 (centre,
right, left), 176 (right), 177 (left); , 117 (left, right
top) ; Trevor Clifford p.28; Cooper-Hewitt
Museum (Zelco calculator) p.112; Corbis pp.14, 16
(top), 98 (top), 108, 116 (top), 118, 119, 203, 209,
222, 244, 247; Craftspace Touring p.204; the Croc
pp.154, 157, 260; Denford pp.235, 239; Design
Council/Manchester Metropolitan University
p.166; Diesel p.218 (bottom); Dorling Kindersley
p.112; Milton Glaser p.34; Chris Honeywell
pp.117, 181 (plastic moulded kettle); Larry
Lee/Corbis p.128; London Transport Museum
p.126, 127, 128 (left); Microsoft p.154; Peter
Morris pp.16 (bottom), 17, 19 (centre right,
right), 27 (right), 33, 34 (top), 41, 113,120, 130,
181 (copper kettle); Photodisc pp.187, 220; Pro-
lok p.246; Science and Society pp.19 (left, centre
left), 125 (all); Science Photo Library p.185;
Stapleton Collection/Corbis p.111; Techsoft pp.158
(both), 162, 163 (both); Trip/H Rogers p.181, (TV
set); Trip/TH Photo p.132; Victoria and Albert
Museum Picture Library pp. 103, 117, 121; 130,
181; Virgin p.35 (all).

The publishers have made every effort to contact
copyright holders. However, if any material has been
incorrectly acknowledged, the publishers would be
pleased to correct this at the earliest opportunity

Cover photographs by: Richard Davies/Foster and
Partners (model of building), Salisbury
Playhouse/Richard Foxton/Nick Withers/Gemma
Mount (set design and poster for production of
'Equus'), Haddon Davies (perfume bottles).

Tel: 01865 888058 www.heinemann.co.uk

Contents

Part 3 Advanced GCE (A2)

Part 1

Introduction

This Advanced Design and Technology book is designed to support the Edexcel Advanced Subsidiary (AS) and Advanced GCE Specification for Product Design: Graphics with Materials Technology.

The book follows the structure of the Specification and is intended to support you through the course. The content of the book will provide you with a great deal of knowledge and understanding and will help you prepare for assessment. As with any Advanced level course you are advised to read around the subject to broaden your knowledge and understanding.

How to use this book

Part 1

Part 1 provides advice on how to use the book, how the course is structured and how to manage your own learning during the course. It includes advice on planning, organising and managing your work. You should read this section before starting your course.

Parts 2 and 3

Parts 2 and 3 provide unit-by-unit guidance on each of the AS Units 1–3 (Part 2) and the A2 Units 4–6 (Part 3). These sections will provide you with the knowledge and understanding to help you through your course. They will also help you prepare for internal and external assessment and increase your chances of success. You should therefore refer to these sections of the book for guidance on the subject content and to help you understand how each unit is assessed.

Parts 2 and 3 also look at the structure, subject content and assessment requirements of the Edexcel Specification. Each of the three AS and three A2 units use similar headings and subheadings to those found in the Specification, and each unit is structured as shown in Table 1.

Tasks, questions and information appear throughout the text as follows:

- The tasks:
  - help you to understand issues such as industrial practices
  - give you practice in some aspect of designing or manufacturing
  - help you practise specific skills, such as how to do market research or test the suitability of materials.
- 'Factfile' boxes contain information which may explain:
  - technical terms
  - illustrate points in the text.
- 'Think about this!' boxes explain different issues, such as industrial practices or 'value issues' which may influence your design decisions.
- 'To be successful you will:' boxes appear at the end of each section in the coursework units. They contain the assessment criteria that you will need to meet, in order to be successful.
- In A2 Units 4–6 you may find 'Signposts' to the AS units. These refer you back to information or topics that have been discussed in the AS course.

Table 1 Structure of book's AS/A2 units

Summary of expectations for the unit	Unit content	Checklists/exam preparation
The first page of the unit summarises: • what you are required to do • what you will learn • how the unit is assessed.	This covers the subject content in detail. It: • explains what you will learn in each unit • helps you understand the assessment requirements • provides tasks, questions and information to guide you through the unit.	The last page of the unit provides checklists and/or exam preparation to help you: • check the progress of your coursework • revise for and prepare for the exams • be as successful as possible.

- The coursework Units 2 and 5 will provide many opportunities for you to generate evidence for your Key Skills portfolio. You may find it helpful to check out Key Skills requirements for Communication, Application of Number and Information Technology at Level 3.

- Technical terms are in **bold** when they first occur and are explained in context in the text. They also appear in a glossary on pages 305–316.

How the course is structured

There are three units at Advanced Subsidiary (AS) and three units at Advanced GCE (A2).

The AS units

The three AS units combine to make the AS course. The AS units:

- build on the knowledge, understanding and skills you developed through the study of GCSE Design and Technology
- provide a discrete course leading to an AS qualification, *or*
- provide the first half of the Advanced GCE course. The AS units contribute 50 per cent of the Specification content. You must follow the AS course before progressing to A2.

The A2 units

The A2 units combine with the three AS units to make the Advanced GCE course. The A2 units:

- build on the knowledge, understanding and skills developed in the AS course, to achieve the full Advanced GCE standard

- provide the other 50 per cent of the Specification content
- enable you to achieve a greater level of sophistication and more in-depth knowledge and understanding.

A summary of the AS and A2 units is provided in Table 2.

How the AS units are assessed

Units 1 and 3 are externally assessed by examination. Unit 2 is the coursework unit. It is assessed by internal marking and external moderation by the Edexcel Moderator (see the summary in Table 3).

How the A2 units are assessed

Units 4 and 6 are externally assessed by examination. Unit 5 is the coursework unit. It is assessed by internal marking and external moderation by the Edexcel Moderator (see the summary in Table 4).

Table 2 Summary of the AS and A2 units

AS 50% of the Specification content		A2 50% of the Specification content	
Unit 1	Industrial and commercial products and practices	Unit 4	Further study of materials, components and systems with options
Unit 2	Product development I	Unit 5	Product development II
Unit 3	Materials, components and systems with options	Unit 6	Design and technology capability

Table 3 Assessment of Units 1–3

AS 50% of the Specification content		Assessed by:	% of AS course	% of A2 course
Unit 1	Industrial and commercial products and practices	External assessment $1\frac{1}{2}$-hour examination	30%	15%
Unit 2	Product development I	Internal assessment Coursework project	40%	20%
Unit 3	Materials, components and systems with options	External assessment $1\frac{1}{2}$-hour examination	30%	15%

Table 4 Assessment of Units 4–6

A2 50% of the Specification content		Assessed by:	% of A2 course
Unit 4	Further study of materials, components and systems with options	External assessment $1\frac{1}{2}$-hour examination	15%
Unit 5	Product development II	Internal assessment Coursework project	20%
Unit 6	Design and technology capability	External assessment 3-hour examination	15%

Unit guidance

The following section guides you through the three AS units and the three A2 units.

Unit 1 Industrial and commercial products and practices

This unit enables you to develop an understanding of industrial and commercial practices through product analysis. Throughout the unit you should investigate the design, manufacture, use and disposal of a range of products. The products you investigate should include both 2D and 3D elements, such as a product and its packaging. You will find information about 2D/3D elements in the sections describing Units 2 and 5 below. The areas of study for Unit 1 include:

- products and applications – the processes involved in product design and development
- materials and components – the range of materials and their working characteristics
- industrial and commercial practice – the processes involved in product manufacture, including manufacturing systems, stages of production, detailed manufacturing methods, service to the customer, energy use in industry and the use of ICT systems
- quality – the importance of quality to the success of products
- health and safety – health and safety at work and risk assessment.

Throughout the unit you should undertake a variety of tasks to enable you to understand the subject content. For example, you could:

- work collaboratively with others on some investigative activities, e.g. when analysing a range of products
- work individually on some tasks, e.g. when developing creative communication skills to record the investigation of products (communication skills can include writing, drawing, sketching, graphics, charts, flow diagrams, systems diagrams, computer-aided design (CAD), modelling etc.)
- work individually on the detailed analysis of one product, to gain a personal understanding of product development and manufacture.

Assessment (written exam)

Unit 1 is assessed through a $1\frac{1}{2}$-hour product analysis exam, which assesses your understanding of the unit subject content. The style of assessment remains the same each time the unit is assessed, but the product to be analysed will be different.

Unit 2 Product development I

Product development I is a full coursework project. At AS level, you are expected to take a commercial approach to designing and making products, to meet needs that are wider than your own. This could mean designing and making a one-off product for a specified user or client, or designing and making a prototype product that could be batch or mass produced for users in a target market group.

One of the key coursework requirements within Graphics with Materials Technology is the inclusion of both 2D and 3D elements. The 2D/3D elements should be linked by the theme or context for design, so that one supports and underpins the other. You will be asked to produce the following:

- A folder (in A3 or A2 format) that summarises the development of the 2D/3D coursework elements.
- A 2D element, developed from traditional or modern graphics media. The 2D element should be linked to and support a 3D outcome.
- A 3D model or prototype product, constructed from modelling materials *and* at least one resistant material (which could include some wood and/or metal and/or

plastic). The 3D outcome should be semi-functioning. This means that it should be testable against aesthetic criteria but does not need to be a functional working product.

For example, a 3D prototype board game could include the playing pieces and a storage system. The 2D element of the project could cover concept ideas for a point-of-sale display which clearly shows how the product relates to the display.

The inclusion in your project of both 2D and 3D elements should enable you to demonstrate fully your potential in Graphics with Materials Technology.

When choosing a coursework project at AS, you should ensure that the project enables you to meet all the assessment criteria. Before you start your coursework you should refer to the section on Unit 2, which explains the assessment criteria in detail. As you work through your AS coursework project, refer to this section as and when you need.

Assessment (coursework project)

Unit 2 coursework builds on the designing and making skills you learned at GCSE level. You are required to submit a coursework project folder and a product with 2D/3D elements for your AS coursework project. The AS coursework is internally marked by your teacher or tutor and externally moderated by the Edexcel Moderator.

Unit 3 Materials, components and systems with options

This unit has two sections, both of which must be studied. Section A is related to materials, components and systems. Section B has two options. You must study one option from Section B. A summary of Unit 3 is given in Table 5.

The two options in Section B have common content for all the materials areas of Product Design. This means you may be taught your chosen option with students from Resistant Materials Technology or Textiles Technology.

During the unit you should undertake a variety of tasks to enable you to understand the subject content. For example, you could:

- work individually on some activities, e.g. when using a database to research materials and components
- work individually on some tasks, e.g. when undertaking practical tasks to develop understanding of working properties and processes
- work collaboratively with others on some activities, e.g. when testing materials.

Assessment (written exam)

Unit 3 is assessed through a 1½-hour written exam which assesses your understanding of the unit subject content. You are required to answer two question papers – Paper 1 Section A and Paper 2 Section B – but both papers will be given at the same time.

- *Section A: Materials, components and systems* There will be six short-answer, knowledge-based questions, with each question worth 5 marks. Examiners will look for concise answers, often a description, explanation or annotated sketch which show your understanding of the topic or process. You are advised to spend approximately 45 minutes on this section of the paper.

- *Section B: Options* There will be two compulsory, long-answer questions, each worth 15 marks. These questions will be the same for all the three areas of Product Design. You should support your written answers to the exam questions with examples found within your own materials area. You are expected to demonstrate understanding of the technology associated with the option studied and to apply your knowledge and understanding to more open-ended questions. You are advised to spend approximately 45 minutes on this section of the paper.

Unit 4 Further study of materials, components and systems with options

Unit 4 has two sections, both of which must be studied. Section A is related to materials, components and systems. Section B has two options. You must study the same option from Section B that you studied in Unit 3. A summary of Unit 4 is given in Table 6.

Table 5 Unit 3 areas of study

Unit	Level	Components	Areas of study
3	AS	Section A: Materials, components and systems	• Classification of materials and components • Working properties and processes • Testing materials
		Section B: Options	• Design and technology in society • CAD/CAM

Table 6 *Unit 4 areas of study*

Unit	Level	Components	Areas of study
4	A2	Section A: Further study of materials, components and systems	• Selection of materials • New technologies and the creation of new materials • Value issues
		Section B: Options	• Design and technology in society • CAD/CAM

During the unit you should undertake a variety of tasks to enable you to understand the subject content. For example, you could:

- work individually on some tasks, e.g. when undertaking practical tasks to develop understanding of the relationship between properties and the selection of materials
- work individually on some activities, e.g. when using the Internet to research information about new technologies and new materials
- work collaboratively with others on some activities, e.g. when developing understanding of the impact that 'value' issues have on the design, development, use and disposal of a range of products.

Assessment (written exam)

Unit 4 is assessed through a 1½-hour written exam which assesses your understanding of the unit subject content. You are required to answer two question papers – Paper 1 Section A and Paper 2 Section B – but both papers will be given out at the same time.

- *Section A: Materials, components and systems* Section A is similar in style to Unit 3, consisting of six short-answer, knowledge-based questions, with each question worth 5 marks. Examiners will look for a more in-depth response when describing or explaining topics or processes at A2 level. You are advised to spend approximately 45 minutes on this section of the paper.

- *Section B: Options* Section B is similar in style to Unit 3, consisting of two compulsory, long-answer questions, each worth 15 marks. You are expected to demonstrate an in-depth understanding of the technology associated with the option studied and to apply your knowledge and understanding to more open-ended questions at A2 level. You are advised to spend approximately 45 minutes on this section of the paper.

Unit 5 Product development II

Product development II is a full coursework project. It is said to be 'synoptic' to the AS project

because it enables you to build on the knowledge, understanding and skills that you experienced in the AS project. At A2 level you are asked to produce the following:

- A folder (in A3 or A2 format) that summarises the development of the 2D/3D coursework elements.
- A 2D element, developed from traditional or modern graphics media. The 2D element should be linked to and support a 3D outcome.
- A 3D model or prototype product, constructed from modelling materials *and* at least one resistant material (which could include some wood and/or metal and/or plastic). The 3D outcome should be semi-functioning. This means that it should be testable against aesthetic criteria but does not need to be a functional working product.

The inclusion in your project of both 2D and 3D elements should enable you to demonstrate fully your potential in Graphics with Materials Technology.

At A2, you are required to work more independently, which may involve using a wider range of people to support you in your work. You should take a commercial approach to designing and making products, to meet needs that are wider than your own. This may involve designing and making prototype products that could be batch or mass produced for a range of users in a target market group. At A2, projects may be developed in collaboration with potential users of your product, or in collaboration with a client (such as a local business or organisation). For example, collaboration could involve consulting with a client or users to develop design briefs and specifications. Consultation may also take place during the development and evaluation of the product. Your A2 project should demonstrate:

- clear progression from the standard achieved at AS level
- a greater level of sophistication and more in-depth knowledge and understanding, than was evidenced at AS level.

At A2, progression and sophistication should be evidenced mainly through your ability to demonstrate a higher level of design thinking. You can evidence this through:

- undertaking research that targets more closely the problem/design brief
- selecting and using *relevant* research information – the folder content should contain a *conclusion* to research information
- making closer connections between relevant research and the development of your ideas
- using a more refined approach to focus ideas, so that they meet more closely the requirements of the Specification
- using a higher level of understanding about materials, components and systems when designing and manufacturing your product
- manufacturing a high-quality product, which may be more complex than at AS level. Your A2 product should 'work' in terms of meeting the design concept. Your A2 product should function well and fulfil specifications more closely than at AS level.

Assessment (coursework project)

Unit 5 coursework builds on the designing and making skills you learned in your AS project. You are required to submit a coursework project folder and a product with 2D/3D elements for your A2 coursework project. The A2 coursework is internally marked and externally moderated, using the same assessment criteria as the AS project. Before you start your coursework you should refer to the section on Unit 5, which explains the assessment criteria in detail. As you work through your A2 coursework project, refer to this section as and when you need.

Unit 6 Design and technology capability

This unit is called design and technology capability because it assesses the knowledge, understanding and skills you have learned throughout the whole Advanced GCE course.

Unit 6 focuses on the knowledge and understanding found within the designing and making process. This includes knowledge and understanding of product development and manufacture. Since this knowledge and understanding is taught throughout the whole Advanced GCE course, you are not expected to learn anything new during this unit. However, you should ensure that you review and revise what you have already learned. This includes preparing thoroughly, using exam practice.

Assessment (design paper)

Your centre will be sent a Design Research Paper at least six weeks prior to the exam. This paper will provide clear guidance about *one* design context that you will need to research.

In the examination, you will be asked to produce a design solution to a given design problem that is based on the context that you have researched. You will be asked to describe how the solution can be manufactured. You should demonstrate the ability to 'think on your feet', not to recall information.

The assessment covers knowledge and understanding related to designing and manufacturing. The style of assessment will remain the same each time the unit is assessed, but there will be a different design context and design problem to solve for each exam.

Managing your own learning during the course

The purpose of this section is to help you take more responsibility for planning and managing your own work.

Accepting this responsibility is an essential feature of any course that you will be undertaking at AS and Advanced GCE level. The ability to manage your own learning and plan your own work is also an essential skill in higher education and is highly valued by employers.

In order that you may take responsibility for your work, you need to be very clear about what is expected of you during the course. This book aims to provide you with such information. It may be helpful for you to begin by taking the following steps:

- Read Part 1 before you start the course, so you understand the course structure and the assessment requirements. It will also give you an overview of the requirements of each unit.
- Get to grips with the coursework projects that you need to produce and with the deadlines that you are required to meet. Investigate the coursework assessment and mark scheme.
- Before you start a unit, read the relevant 'Summary of expectations' in Part 2. This will give you an understanding of the unit requirements and provide information about how each unit is assessed.

Taking responsibility for your own learning

Once you have a clear understanding of the course requirements, you can plan your time and your work. Taking responsibility for this will enable you to increase your level of autonomy. This means being more independent and being more responsible for your own decisions. Being independent will require you to manage your own learning, develop project management skills and use a wide range of support beyond your teacher or tutor.

Project management

Learning to manage your work is called project management. This is an essential skill in higher education and in employment.

Project management means knowing:

- what is required
- planning and setting targets
- managing time, resources, budgets and people
- monitoring progress
- getting it right.

In order that you manage your own projects successfully, you will need to develop the following:

- research skills
- communications skills
- understanding industrial and commercial practice
- ICT skills.

Research skills

You will need to use both primary and secondary research to help you design, develop and manufacture your product.

You can use primary research to identify:

- user preferences – such as buying behaviour, taste and lifestyle of target market groups
- market trends – such as style, design, colour and product price ranges
- existing products – by collecting information about the design and manufacture of products through product analysis.

You can use secondary research, including information from books, magazines, catalogues, databases, CD-ROMs or the Internet, to identify:

- existing products and price ranges
- information about materials, components, systems, processes, technology, production, quality issues, etc.
- the work of other designers.

Communications skills

During your Design and Technology course you will need to apply a range of communication skills, such as:

- talking and listening to others
- reading, analysing and recording research information
- developing an understanding of form, function and design language
- drawing and sketching
- using professional practice, e.g. writing reports and presenting ideas to peers or 'clients'.

Understanding industrial and commercial practice

You will be expected to develop an understanding of industrial approaches to design and manufacture during your course and you will be expected to evidence them in your coursework. You will learn about industrial practices in Unit 1, but there are other approaches you could use to enhance this understanding. For example, you could:

- investigate work-related materials produced by a business
- use the expertise of visitors from businesses
- make an off-site visit to see a business at work
- use work experience to enhance your understanding of how organisations work
- use modern contexts to develop products for real/imaginary companies, i.e. a brand image appropriate to a specific company or retailer
- use ICT to research information about products, i.e. finding out about materials, components, systems, processes, the way products are marketed and company values.

You should provide examples of industrial practices in AS and A2 level coursework *where appropriate*. It is not appropriate simply to add a concluding page to explain what changes would be made to the product if it were to be manufactured in high volume. Instead, you should evidence industrial understanding through the designing and manufacturing activities that you normally use to develop products. These may include:

- developing design briefs and specifications
- undertaking market research and product analysis
- modelling and prototyping prior to manufacture
- producing a production plan
- testing against specifications.

In order to evidence industrial practices, you should use industrial-type terminology. You should make use of the glossary at the end of this book and refer to the coursework units in Parts 2 and 3.

ICT skills

You will need to investigate how you can use ICT in your coursework projects and use it *where appropriate and available*. The use of ICT is assessed in the coursework projects, when developing and communicating design proposals, in planning manufacture and in product manufacture.

You will not be penalised for non-use of ICT. There are many occasions when developing hand and eye coordination should be encouraged. When thinking through the development of an idea, it is often more appropriate to use hand-drawn techniques. You should develop both hand and computer skills.

In order to enhance your design and technology capability, you could investigate how you might use:

- ICT for research and communications – such as using the Internet, email, video conferencing, digital cameras or scanners
- word processing, databases or spreadsheets for planning, recording, handling and analysing data
- CAD software to model, prototype, test and modify design proposals in 2D/3D
- computer-aided manufacture (CAM) for computer control, using computer numerically controlled (CNC) machines.

Part 2
Advanced Subsidiary (AS)

Industrial and commercial products and practices (G1)

Summary of expectations

1. What to expect

This unit will help you to develop your understanding of products and the industrial and commercial practices by which they are designed and manufactured. The principle means of developing this understanding is through the process of 'product analysis'.

2. How will it be assessed?

The work that you do in this unit will be externally assessed through a 1½-hour Product Analysis exam. The assessment criteria are set out in Table 1.1. You should use appropriate specialist or technical language and correct spelling, punctuation and grammar.

Table 1.1 Assessment criteria for Unit 1

Assessment criteria	Marks
a) Outline the product design specification for this product.	7
b) Justify the use of: 　i) material or component 　ii) material or component.	3 3
c) Give FOUR reasons why this product is one-off, batch or mass produced.	4
d) Describe the stages of production for this product. Include references to industrial manufacturing methods.	16
e) Discuss quality issues for this product.	8
f) Discuss product safety issues associated with this product and its production.	8
g) Discuss the appeal of this product.	8
h) Quality of written communication.	3
Total marks	60

3. What will be assessed?

You must demonstrate your understanding of the unit content through the analysis of a product in the Product Analysis exam.

4. Content

Product analysis is based around knowledge and understanding of the following areas:

- Products and applications:
  - the processes involved in the development of a range of manufactured products
  - the **form** and **function** of different products
  - trends, styles, new technical capabilities, and social, political and ethical influences on the design, production and sale of products.

- Materials and **components**:
  - the range of materials and their potential application
  - working characteristics of materials: physical, chemical and composite.

- Industrial and commercial practice:
  - manufacturing systems, including one-off, batch, high-volume, bought-in parts
  - stages of production
  - detailed manufacturing methods, when combining or processing materials
  - service to the customer, including legal requirements and the availability of resources
  - the forms of energy used by industry and this energy's impact on design, manufacturing and the environment
  - the use of ICT by industry in the design and manufacture of products.

- **Quality** in terms of the product:
  - **fitness-for-purpose**
  - meeting the criteria of the specification
  - accuracy of production
  - appropriate use of technology
  - how it looks and feels.

- Health and safety:
  - the regulatory and legislative framework related to materials and equipment
  - standard risk assessment procedures in product design and manufacture
  - safe working practices.

5. How much is the unit worth?

This unit is worth a total of 30 per cent of your AS qualification. If you go on to complete the whole course, then this unit accounts for 15 per cent of the full Advanced GCE.

Unit 1	Weighting
AS level	30%
A2 level (full GCE)	15%

1. Products and applications

Figure 1.1 *The first designers were our early ancestors – this Chinese cave painting of a horse dates back to 13 000 BC*

The processes involved in the development of a range of manufactured products

Except for nature itself, everything around us has been 'designed' at some point. Someone has taken conscious decisions about the products, places and communications we use everyday – their shape, size, materials, colour, weight, and so on.

The very first designers, although they were not known as such, were our early ancestors who shaped stones to make simple tools, daubed the walls of caves with images (see Figure 1.1) and used animal skins to create simple garments to provide protection from the weather.

Today's designers and manufacturers are responsible for creating the things people need and want in order that they can survive physically and psychologically. Products not only need to perform in the way we expect them to, they also need to make us feel satisfied and valued. At the same time a product has to be commercially viable – in other words, it must be suitable for manufacture (or reproduction) and distribution at a price that makes a profit for the company that invests in it.

The role of the designer

Technology exists to enable us to extend the physical capabilities of our bodies and minds. It enables us to run faster, reach further, live longer, eat what we want when we want, send messages further and quicker, and so on. The invention and development of new materials, manufacturing processes and mechanical and electronic devices has enabled us to produce some amazing things. In itself, however, a particular invention may prove to be difficult to use – it might be awkward to lift or hold, or contain parts that break easily. Not knowing which switch or dial to turn, and when, can be confusing.

An important role of the designer is, therefore, to help make technology easier to use. Often this involves taking the components of a device and creating a 'container' for them which is shaped for the human hand or other part of the body. The container will help protect both the components from damage, and the user from exposure to sharp edges or dangerous electrical connections. The position and type of displays and controls can help make it easier to carry out the right actions in the right order to make the product work properly.

Meanwhile the designer of a printed product needs to ensure that the graphic is easy to understand by using legible **typography**, informative images and a clear, well-structured and consistent layout.

Meeting needs

Beyond our basic requirements for food, warmth and shelter come our needs to feel secure, confident and happy – a general sense of well-being. The experience of using a product can help make us feel satisfied, positive and mentally enriched. We may have been stimulated by the sleek, stylish appearance, reassured by the solidity of materials and construction, or delighted by the touch and feel of a textured surface and the resistance and click of a control dial. Photographs and dynamic use of typography can excite and stimulate us into wanting to read further. These experiences are far from accidental: they have been deliberately designed to arouse and please us.

Not only do individual products help fulfil our physical and psychological needs, but a range of products and services combined together can create a highly satisfying experience. For example, a meal served in a good restaurant initially satisfies our physical needs for food, water, and shelter. A well-designed menu stimulates our desire for interesting combinations of flavours and 'mouth feel' textures.

Designing might be a lot easier if we all had the same physical and psychological needs to satisfy. Unfortunately, we don't; different people have very different needs. In terms of physical needs, there is a wide range of capability in what we can hold, lift, turn, etc., and in the use of our senses to see, hear and feel things. In psychological terms, there is an even greater range. Apart from individual preferences for certain colours, patterns, materials and so on, people from different cultural backgrounds live and work in very different ways – what can be a positive experience for one group of people can be a negative experience for another:

- There are social issues to consider. A new device that does something in half the time with no human operator may represent a significant saving of money, but it could put people out of work and have a major impact on the prosperity of a particular local community.
- There are moral implications to consider too. Some products are designed to invade privacy, or potentially to cause harm. Even if this is not the main purpose of the product it can be an unexpected and unwanted side-effect.

- Finally, there are environmental issues. The use of natural resources and the production of many synthetic materials depletes the world's **finite resources** or can easily damage the delicate balance and life cycles of nature. Products need to be designed to minimise their impact on the environment – is all that packaging really necessary?

The above considerations have a major impact on what the designer can do. There are strict performance specifications that must be adhered to in order to ensure that a product is safe to use and safe to make. There are also a range of legal requirements concerned with the originality of the product, for example copyright issues.

Will it make a profit?

The work of a designer is usually commissioned by a client. The client might also be the manufacturer, but this is not always the case. Large-scale manufacturing requires considerable investment in setting up a factory, equipping it with expensive production tools, buying in bulk stocks of materials, employing and training a workforce, setting up a distribution system and launching a sales and marketing campaign. At the end of the day, the original investor wants to see a significant return on the outlay and, therefore, needs to be confident that the design is marketable and the production costs have been kept to the minimum.

Designing is, therefore, a very demanding activity. There are so many different people to try to satisfy, and the demands they make can sometimes be contradictory. To be successful in design work designers will need to be curious about how people behave and how things work and are made. The reward comes in knowing that in some way people's quality of life has been improved, within the constraints of financial viability.

Task

Evaluate the packaging for a range of food products. For each explain:

- who the product is aimed at – the target market
- how the packaging design attracts this target market
- any moral, social or environmental issues related to the packaging
- legal issues related to the packaging.

What are graphic products?

The term graphic products covers a wide range of design activity. Your choice of products to analyse needs to include both 2D and 3D elements. For example, a product related to a corporate image could include a 3D product, supported by a 2D element, such as point-of-sale advertising or marketing materials. The 2D element is concerned with the communication of ideas, information and messages. Any 3D product that you analyse needs to be constructed from a range of modelling materials *and* resistant materials (which could include some wood and/or metal and/or plastic). A simple example of the type of graphic product you could analyse would be a 3D product, such as perfume bottle, a calculator or tube of toothpaste, together with the product's packaging. Typical examples of graphic products might include:

- a product with its packaging or with advertising materials
- a point-of-sale display with marketing materials
- a corporate identity system with 2D and 3D elements
- an architectural model, together with 'concept' drawings.

Figures 1.2 and 1.3 illustrate a 2D and a 3D graphic product.

Design and development

The product design cycle

Designing in the more economically developed world inevitably involves money, maximising **profit** and reducing costs. Design and development are expensive activities and must be paid for by the earning potential that they create. This means that the manufacturing industry requires all its functions to be profitable, including design.

The **product design cycle** starts with the perception of need. Research may show that there is a 'pull' from the market that is creating demand or that competitors' products are likely to gain a bigger market share, so stimulating a response. Alternatively, technological or productivity opportunities become available to encourage the development of new products. This is called a '**technology push**'. There is also the situation where surplus capacity becomes available through changes in the market; this will also stimulate product development.

Figure 1.4 is a model of a product design cycle and demonstrates the relationship between these various aspects and their influences. In addition, it highlights the influence of shareholders and the demands of the retailer. Many large retail chains have more power and influence than manufacturers through advertising and promotion and it is they who have become an influential part of the product design cycle. It should be remembered that it is not always the consumer who influences this process.

Task

Look at a household product such as a clock or radio. Consider whether the design has changed over time through need, changes in style or technology push.

Figure 1.2 A 2D graphic product

Figure 1.3 3D graphic products

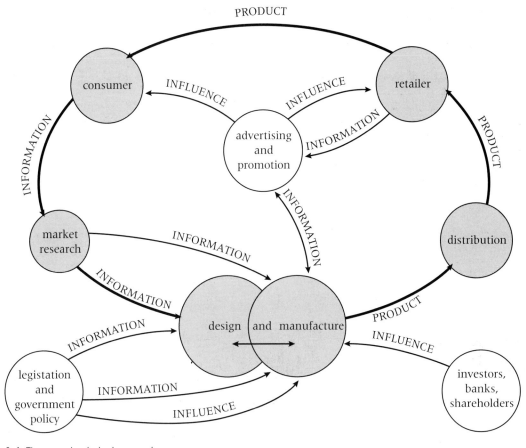

Figure 1.4 *The product design cycle*

How ideas are conceived and developed

On some occasions it is the designer who identifies the need for a particular product. One of the most interesting recent examples of this is Trevor Bayliss who came up with the idea for a clockwork radio for use in less economically developed countries where batteries are extremely expensive (see Figure 1.5). After developing a series of working **prototypes**, he

Figure 1.5 *The BayGen wind-up radio*

finally found a financial backer to set up a production factory.

Identifying problems that need solving is not easy. In most situations, the designer begins work in response to an outline brief provided by the client. This might well be based on a specific market need identified by a marketing organisation. For example, 'People are interested in purchasing electro-mechanical pet animals for their children'.

Investigation

A simple statement such as this does not provide enough information for the designer to go away and start sketching ideas. There are some immediate questions to ask, such as the following:

- What sort of pet?
- What age of children?
- What sort of cost?
- How many are to be made?
- What are the appropriate safety regulations?
- What is already on the market?
- How long have I got?

To find the answers to these questions, research is needed and the designer will need to use a wide range of sources of information.

The skill of the designer lies not so much in gathering information, but in deciding which is the most appropriate information to use in that situation.

Research information is used to produce a design specification. This provides a detailed series of statements about the product that can be agreed with the client. It will identify the following:

- *Fixed criteria* (for example, the toy must be a pet and be suitable for children aged 3 years and above) and criteria that allow the designer to develop ideas and take decisions about, for example, the type of pet and the sort of movements it makes.
- *Specification statements* which follow on directly from the conclusions that have been derived from research information. For example, the discovery that parents of children under 3 years did not think such a toy would be suitable, and that more rigid safety requirements are enforced for toys for younger children, leads to the statement that the toy should be for children over 3.

Product design specification

A **product design specification** sets out the known requirements for the final outcome. It provides a checklist against which the product design can be usefully compared during design development and at the end of the design process. It should be remembered, however, that the design specification is not necessarily fixed forever at this stage. During the development stages new information may emerge that suggests that modifications to the specification are needed, and these will be discussed and agreed with the client.

The nature of a particular product will determine the overall content of a specification. Typically, a specification might address the following points:

- *What is the aim of the product?* What general need is being satisfied, and in what particular situation or circumstance? For example, 'The point-of-sale display is intended to advertise a new confectionery product. It will be placed on the counter top.'
- *Function.* What should the product do to achieve its purpose, and are there any specified ways in which it should do so? For example, 'The display must contain at least 50 chocolate bars. The name of the bar must be clearly visible when the display is fully stocked.'
- *Materials and components.* Where specified, what materials, **components** and manufacturing methods, or processes must be used? The materials, the manufacturing process and the quantity to be produced will influence many design decisions. For example, 'The stand must be made from corriflute, to be supplied from existing stock'. Or, 'The power source should be a P53 battery that the manufacturer currently has in stock.'
- *Market.* Who is to be the target purchaser and the user? Are their requirements the same? For example, 'The display is intended to attract children between the ages of 10 and 14. A contemporary computer graphic cartoon image is needed to identify the confectionery with the pre-/early teenage market.' What are the physical and psychological needs of the intended user in relation to the product and the environment in which it will be used? For example, 'The stand must be easy to assemble by the retailer. Instructions must be as visual as possible. It must look sturdy and durable.'
- *Aesthetics.* What particular sensory qualities (form, colour, texture, sound, smell) are required? For example, 'An electronic circuit should be incorporated to produce a sequence of flashing lights and pleasing electronic sounds in order to attract attention.'
- *Quality and safety standards.* Reference should be made to appropriate **BSI (British Standards Institution)** and other standards. It is essential to remember that safety should never be compromised, but quality is often compromised for the sake of cost. For example, 'The product must conform to BS—. There must be no sharp edges. The stand must be well balanced.'
- *Performance.* How often and by whom should the product be maintained? How long should the product last in normal operational conditions, in terms of time, and/or frequency of use? For example, 'The display stand is expected to last for the duration of the six-week promotional period.'

Other criteria may be relevant depending on the product to be designed. For example, environmental standards.

Task

Product analysis is used extensively in many industries to develop specifications and ideas for new products. Choose a graphic product such as a point-of-sale display, a corporate identity system or promotional material. Develop a product design specification for your chosen product, taking into account: function, purpose, performance requirements, market and user requirements, aesthetics, quality standards and safety.

1906 – Strowger
Calling Dial,
Stroger USA

1931 – Siemens telephone,
designed by Jean Heiberg

1970s – standard GPO
British telephone

2000 – a modern
mobile phone able
to access the
Internet and e-mail

Figure 1.6 *The evolution of the telephone*

Form and function

It is interesting to look at a product such as a telephone to see how it has changed over the 100-year period of its existence and consider what has driven the development through the various stages (see Figure 1.6):

- Advances in plastics in the 1950s and 1960s.
- **Miniaturisation** of electrical and electronic systems that began in the 1970s.
- Development of digital communications networks and the advent of mobile phone technologies.
- Internet and email telephone access.

It is only in recent years that consumers have had a wide choice. This has now become evident through the demand for mobile phones with a degree of style, colour and 'street cred'.

Task

In relation to progression in telephone design, consider at what point did size and weight become important? Will the next generation of phones with video screens be progress or a novelty? Will the consumer be persuaded that this is what one needs?

Alternative ideas and their feasibility

When the specification has been agreed the designer can get down to thinking about possible solutions to the problem. In reality, however, it is likely that a number of ideas will already have been generated, ready for further consideration.

A popular technique is exploring analogies – thinking about things that the problem is similar to. For example, the answer to a structural problem might be found from studying the skeleton of an animal, or the structure of a vegetable.

At some stage, reference back to the design specification needs to be made to see which ideas best fit the requirements. Often two or three approaches are identified as being suitable for further development. Although the next stage will involve a great deal more focus on the detailed choices of shapes, sizes, materials, etc., it is sometimes necessary to discover completely new approaches, or to solve a particular problem by using an element of a completely different idea that was rejected earlier on.

Performance modelling

At the heart of the activity of designing, and what makes it unique from any other discipline, is the process of **modelling**: creating, making and testing ideas. This detailed development stage is usually the longest part of the design process. Eventually, the design reaches a stage where sketching it on paper is not sufficient to model how it would be in reality. Again, the skill of the designer lies not so much in the modelling technique used, as in the decision as to which technique is the most appropriate.

Computer technology is often used in the modelling process. 3D designs can be represented on screen and turned through any angle, and adjustments made easily and quickly. These data can be used to produce real objects on computer numerically controlled (CNC) machines, called 'rapid prototypes'. Spreadsheets can be used to manipulate data and explore the impact of changes to volumes, quantities and costs of materials and components.

Before a more detailed model of an idea is made, it is essential to be clear as to what specific aspect of the design will be developed further as a result. For example, your aim might be to resolve an issue regarding stability, or the positioning of displays and controls, or to learn more about how potential users react to the overall design concept. The purpose of creating the model will determine what sort of model is appropriate. Sometimes it might be in the form of a test rig to help determine the measurements for a particular mechanical component, or a non-working appearance model.

Creation of mock-ups and prototypes for testing products and systems

Performance tests involve simulating the use of the product in some way. For example, a poster could be exposed to the sunlight for different periods of time to see how quickly the inks faded, or a pop-up mechanism cold be tried, say 100 times, to discover if the elastic band snapped or the card became torn.

Where the product is to be tested to simulate use over a much longer period of time, a special test rig may be constructed and data about the rate of deterioration of performance of the product can be collected using sensors and a computer system to record and analyse the data.

Some tests can be easy to set up, such as dropping a package from a certain height to see if the contents remain intact. The reliability of other tests can be more difficult to establish and external factors can easily skew a set of results.

Throughout this development process designers draw extensively on their knowledge and understanding of the properties and characteristics of materials and components and the possibilities of manufacturing processes.

Designing for manufacture

The design of a 2D or 3D product needs to be developed from the start with the eventual production process clearly in mind. Considering the requirements of manufacture is not something simply done at the end of the development of the product: the costs and availability of the methods of production will influence many of the developmental decisions taken as the final design takes shape. The number of items to be printed or made is also an important factor to be kept in mind at all times.

For example, to facilitate the manufacturing process a high-volume run of a product made from moulded plastic will need to include shapes and forms that are well rounded. Meanwhile some parts and fittings (for example motors, hinges, etc.) may need to be specified as standard components from stock to help reduce production costs. Using stock sizes of materials such as timber, metal, plastics and paper facilitates larger and more cost-effective bulk purchase. In addition, the available quality of paper and method of printing to be used can make a great deal of difference to the legibility of type and image. For example, it is difficult, although not impossible, to control colour

consistency and the sharpness of lettering using conventional screen printing processes.

Underlying these decisions about design and manufacturing are the requirements for the ultimate quality of the product, in terms of its appearance, functionality, reliability and consistency. Many products (for example a newspaper) are not intended to 'last a lifetime' and so do not need to be manufactured to high performance specifications. Indeed, to do so would unnecessarily increase their production costs. Similarly, a product with an anticipated working life of, say, ten years should not contain materials and components that are likely to fail within six months, as this would increase the manufacturer's repair and replacement costs and give the company a poor reputation for reliability.

Before production can begin, a manufacturing specification is produced. Accurate drawings and other information about the product to be made are needed to ensure identical products can be consistently produced. This will include full details of the materials and components, the shapes and sizes, the finishes to be applied, the number to be made and so on.

Designs also need to be officially approved in terms of safety, and the appropriate documentation will need to be passed to the relevant authorities.

Depending on the product and its novelty, the designer or manufacturer may wish to take out a patent to prevent other companies copying the design without permission. Designers sometimes work in conjunction with a specialist production engineer to plan the mass production of the product, or the printing of multiple copies of a 2D graphic product. The design can still change at this stage to accommodate the particular requirements of the available materials and manufacturing technology. Extensive flow diagrams need to be produced to develop and record the most efficient flow of materials, components, products and people around the factory floor. Techniques such as 'just-in-time' help ensure that, while stocks can be kept to a minimum, there is always a steady supply available when and where needed (see page 25).

Task

Imagine you have been asked to evaluate one of the following products. Identify and describe the range of methods you might use:

- a domestic cooker
- a computer scanner
- a sewing machine
- a corporate identity system and promotional material for a bank or building society.

2. Materials and components

The range of materials and their potential application

In the design of graphic products there is a wide range of materials available, but the type of materials that the graphic product designer will be most familiar with will be papers, cards and model-making materials.

Papers, cards and model-making materials

There is a wide range of different sizes, weights, colours and finishes of papers and card, or board as it is often called. It can be plain white or coloured, or specially coated. Any size or colour you like can be specified!

Size

There are many standard sizes, such as the 'A' range (for example A6, A5, A4, A3, A2, A1 and A0), but special sizes can be cut.

Weight and quantity

Paper and card are also measured in weight. A standard piece of paper used in a computer printer or as a letterhead tends to be between 70 and 90 grams per square metre (gsm). The cover of this book is 220gsm and the pages of this book are 90gsm. Newspapers use a much thinner paper – around 45gsm. Card and board are measured in micrometers, or microns for short.

Paper tends to come in standard quantities. For example, most packs of A4 'office' paper come in quantities of 500 sheets, known as a ream, although smaller quantities of special finish paper and card are available. In the printing industry papers are purchased in much larger quantities and may be sourced from anywhere in the world, depending on the cost, quality and availability. Paper is one of the world's basic commodities.

There are many types of paper, each with particular characteristics. For example, bond has high strength and durability, takes ink well and can easily be erased. Some papers come with a ready-printed background, such as isometric and perspective paper, or ready for use to give a coloured background to a black and white laser print.

Other materials

Providing it is not the material itself that is being modelled, 3D models can be made out of just about anything! Parts from existing products can be ingeniously re-used and painted appropriately. There are, however, some materials that particularly lend themselves to the construction of models, such as plywood, MDF, balsa wood, papier mâché, foam board and corriflute.

Some graphic products will involve the use of other materials such as:

- plastics, for example thermoplastic and thermosetting polymers
- woods, including softwoods, hardwoods and manufactured boards
- metals and alloys
- composite, synthetic and laminated materials
- ceramics, glass and rubber.

Standard components

Components are standard, ready-made items that are used in the manufacture of a 3D product, such as nuts, bolts, rivets, hinges, clips, fasteners, gears, brushes, bearings, cams, plastic and electrical wire. These are usually **bought in** from suppliers or subcontractors. In some situations a **standard component** may need to be modified. Some component parts, such as a circuit-board, may be specially created and used across a range of the manufacturer's products.

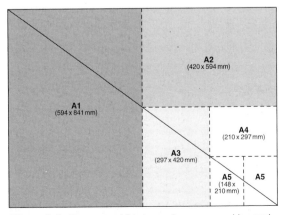

Figure 1.7 Common 'A' sizes of paper and board

Task
Choose a graphic product and identify the materials and components used in its manufacture. Explain why the properties and characteristics of these materials make them appropriate for the performance requirements and appearance of the product.

Working characteristics of materials
You need to be aware of the properties and characteristics of materials and components in order to enable you to make informed decisions, whether you are carrying out product analysis or

deciding upon product design and manufacture.

There is an enormous range of materials and components available and the problem is knowing which ones to choose. Each one has its own particular performance characteristics – the way it behaves. For example, paper tears easily but plastic is difficult to break. It all depends on what you want the material to do when in use and/or in manufacture. For example, should plastic or metal be used in the production of a kettle? The manufacturing options will be governed by the material used. Plastic is far more versatile and can be moulded into 'jug' shapes, but to create tall shapes in metal is expensive. The advent of plastics has enabled many jug designs to come on to the market.

As well as the familiar, traditional materials there are new specially created composite materials that have their own unique performance characteristics. Indeed, today, rather than having to make do with whatever happens to be available, designers can ask a materials technologist to develop a composite material that will perform exactly the way they want it to, although this, of course, adds considerably to the development costs of the product.

Finding out about materials through product analysis

When looking at materials and their characteristics from the perspective of a made product, it is necessary to turn the design and realisation process on its head and ask why and how using that material has enabled a product to come into being. The examples below are all drinks containers. They are common products that enable us to focus on the materials used in manufacture.

In 1884, Coca Cola was first sold in glass bottles; 21 years later, the unique Coke bottle shape was created by the Root Glass Company of Indiana. Glass bottles are used to this day, but in the 1970s a new material for drink packaging began to take hold of the market. This was polyethylene terephtalate (PET), and by the mid-1990s, most of the drinks on the supermarket shelves were packaged in PET bottles. What is so special about this new material that has enabled it to replace glass in this area of the market? By focusing first on the requirements of the supermarket and the consumer, it is possible to begin to see why the particular properties of this material are well suited to its function:

- It is possible to buy and sell drinks in large quantities, but large glass bottles are heavy for both distribution and taking home. A 330 ml PET bottle weighs only 10 g.

- Breakages are minimal with PET bottles, whereas they are a serious issue with glass, particularly large bottles when people are serving themselves.
- PET bottles are cheaper to produce. The Coca Cola company has even carried the Coke bottle shape into PET.
- Health and safety are improved; glass bottles are reusable, but collection, sorting and cleaning are often not cost effective. Doorstep milk delivery can be the exception. PET bottles have a very low failure rate and are easier to seal, even with gas release slots in the neck.
- Both PET and glass can be recycled, but collection and sorting of PET bottles coupled with the small amount of material in each makes glass the most cost-effective drinks package to recycle.

The other popular form of carbonated drinks and beer packaging is the aluminium can. This material has a number of properties that lends itself to this particular application:

- Aluminium is a plentiful resource, although the initial processing costs are higher than many other metals. It is, however, easy to re-melt, making it potentially cost effective to recycle.
- It is very malleable and can be formed into cans with a wall thickness of just 1 mm, thus reducing the material costs and the weight. The negative aspect to this is that as each can contains so little aluminium the incentive to recycle is reduced.
- Aluminium is corrosion resistant, although drinks cans are coated internally to withstand the acidic nature of many drinks. The malleability of the material enables the formation of complex flanging for the top and for the ring-pull rivet which is actually formed from the material on the top of the can.

Tasks

1 Carry out a comparative product analysis of glass, PET, paper and aluminium drinks containers. Focus upon these areas:

- applications, including carbonated, fresh juice and alcoholic drinks
- the screw top versus the ring pull
- material choice in relation to the type of drink and the volume of the container.

2 Juice containers are made from coated paper or card. Why is the rectangular shape advantageous, but not one that could be applied to other drink products?

The relationship between structure, properties, working characteristics and materials

By looking at drinks cans, it can be seen that the characteristics of the material are major factors when making a choice for particular product applications. The best material, however, will be the most appropriate after balancing all of the aspects such as cost, availability and working characteristics.

The look and feel of a product is most important. Depending on the purpose of the product, decisions about the material and finish to be used may well be influenced by how attractive its appearance and surface texture will be. The weight and smell of materials and finishes is also important. End users may be influenced simply by the colour of an item or by a recent advertising campaign that appealed to them; the idea that having that particular brand in a particular colour will make a statement about the kind of individuals they are.

Below is a list of some terms commonly used to describe the physical performance characteristics of materials:

- fusibility – the ability of the material to change into a liquid state when heated
- density – the amount of matter within an object
- electrical resistance – the extent to which the material resists or conducts electricity
- thermal resistance – the extent to which the material resists or conducts heat
- environmental resistance – the extent to which the material decomposes as a result of outside forces, such as water or light
- optical properties – the extent to which the material is opaque, translucent or transparent

- strength – how well the material withstands force without breaking or permanent bending
- hardness – the material's resistance to abrasion and indentation
- toughness – the ability to withstand sudden forces or shocks
- elasticity – the ability to bend and flex, and then return to its previous shape
- **plasticity** – the ability to change permanently when a force is applied
- **malleability** – the extent to which the material deforms under pressure
- ductility – the extent to which the material deforms through bending, twisting or stretching.

Finishes

Materials often have surface finishes applied to them to improve their performance characteristics. For example, paper and card can be:

- polished, laminated, or embossed
- coated, using a gloss or matt varnish.

There are four main areas that influence the choice of material and finish:

- function – what the product is to be used for
- scale of production – how many are to be made
- economics – the manufacturing costs
- availability – the ease with which the necessary quantity can be obtained.

So one particular material might have just the right strength and finish and be well suited to the particular manufacturing process needed to make tens of thousands of copies, but involve the use of very expensive tools, and currently be in very short supply from a foreign country.

3. Industrial and commercial practice

Manufacturing systems

Making a product costs money. The more it costs, the higher the price the product will need to sell at in order to make a profit. If the selling price of a product is high, then only a few people will be able to afford to buy it. Manufacturing products that the majority of people can purchase requires the use of different manufacturing systems.

There are many elements that contribute to the cost of making a product. The materials and components are often one of the smallest expenses. Labour, production equipment, the costs of marketing and retailing and tax account for much of the final selling price. The balance between costs varies with different products – some may require more to be spent on, say, materials, but less on marketing.

Minimising the costs of production is a major concern of manufacturing industries. Two of the main ways they can achieve this is by:

- making bulk purchases of materials and components
- reducing labour costs by simplifying the production process.

How products are manufactured

The level of production, that is the number of products that need to be made, determines the method of manufacturing to be used. 3D products and printed graphic products are made using slightly different manufacturing systems.

Short-term runs

Few 3D products are one-offs, as they are relatively expensive to create. A bridge to span a specific river, a made-to-measure fitted wardrobe or a commemorative item of jewellery would come under this heading. Often these products will use standardised components (see page 21) to reduce production costs. Even a craft item, although made by one person, is likely to be made in a small batch, with perhaps minor individual changes to make each piece unique.

A graphic product created to communicate information about a proposed design (for example an architectural model or a rendering of a car) would be classed as a one-off. A graphic product that is printed or reproduced in some way cannot, by definition, be described as being a one-off. However, a mural or some other handwritten, drawn or painted artefact could be.

Batch production

In batch production, a specified quantity of a product will be made, enabling some bulk purchase of materials, and some potential savings in labour costs if the production process is broken down between a number of people. For example, it might be decided to screen print 500 T-shirts for special sale at a local concert. Only one stencil will be needed, and there will be savings on buying larger volumes of inks and plain T-shirts. Two people who work together will be able to produce them in less than half the time it would take one person.

Many printed products are produced on a batch production basis. For example, 10 000 copies of a book might be printed initially. Six months later, when all these have been sold, a further batch of 10 000 (or some other quantity) might be reprinted, using the same plates or film, or electronic files. In the interim, many other different books will have been printed in different quantities on the same presses.

Many 3D products are batch produced. For example, 2000 white hair-dryers might be manufactured, and then 1000 blue ones, followed by 500 green ones that use a more powerful motor and have more features.

Each time a new batch is started, a number of elements in the production process will need to be changed – different coloured inks, different mouldings, for example, or even placing machinery in different locations in the production line. Successful batch production involves a high degree of flexibility and must be carefully planned to reduce downtime, that is when the production line is not actually making anything.

High-volume (mass) production

High-volume, or mass production differs from batch production only in that the production line is completely dedicated to producing one product for an extended period of time, which is unlikely to be known when the line is set up. A new model of a car, for example, may come to be produced for many years if it is successful, while the production of another might cease after only six months if it fails in the market place.

Few graphic products could be said to be truly mass produced. Although large quantities of national newspapers are printed every night, the content changes through the day, and many printing presses fit other short-run local newspapers in between.

Continuous production

This term is used to indicate products in which the manufacturing process is continuous, and never changing. It often applies where stopping the production line would be difficult, as with steel production. Some food and chemical production is continuous.

Standardisation and use of modular sub-assemblies

In all the manufacturing systems described above, component parts may be produced using a different system. For example, the electronic circuits for a car may be made in another part of the factory in smaller batches, as required. **Sub-assemblies** such as these are often bought in ready-made to the required specification from other specialist companies.

Some products use **standardised components** to reduce manufacturing costs. For example, a car may be fitted with a 'standard' CD player mechanism, which is also used by other manufacturers. Other components might be standardised within the product (for example using the same sizes of bolts throughout).

Task

Choose an example of a graphic product and say if you think it has been one-off, batch or mass produced, or has used a mixture of these processes. Explain the benefits/disadvantages of each method of production.

Stages of production

The production of most products goes through a series of stages as outlined below.

Preparation

This refers to the preparation of the raw materials to be used. In many cases the materials will be supplied ready-processed. For example, a printing works buys in paper rather than attempting to make it from wood pulp. However, it might wish to prepare the paper in a particular way, or to cut it ready for use on a particular press. With 3D products, materials might need to be treated or shaped to prepare them for processing.

Some materials might be locally available, but it is not uncommon for sources around the world to be used, and fluctuations in global currencies, variations in shipping times and costs and availability can make a significant difference, particularly with bulk purchases where ordering large quantities may be a more cost-effective way of operating.

A printed product might be designed in the UK, printed in Europe or the Far East, on paper delivered from Scandinavia, and then sent back to the UK for sale.

Processing

This is where the product is given its final shape or form, or printed surface. A printed graphic product will result from the inks being applied to the paper. A 3D product may be moulded, drilled, pressed, etc., into the correct shape.

Assembly

The pages of a book or magazine are not printed one by one. Each plate might carry 16 or 32 pages. These need to be cut, folded and bound to make the final product. A 3D package is likely to need to be stamped out, and any transparent plastic 'windows' added. The package will remain flat, however, until needed on the factory floor at the point of assembly.

A 3D product may consist of many components that need to be joined together in a particular sequence.

Finishing

Printed products are often given a special protective coating, such as a varnish. On 2D products these are likely to be added before assembly. A 3D product is more likely to be painted, plated or polished after assembly.

Recycling resources

Wherever possible manufacturers are keen to maximise the use of a material or energy resource by minimising its waste. In the printing industry, waste paper can be re-pulped, although it produces a lower grade of paper. Some finishing processes that involve toxic chemicals can cause environmental problems, so these need to be carefully controlled. There is also a social issue in recycling resources in that minimising waste and recycling can have an impact on the environmental issues that affect the world. Recycling can be extremely efficient if the waste is readily available. The main costs of recycling are collection, transportation and sorting. Recycling is beneficial in many ways:

- It helps to conserve **non-renewable resources** and the site from which they are extracted.
- It reduces energy consumption and greenhouse gas emissions that result from processing.
- It reduces the levels of pollution that stem from manufacture and disposal processes.
- It reduces dependency on raw materials.

Stock control and just-in-time

Having the right materials in the right place at the right time is essential for an efficient and economic production line. When materials are ordered in large quantities to reduce costs, they need to be stored properly. Storage space and maintaining the right storage conditions can be costly, so it is important not to overstock. At the same time, it is essential to avoid a situation in which the materials and components that are needed to produce a product have run out, or have been delayed in their delivery to the factory.

Delivery from the factory's own stock to the production line raises similar problems: too much stock at the point of production will get in the way and might deteriorate or become damaged before use. At the same time, halting the whole production line while one stage in the process waits for a batch of components to arrive will be highly costly in terms of lost and delayed production.

The aim is to get the right stock in the right place just in time. This is the idea that production is 'pulled' along in response to customers' orders and the requirements of the manufacturing process – in other words, as close as possible to when it is actually needed. This needs sophisticated planning, monitoring and stock control systems.

The just-in-time (JIT) stock control system was developed in Japan and is based on production being directed in response to customer's orders and the requirements of the manufacturing process. The aim of this is to rid the system of stock at every opportunity. Stock is material and components wasting investment and taking up space. This technique is dependent upon the supply of components and material arriving just in time.

Planning the production

Any production line needs to be organised as efficiently as possible to ensure materials, tools, operators and inspection processes are all in exactly the right place at the right time. Many production processes are broken into a series of sub-assemblies, so that a series of operations will be going on in parallel before everything is finally assembled.

There are various techniques and tools that can be used to help achieve the optimum production layout. The most common is a flow process chart approach that uses various graphic symbols to indicate movement, storage, processing, checking, etc. In large, modern factories use will be made of a computer planning system that will work out the most efficient stages of production and the best layout of machines. When a particular batch of products or a print run has been completed the computer system is used to quickly re-plan the layout for the next, completely different product run.

The final stages of production inside the factory, after any special finishing and final testing processes, are usually labelling, packaging, and storage prior to departure from the factory.

If the product has been made for an external client, it will be delivered as soon as possible after production. Marketing, retailing and distribution to the point of sale becomes the client's responsibility. If, however, the client is also the manufacturer, then the final products will need to be transported to external distribution centres – the aim being to make goods to order and avoid storage.

Manufacturing methods

Printing processes

There a number of basic printing and manufacturing processes used to reproduce graphic products. The choice of which to use will primarily depend on the length of the production run, the unit costs and the quality required.

One of the most common processes used for medium to long colour print runs (that is between 1000 and a million copies) is called offset lithography (see page 100). If, however, the product is text-based and black and white, letterpress (see page 102) may be used.

For high-quality, long print runs (for example for glossy magazines, 'art' books, mail order catalogues, stamps), gravure will be used. If the printing surface is unusual, for example plastic, corrugated card, then flexography becomes the ideal choice.

Finally, screen printing (see page 100) is best for short print runs where low cost and basic quality are the main requirements. Shop window posters and T-shirts are typically produced this way. Longer run, higher quality screen printing processes are available for fabrics and wallpapers.

Colour printing

An essential consideration in terms of cost and processes is the number of colours being used. Full colour is usually achieved through the use of four printing colours: cyan (blue), magenta (red), yellow and black. The application of each colour requires additional processing, so reducing the number of colours used can reduce the cost. Until recently, full-colour printing was significantly more expensive than using just one colour, but recent advances in printing technology have made full colour relatively inexpensive.

Using just one or two colours can still reduce costs, however. A two-colour print, with the creative use of a range of tones and overlays can produce some surprising results. For example, certain colours (such as particular tones of cyan and yellow) can be overprinted to produce black, so a book or CD cover could be created with dark blue text, bright yellow graphics (such as keylines), light blue and yellow background tints, and black and white photographs.

The use of 'spot colours' – very accurate single colours, used perhaps for a corporate logo – and special inks, such as silver or gold, add to the overall printing cost. Unusually shaped papers (for example with a curved edge or cut-out shape) add further production processes that involve extra equipment, labour and production time. Embossing requires a special die, and a further manufacturing process.

Finishing off

The cost of finishing processes can add considerably to the cost of production. Book covers and packaging often have a protective varnish applied, or can be laminated with a plastic film.

Assembling printed sheets together involves yet another cost. Magazines, for example, may need to be folded and stapled – one of the simplest and cheapest methods. Paperback books are glued together in sections. The most expensive books are literally 'sewn' and glued together in sections for maximum durability. These days all such operations are done using automated machinery.

The manufacture of a 3D product has a different set of considerations to a 2D product, but ultimately the process of choosing the methods used comes down to a series of decisions about the length of the production run and the complexity and quality of product required.

Learning about manufacturing through product analysis

The starting point is a decision on what design requirements must be in terms of the functional properties of the component or product. Then the focus will be on the manufacturing process and likely scale of manufacture. With many products, it will be clear whether they have been manufactured on a large or small scale, or are in fact one-off items. When carrying out a product analysis, it is important to be clear what the product does and its purpose. You will need to answer the following questions: What was the designer's intended purpose for this product? What is the intended function? What is the intended market, and crucially, how large is the market? Once you have answered these, you are then in a position to look at how the product was made.

Containers and casings

Figure 1.8 illustrates an example of a container and a casing – a bucket and a television set. Containers and casings of many types have similar design requirements whether they are tamper-evident containers for health care and pharmaceuticals products, casings for television sets, or buckets and bowls. Casings in particular are complex and manufactured to fine **tolerances**.

Design requirements:
- Cheap to produce in high volumes.
- Self finished to a high standard with flat and/or textured surfaces.
- Rigid enough for their intended purpose. A television set is typically more rigid than a bucket. The bucket's flexible nature may be more of a result of the manufacturing process than a design intention although a rigid plastic bucket would be brittle and easy to crack.

Process – injection moulding:
- The most common manufacturing process for household consumer goods. It is cheap, quick, efficient and suits a wide range of plastics and plastic products. Granules of unpolymerised plastic material are heated and then forced under pressure into a mould cavity called a die.
- Suitable for complex shapes with holes and screw fixing bosses and for multiple dies for small products such as containers and lids.
- Mould design is critical and affects the product design. Products are usually designed with tapers for removal although multi-part dies allow even screw threads to be moulded.
- Production rates are high and are highly automated – typically, between an eight and 50-second cycle time.
- Equipment and tooling costs are high. Dies are usually made from tool steels as one-off or small batch products.
- Labour and finishing costs are low with little finishing required other than some removal of runners and occasional excess plastic. With most thermoplastics, all removed material is recycled back into the process.
- Integral hinges and tear-off portions can be moulded into the product by appropriate thinning of the material. Materials that can be injection moulded are mostly thermoplastics. Some composite materials can also be processed using **injection moulding**.

Figure 1.9 shows an injection moulded product. In the injection moulding process, the granules enter via the feed hopper and are melted within the barrel. The molten plastic material is fed to the die by a combined process of an extrusion screw and a final injection. The mould then opens and the product is ejected (see Figure 3.1.5 on page 74 for a detailed diagram).

Drinks packaging

On page 22 we looked at a product analysis exercise related to drinks packaging from the perspective of the properties and characteristics of the material used in the manufacture. It is

Figure 1.8 Casing and container

Figure 1.9 Snap on covers for mobile phones are injection moulded

appropriate now to return to two of these products – PET bottles and aluminium drinks cans – and look at them from the perspective of manufacturing processes.

PET bottles – design requirements:
- Cheap to produce in very high volumes.
- Pressure vessels with a self-finished sterile surface.
- Hollow, rounded shapes with thin walls to reduce weight and material costs.
- Increased material and definition around the neck to facilitate screw top and pressure sealing.

Process – blow moulding:
- It is cheap, uses very little plastic and can produce up to 2500 bottles per hour. Highly automated production means that one operator can manage several **blow moulding** machines. This means labour costs are low.
- Products need not be symmetrical and it is possible to integrate handles for low-pressure containers such as those used for milk, cooking oils and disinfectants.
- Screw threads and undercut features are possible, though holes are not.

Figure 1.10 shows a blow moulded product. See Figure 3.1.6 on page 74 for a detailed diagram of the blow moulding process, the stages involved in blow moulding. A hollow tube of thermoplastic called a parison is extruded downwards between two halves of a mould which closes around the neck and the base. Hot compressed air is then blown into the parison forcing the plastic to expand and line the walls of the mould. The bottle is then cooled and ejected.

Aluminium drinks cans – design requirements:
- Cheap to produce in high volumes.
- Printable outer surface.
- Hollow, rounded shapes with thin walls to reduce weight and material costs.
- Wide neck capable of secondary forming, after filling, to form airproof sealing joints to similar material.

Process – cold forming:
- Various processes are classified under this heading that combine forward and backward extrusion processes and cold drawing. The common features are high pressure and large forces.
- Suitable for any ductile materials including aluminium, copper, tin alloys and some low carbon and alloy steels.
- Good material utilisation and production – up to 2000 units per hour. Highly automated with low labour costs.

Figure 1.10 *A blow moulded product*

- Tooling and equipment costs are high, therefore this process is best suited to high-volume production runs.
- Undercut shapes are not possible.

Cold forming of drinks cans is a multiple process involving 'drawing' and 'ironing'. A disc of aluminium is punched into a die to form a shallow dish shape. This is then drawn down into a deeper die while being 'ironed' by hardened steel rings that thin and polish the material.

Service to the customer, including legal requirements

The designer and manufacturer's responsibility does not end when the product leaves the factory. A consumer may have a particular difficulty using the product – maybe he or she has not been able to follow the instructions, or the product fails to do what the consumer thought it would. Alternatively, it might be faulty. Dealing with these sorts of issues is extremely important in terms of long-term public relations, but can prove costly in terms of staffing and replacement. Manufacturers, therefore, strive to provide goods that are easy to use, perform as expected and are not damaged, either when they leave the factory or while in transit.

At the same time, safety in use is extremely important. A product has to pass a range of tests set against nationally recognised safety standards. These are produced by the British Standards Institution (BSI). There are also various legal requirements that must be observed, as well as European Union directives. These help provide a consumer with reassurance that the product will be safe to use.

However, a particular product may be hazardous in use or storage in other ways not covered by the standards or legislation, which manufacturers remain responsible for. Thorough product testing is needed to ensure that products supplied are as safe as possible.

Forms of energy used by industry

All production demands the use of energy. Energy is used in the initial preparation of any raw materials, in the manufacturing processes and in distribution. Until recently, little attention was paid to the way in which energy and the Earth's natural resources were being used. Levels of consumption and pollution have increased dramatically during the last 50 years. Unfortunately, it is not that easy to stop! For example, as far as a manufacturer is concerned:

- non-renewable resources are cheaper
- durable products mean that there is less replacement purchasing, and less market for spare parts
- regularly changing styles and fashion promotes a throw-away approach

… while as far as consumers are concerned:

- cheaper products are usually preferred
- high levels of packaging are expected
- recycling is often too time-consuming and inconvenient.

Life cycle assessment and environmental issues

Designers and manufacturers do take environmental issues seriously. Long-term investment in increasing the efficiency of use of energy and reducing waste can result in cost savings. A life cycle analysis can be carried out to analyse the use of energy and resources at each stage of production, use and disposal (see Figure 1.11).

There are three basic approaches that can be used to make a product more environmentally friendly. The first is to aim to reduce the amount of materials and energy used in a product. The second is to use recycled materials and/or to make sure that the materials used can in turn be recycled themselves. Finally, component parts can be designed to be re-used at a later date in other products.

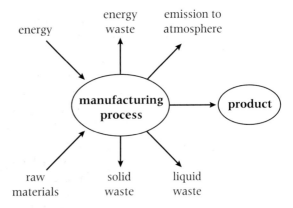

Figure 1.11 Life cycle analysis

Look out for information about how the product you are analysing has been manufactured with respect to the environment:

- How will it be disposed of when finished with?
- Does it include the recycling symbol?
- If so, in what way can it be recycled?
- How easy is this for the consumer to do?
- Has it been made from any recycled materials?

The use of ICT by industry in the design and manufacture of products

At home, and often in school, computers are described as being stand-alone. In industry, they are part of large networks in which a wide range of functions are integrated together.

Computers and other electronic devices have revolutionised the ways in which products are designed and manufactured. Computers and forms of electronic communication (information and communication technology – ICT) are not as new as you might think, but it has only been in recent years that the speed of processing has become such that they have had a fundamental impact on the way in which industry and commerce work.

Every type of software package has an application somewhere in the design and manufacturing process:

- Spreadsheets are used to calculate costs, project the consequences of variation, and for stock control.
- Databases have become a fundamental market research and sales tool.
- Word processors and **desktop publishing** programmes have changed the way text and images are created and exchanged.
- Multimedia software is used to create sophisticated presentations to clients and customers.
- The Internet is used for product development

research and to sell goods directly via a website.

- 2D and 3D graphics packages enable images and 3D products to be modelled on a computer.
- Sophisticated control systems have enabled the widespread automation of many manufacturing and testing processes.

When these last two applications are linked together they are known as computer-aided design/computer-aided manufacture – CAD/CAM. The data created by the CAD program are used directly to instruct a machine how to operate – to print out an image or to create a 3D form. There are many advantages in using CAD/CAM, but the most important are for speed and accuracy, and the ability to rapidly experiment with different ideas.

2D CAD systems

CAD systems are now used in the production of virtually all graphic products. There are essentially two types of software packages – bitmap and vector:

- **Bitmap** packages often have the word Paint or Photo in their titles. They store pictures by recording the colour and position of each pixel in the image. It is easy to make changes to the whole image, and parts of the image can be selected. Bitmap file sizes tend to be large when saved.
- Vector packages usually include the word Draw in their title. They store information about the length, angle and thickness of each line.

Many modern packages include facilities to work with bitmap and vector images, and to convert from one to the other. Alternatively, they may be created in separate packages and brought together in another. For example, the photographs on the cover of this book would have been scanned in (or taken by a digital camera) and manipulated in a bitmap package. The publisher's **logotype** (shown in the bottom right-hand corner) would have been created directly in a vector package. The final layout would then have been developed in a desktop publishing package, where the lettering would have been created and the photos and logo added. The final composite file would have then be sent directly to the printing company (either on a CD or over a high-speed Internet connection).

3D CAD systems

3D CAD packages tend to be more complex than 2D packages, in that information about the length, height and depth of an object is needed. Once entered into the computer, however, it is possible to rotate an object in any direction, enabling the designer to get a much better idea

of what a product will look like in reality. Textured surfaces, lighting and shading can be applied to the surface for added realism. Graphics produced in a 2D program can be quickly 'wrapped round'.

The resulting data can be used to create a 'rapid prototype', using a sophisticated computer numerically controlled (CNC) lathe, driven by the computer, and then, finally, for the construction of the tools that will make the production model. Again this information can be sent electronically, anywhere in the world.

A new area in 3D packages is **virtual reality**. This combines computer modelling with simulations in order to enable a person to interact with or be immersed in an artificial 3D visual or other sensory environment.

CIM systems

CIM stands for **computer integrated manufacture**, and describes a system in which many stages and processes are directly linked together electronically. For example, the original CAD files may be used to generate costings and stock requirements, and be re-used by the marketing department to show to potential customers. Quality control, production flow monitoring and consequent delivery times are all handled by the same system, which can be accessed globally, as needed.

Tasks

1 a) Choose an example of a graphic product and suggest what its main stages of production were, what manufacturing processes might have been used and what environmental issues would need to have been considered. How and when might ICT have been used?

b) Can you spot where sub-assemblies might have been used?

c) What finishing processes seem to have been applied?

2 The examples of product and principle manufacturing processes that have appeared on these pages do not make a comprehensive list. Choose some other product examples and look at the design requirements and principle manufacturing process in a similar manner. Always investigate from an industrial manufacturing perspective. Suitable products could be:

- toothpaste tube and packaging
- DVD and casing
- child's game and packaging.

4. Quality

Quality in terms of the product

Product analysis will enable you to understand why one particular product should be chosen in preference to another. The following are issues that influence consumer choice:

- Design – we are influenced greatly by appearance.
- Manufacture – build quality.
- Performance – if it functions as expected.
- Customer satisfaction – cost and quality in harmony.

Quality assurance and total quality management

All of the above are aspects of **quality assurance (QA)** which stretches from the design concept through to delivery to the customer. Organisations that demonstrate the highest possible standards throughout their company are awarded **ISO 9000**, the international standard of quality.

Total quality management (TQM) is the goal of companies which seek to establish the highest possible standards of quality at all levels, not just within their manufacturing sectors. TQM is about the attitudes that people within an organisation have. It stems from being valued as part of a team and sharing the benefits of that team's successes.

Quality control

Quality control (QC) is a part of the achievement of quality assurance. QC is concerned with monitoring and achieving agreed standards as a result of inspection and testing.

Quality control is applied at critical control points at all stages of manufacture, and typically starts with an inspection of the raw materials and components that have been delivered by outside suppliers. In a printing works, for example, the paper and inks will be checked to ensure they meet the expected specification. Electronic files, photographic film or printing plates will also need to be approved.

Tolerance is the degree to which a component is acceptable in order to function in accordance with its design specification. Precision is extremely costly to achieve. Tolerance describes the acceptable level of variation from a nominal size or standard. For example, the diameter of a bottle must be such that it fits within the bottle-filling process and also contains the required volume of liquid. The dimensional tolerance is likely to be described as + or –0.8 mm, i.e. any size between 53.2 mm and 54.8 mm would be acceptable.

Inspection is the examination of the product components or material to determine if specified standards and tolerance limits are being met and adhered to. Inspection incorporates the monitoring of:

- material used
- dimensional accuracy
- quality of appearance and finish.

Testing is concerned with the product's performance and can be either non-destructive or to the point of failure. Testing procedures seek to determine:

- function
- life expectancy
- durability under adverse conditions.

Statistical quality control

Throughout the process of production, materials, components and the product in the process of being completed are subject to checks for quality. In a print run of, say, 100 000 CD covers it is not viable to inspect every one for flaws. Instead, a sample is checked – for example one in every thousand. It will be checked mainly for accuracy of colour matching and registration (the positioning of the different colours). The variations will be recorded, and the data from a range of samples analysed statistically. If a trend in variation is observed (for example there is a consistent deterioration in the quality of a certain coloured ink), then the printing press will be adjusted (or if necessary repaired) before the quality of the print reaches its pre-set tolerance limit and would need to be rejected.

Processes may go out of control for a variety of reasons, for example as a result of:

- human error
- machine or process failure
- tool wear
- material and component variations
- environmental changes, temperature fluctuations, vibration and so on.

Task

Choose a graphic product and suggest a number of specific points in its manufacture where quality control checks are made.
For the same product, list a range of internal and external quality standards that the manufacturer might have needed to comply with.

5. Health and safety

The regulatory and legislative framework related to materials and equipment

Manufacturers have a responsibility to ensure that the workers they employ are not exposed to potential injury or illness. Beyond this basic responsibility, many companies have discovered that taking a positive interest and concern in the well-being of their employees contributes to improved attendance and productivity.

All production exposes the workforce to risk. There are various Acts and regulations that manufacturers are required to follow, including the following:

- the Factories Act (1961)
- the Health and Safety at Work Act (1974)
- COSHH (Control of Substances Hazardous to Health).

While these cover general circumstances, it is necessary to consider the particular risks of each production process.

The risks involved at each stage of production need to be identified and assessed, and then minimised as far as possible. This is known as **risk assessment**, which usually covers things like:

- adequate training
- suitable clothing
- adequate lighting and space

- inspection and maintenance of equipment
- clear safety and emergency procedures.

The 1992 Management of Health and Safety at Work Regulations requires a risk assessment be made in all workplaces to determine the hazards that may cause hurt or damage.

A hazard is defined as anything that has the potential to cause hurt or damage, such as:

- machinery that could be faulty
- walkways that could be overcrowded or slippery
- storage of dangerous substances
- personnel undertaking unfamiliar procedures.

The risk must be assessed and evaluated through the analysis of the chance of the potential degree of hurt or damage that could occur. The hazard should then be eliminated or controlled at an acceptable level.

Task

Choose a graphic product and identify a number of health and safety issues that the manufacturer might need to have taken into consideration. Discuss the safety of consumers in their use of the product, and the likely safety precautions needed in production.

6. Aesthetic qualities

Form and function

To a certain extent, it is true to say that the form (that is, the three-dimensional or 3D shape) of a product needs to follow its intended function.

For example, a teapot that looks elegant, but fails to pour the tea properly without spilling it can hardly be classed as good design. At the same time, products need to be created with a great deal more than just their physical function in mind.

A design team needs to ensure first and foremost that a product is fit-for-purpose. All products have a function – the thing they are intended to do.

For example, they may intend to:

- entertain
- inform
- contain something else
- move someone or something from one place to another.

Indeed, many products have more than one function. For example, a package is rarely just to protect its contents, but also serves to provide information about what it contains, often provides a means of display in a shop, and sometimes a device to make it easier to carry. Some packages may be thrown away, while others will be re-used during the life of the product, or even be capable of being re-filled. Making a product easy and safe to dispose of is also a 'function' of the design.

Aesthetic and cultural influences

Beyond these physical requirements are the emotional requirements. People respond to images, objects and their surroundings which can make them feel stimulated or displeased. We don't just respond to visual images, but sounds, smells, textures and tastes as well.

These responses are known as **aesthetics**. Different people respond to different things in different ways. Our feelings about things tend to be influenced by our expectations and memories of previous events.

While to a certain extent aesthetic response can be simply described as personal taste, there are some responses that particular cultures tend to make. It is these responses that the designer can anticipate.

For example, in the western world, the colour red is generally thought of as being stimulating or alarming, hence it used to draw rapid attention. Meanwhile, green is considered to be a calm colour. Green is used to suggest relaxation and continuity. Likewise, certain patterns, shapes, textures, lines, colours, compositions, sounds, smells and tastes can all evoke different responses.

Beyond the primary function of a product, and the way it makes us respond aesthetically is a third, social level – what it says about us as a person. Buying a certain designer brand or other well-known name of product reinforces our association with the qualities, attitudes and values of a particular group of people we aspire to be like.

Trends and styles

Fashion, or style, is a major consideration in the design of products. Style defines a particular set of shapes, colours, motifs, typographies, etc., that are recognised as being of a particular time and place.

In the specific terms of graphic products, one of the first styles it is possible to identify dates back to the late Victorian and Edwardian era. This is when the mass reproduction of the printed image became possible (see Figure 1.12). The work of the Arts and Crafts movement, the pre-Raphaelites and the Japanese print were highly influential, followed by the more organic Art Nouveau shapes and colours.

Figure 1.12 *A Victorian theatre programme*

The development of mass production and the fascination for speed during the 1920s and 1930s is echoed in the highly rectilinear shapes, bright 'sunburst' colours, and the streamlined qualities of the graphic styles of the time.

The streamlined style continued into the 1950s, combined with images drawn from the world of science and new technology (see Figure 1.13).

Mid to late 1960s' graphics are characterised by references to 'Swinging Britain' and to 'psychedelic' free-form, swirling shapes and bright fluorescent colours (see Figure 1.14).

The use of the computer had a major impact on the graphic styles of the last 20 years of the twentieth century, enabling designers to create significantly more complex images than before, combining text, images and graphic devices in new creative and exciting ways.

At the start of the twenty-first century the world wide web is providing new challenges for designers to create layouts that are animated, interactive and in the control of the 'reader'. The recognisable characteristics of such styles are only now starting to emerge.

Achieving good quality design is, therefore, a challenging task. It requires a thorough understanding of what a product must do physically, how people will respond to it emotionally and what it will say about them as a person. All this must be achieved at a price they are willing to pay for the product.

Task

a) Choose a graphic product and identify its primary functions – what it is intended to do.

b) Go on to indicate the decisions the designer has taken in terms of aesthetics to make the product more acceptable and satisfying to people.

c) Finally, discuss the trends and styles of the time it is reflecting.

Figure 1.13 *A car advertisement from the 1950s*

Figure 1.14 *An image of the swinging '60s – this poster was inserted into a Bob Dylan album, 1966*

7. Developing product analysis skills

In order to become proficient in product analysis, you need to undertake a wide range of product analysis activities. Earlier in this unit, we looked at information about product design and manufacture intended to help you to understand the issues that are most relevant to different types of products. The issues are not always the same. Quality and fitness-for-purpose will always be issues but safety in relation to a book, for example, is quite different from safety in relation to a household electrical appliance.

You are surrounded by manufactured products, far too many to cover within any textbook; you need, therefore, to develop a strategy to assist you with this process. This is best achieved by practice and by using the framework that forms the structure of the assessment process that is set out in Table 1.2.

It is also important to investigate different types of products, and products that have been manufactured using a range of production processes, levels of production and from a variety of materials. It is best to start with simple products, that is those that are not made up of a large number of components, and then move on to more complex ones as your ability develops.

The Virgin **brand** is a complex and sophisticated **corporate identity system** (see Figure 1.15). It bestows a series of product expectations and lifestyle aspirations on to a wide range of everyday goods and services, ranging from financial products to mobile phones, from CDs to cola drinks and from airlines to railways. It achieves this though a

Table 1.2 *Unit 1: product analysis assessment grid*

Assessment criteria	Marks
a) Produce a basic product specification with regard to: • function • purpose • performance • market • aesthetics • quality standards • safety.	7 marks
b) Justification of how properties/characteristics, materials, components, sub-systems are important to the product and its function.	6 marks
c) Explaining the level of production.	4 marks
d) Stages of production including: • preparation • processing • assembly • finish.	16 marks
e) Product quality issues: • quality control in production • quality standards.	8 marks
f) Health and safety issues including: • safety in the use of the product • safety procedures in production.	8 marks
g) Aesthetic qualities including: • form • function • trend/styles • cultural.	8 marks
h) Spelling, punctuation and grammar.	3 marks
Total marks	60 marks

Figure 1.15 *The Virgin brand*

carefully controlled, unified presentation of a distinctive use of typography, colour, packaging, staff uniforms and marketing messages. Virgin's products are extensively promoted through coordinated programmes of press and trade advertisements, high street posters, TV and radio advertising, special offers and publicity stunts.

Undertaking an analysis of the Virgin brand would be a challenging and possibly daunting task, but one that could easily and effectively be broken down into a series of smaller elements, each focusing on different products and services. What is the target market for Virgin Cola, for example? How is the company's corporate identity scheme adapted and applied to its packaging? What media are used to promote it?

The assessment process

The main purpose of the assessment grid, set out in Table 1.2, is to enable the process of assessment to take place. It is, however a useful framework and checklist that can be used to support the process and understanding of product analysis. If you use this for all of your activities, then you will be familiar with it when it comes to your Product Analysis examination.

Notice also the distribution of the marks. This indicates the areas that are considered to be most important and, therefore, most demanding of your time and effort. The relationship or 'weighting' between the various elements of the assessment grid will become clearer as you work through your course of study.

The Unit 1 examination is a single product analysis question. It will follow the same format and be marked out of 60 using the assessment grid shown in Table 1.2.

The examiner will be looking for:

- analytical skills – this is about product analysis; get to the point and address those aspects that are in the assessment grid
- your ability to draw upon and apply your knowledge – take the opportunity to demonstrate your knowledge of the unit content in relation to materials and processes, etc. through your answer
- clear thinking – stay on task; you will not be awarded marks for comments that are not relevant to the product, materials or process in question.

Within the examination a further 3 marks are allocated to your ability to communicate. This includes the appropriate use of specialist/technical vocabulary and your ability to spell, punctuate and use grammar. The paper is designed to give you the opportunity to demonstrate your capability.

Throughout your course you should make every effort to become accustomed to the correct vocabulary. Always use the correct names for tools and processes. If you don't know what a specific tool is called, then look it up or ask. Be precise; use the word 'hacksaw' or 'tenon saw' rather than just 'saw'. The process of removing wood, plastic or metal using a lathe is called 'turning' not 'lathing'. The terms 'shape' and 'form' are not interchangeable. Do not use 'soft' to describe a material when you mean 'malleable', or 'hard' when you actually mean tough. Glass is hard but it is not tough, although it can, of course, be toughened.

Exam preparation

Practice exam question

Below is a specimen examination question. It is similar to the one you will get in your examination. Spend 1½ hours attempting your answers. Use the assessment grid in Table 1.2 on page 35 as a checklist.

Figure A

Figure A shows a point-of-sale stand. The stand is made from thick card which is scored, ready to be folded and slotted together. It displays videos with their thermoplastic box covers.

a) Outline the product design specification for the point-of-sale stand and video box cover. (7)

b) Justify the use of:
   i) thick card (3)
   ii) thermoplastic. (3)

c) Give four reasons why this point-of-sale stand would be produced in a batch. (4)

d) Describe the stages in production of the point-of-sale stand. Include references to industrial manufacturing methods. (16)

e) Discuss the quality issues for the point-of-sale stand. (8)

f) Discuss health and safety issues associated with the point-of-sale stand and its production. (8)

g) Discuss the appeal of this point-of-sale stand. (8)

UNIT 2

Product development I (G2)

Summary of expectations

1. What to expect

You are required to submit one coursework project at AS. This project should build on the knowledge, understanding and skills you learned during your GCSE course.

Your project should comprise both 2D and 3D elements, together with a coursework project folder. You can find more information about the 2D/3D coursework elements in this unit. It is important to undertake a coursework project that is of a manageable size, so that you are able to finish it in the time available.

The project should represent an amount of work that is proportional to the time taken and the marks awarded – so understanding how your project is assessed will help you plan your work.

2. How your project will be assessed

Your coursework project covers the skills related to designing and making and will be assessed using the criteria given in Table 2.1.

You must attempt to cover all the assessment criteria. If you take account of how the marks are awarded when planning your work, you will be able to spend an appropriate amount of time on each part of your coursework project. This will give you more chance of finishing in the time available and increase your chance of gaining the best possible marks.

Your coursework project will be marked by your teacher, using the assessment criteria in Table 2.1 and then sent to Edexcel for the Moderator to assess the level at which you are working. It may be that after moderation your marks will go up or down. The Moderator also gives you a grade for your coursework project.

3. Choosing a suitable project

As a student designer, your choice of AS coursework project is very important because you need to balance designing and making a successful product with meeting the needs of your coursework project assessment.

At AS level, you are expected to take a commercial approach to designing and making products that meet needs that are wider than your own. This could mean using one of two different approaches to product design:

- *either* designing and making a one-off product for a specified user or client
- *or* designing and making a prototype product that could be batch or mass produced for users in a target market group.

The key to success is to choose a project that is both enjoyable and challenging, so that you will feel inspired to finish it in the time available.

Table 2.1 AS coursework project assessment criteria

Assessment criteria	Marks
A Exploring problems and clarifying tasks	10
B Generating ideas	15
C Developing and communicating design proposals	15
D Planning manufacture	10
E Product manufacture	40
F Testing and evaluating	10
Total marks	100

4. The coursework project folder

The coursework project folder should be concise and include clear information that is appropriate to the project. This means that you will need to be very selective about what to include, so that you target the available marks.

For example, your research needs to be planned so that you find out useful information that will help you make decisions about what you intend to design and make.

Do not be tempted to waste your valuable time finding out information that has no relevance to your project because you will not gain any more marks for it.

Your coursework project folder should include a contents page and a numbering system to help its organisation. The folder should comprise around 20–26 pages of A3 or A2 paper. The title page, contents page and bibliography are included as extra pages. Table 2.2 gives a guideline for the page breakdown of your coursework project folder. *Please note, however, that this is only a suggestion and you may find that your folder contents vary slightly from this, because of the type of project that you have chosen.*

Table 2.2 Coursework project folder contents

Suggested contents	Suggested page breakdown
Title page with Specification name and number, candidate name and number, centre name and number, title of project and date	extra page
Contents page	extra page
Exploring problems and clarifying tasks	5–6
Generating ideas	3–4
Developing and communicating design proposals	3–4
Planning manufacture	3–4
Product manufacture	3–4
Testing and evaluating	3–4
Bibliography	extra page
Total	20–26

A Exploring problems and clarifying tasks (10 marks)

1. Identify, explore and analyse a wide range of problems and user needs

Your choice of AS coursework project is very important, because you are expected to design and make to meet needs that are wider than your own. This could mean designing and making a one-off product for a single specified user or client, or designing and making a prototype product that could be batch or mass-produced for a range of users in a **target market group**. Whichever type of product you design and make, you will need to explore and identify a realistic need or problem through investigating and analysing the needs of people in different contexts.

2D and 3D elements

Your choice of project is important because it must include both 2D and 3D elements. The 2D/3D elements should be linked by the theme or context for design, so that one supports and underpins the other. For example, if your theme is corporate image, the 3D element could be a prototype product, supported by the 2D element. This could be point-of-sale advertising or marketing materials, both related to the 3D prototype product.

For your coursework project you are, therefore, asked to produce the following:

- a folder (in A3 or A2 format) that summarises the development of the 2D/3D coursework elements
- a 2D element, developed from traditional or modern graphics media – the 2D element should be linked to and support the 3D outcome
- a 3D model or prototype product, constructed from modelling materials and at least one resistant material (which could include some wood and/or metal and/or plastic) – the 3D outcome should be semi-functioning.

A simple example of the coursework requirement would be the development of a 3D prototype product, together with its packaging. The product, such as a perfume bottle or calculator, could be modelled from modelling materials. Although this would not have to 'work' as a functioning product, it would still need to be a good quality model, testable against aesthetic criteria. The 2D element – the packaging – should be developed from graphics media, as a separate outcome from the coursework design folder.

Don't forget to consult with other people, including your teacher or tutor to ensure that you choose a suitable project.

Exploring and identifying the needs of users

The needs of users change according to **demographics**, expectations and lifestyle. In other words, the kind of products that people want to use change according to their age, and their expectations change according to the kind of lifestyle they aspire to. Many manufacturers and retailers target potential customers and attempt to match their lifestyle needs with products. They often do this by using **market research** to identify the **buying behaviour**, taste and lifestyle of potential customers. This can establish the amount of money they have to spend, their age group and the types of products they like to buy. New products can then be developed to meet customer needs. (Consider the questions in Figure 2.1.)

There are many starting points for investigating the needs of users. You may, for example, decide to investigate trends in leisure pursuits, lifestyles or demographics. This might require you to find out where people shop, their gender, age group, their level of disposable income, their **brand loyalty**, the newspaper and magazines they read, the music they like, TV programmes they watch, their leisure interests and where they eat out. This kind of in-depth investigation can provide numerous contexts for design – perhaps you could even find a 'gap in

Figure 2.1 *What problem does this product solve? What is the target market for this product? What are its key design features? How does this product meet the needs of its users?*

the market' for a new product and devise a marketing campaign for it. Maybe there is the need for a new leisure centre or the traffic flow in your town centre needs re-designing – could you produce an architectural model and interior/exterior **presentation drawings**?

Task

Finding out about the needs of users
The needs of your users, or target market group, can provide a good starting point for design. Manufacturers often build up **customer profiles** to help establish the characteristics of products that customers want. You can use a similar practice. Write a customer profile to describe target market groups for the following products and their packaging. In doing this you could think about customers' lifestyles, values and tastes.

a) board game   c) football kit
b) toothpaste   d) calculator

Identifying and analysing a realistic need or problem

The characteristics of many products need to change gradually over time, if they are to fulfil the developing needs of users and sell into a market at a profit. Once the requirements of users are known, they can be analysed and problems which give rise to the need for new products identified. Often a new product may be developed from an existing one by modifying its aesthetic or functional characteristics to meet changing user needs. This kind of product development often involves redesigning to add value to a product – to improve its performance, function or appeal.

Updating existing products through colour and styling to create a more modern or fashionable look is sometimes used to increase the market share of products that have been on the market for some time. The redesign of products with a strong brand image, such as sweets or DIY tools, needs to be carefully managed. Any changes must be gradual so that the recognisable brand image is maintained.

One of the most important areas of product development is the creation of new brands that target specific market groups. The young sports market is a prime target for this type of **marketing**. There are many views about the morality of marketing brands because some young people can be pulled in to a cycle of needing to be seen to be wearing the latest 'logo' or **brand**

name. What do you think? Are young people under pressure to be seen in the latest styles? Is this kind of brand marketing ethical?

Clarifying the task

Once you have identified user needs and a problem to be solved, you can start to focus the problem by asking questions about the kind of product that you could design and make to solve the problem (see Figure 2.2). You can think about the purpose and potential for the product: How will it benefit users? Will the product be a prototype product that could be mass produced or a one-off specialist product like a scale model? Are there any products currently on the market that fulfil a similar need? What kind of materials do they use? What kind of materials and processes could you use? Asking questions like these will help you clarify what you are trying to achieve and enable you to write a design brief.

> **To be successful you will:**
> - Identify clearly a realistic need or problem.
> - Focus the problem through analysis that covers relevant factors in depth.

2. Develop a design brief

Developing a design brief will focus your mind on what you want to do. The design brief should develop from your exploration and analysis of user needs and problems and from your identification of a potential product that is feasible for you to make.

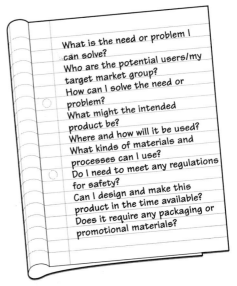

Figure 2.2 *The type of questions you could ask about a product*

Your design brief needs to be holistic – in other words, it needs to encompass the development of both the 2D and 3D outcomes. It should also be simple, concise and should explain what needs to be done, but it should not include the solution to the problem. It is intended to give you direction but not to be so precise and specific that you are not left with any room for development.

Task

Developing a design brief

Your design brief should be simple, concise and explain what needs to be done, without going into too much detail. For example 'Design and make a poster to advertise a new perfume'. This brief could encompass an advertising campaign, together with a prototype model of a perfume bottle or maybe a promotional gift.

a) For the design brief above identify the client, the 2D and 3D elements including at least one resistant material and the range of potential users.

b) Choose three different contexts. Write a design brief for each, identifying the 2D/3D elements, their purpose and the target market group.

Try not to be too ambitious when you write your design brief because even the simplest idea has a habit of becoming more complex as it develops! Use your design brief to help plan your research, so that you can target what you need to find out.

To be successful you will:
- Write a clear design brief.

3. Carry out imaginative research and demonstrate a high degree of selectivity of information

Research

Your research needs to be targeted and it needs planning in order to find out useful information that will help you make decisions about what you intend to design and make (see Table 2.3). Read your design brief and your analysis of the problem – this will help your planning. For example, the design brief 'Design and make a poster to advertise a new perfume' may lead to research into trends in advertising, the needs of the target market group (male/female?), buying behaviour, the investigation of existing perfume products or the marketing of promotional gifts. This would include research into their design, possible materials, modelling processes, quality and safety needs.

Task

Market research

For a product of your choice, produce a market research report that identifies the buying behaviour, taste and lifestyle of potential consumers. Find out the size of the target market group, their preferred brands and the competition from existing products.

You should undertake both **primary** and **secondary research** using a range of sources, such as the ones listed in Table 2.3. Use these research ideas and your research planning to target the research you need to undertake. This will depend on the type of product you intend to design and make – you are not expected to research everything listed in the table.

Table 2.3 Research ideas and where to find information

Research ideas	Information sources
Market trends, design and style trends, **niche markets**, new product ideas, user requirements, buying behaviour, lifestyle, demographics	Market research, window shopping, shop surveys, visits, the work of other designers, art galleries/museums, user surveys, questionnaires, product test reports, consumer reports, people, the Internet
Materials, components, corporate image, advertising trends, systems, processes, scale of production, product performance requirements, product quality, product price ranges, value for money	Analysis of existing products, materials testing, books, newspapers, electronic media, CD-ROMs, databases, the Internet, exhibitions, reports, magazines, catalogues, libraries, local colleges, industry, people
Quality control and safety procedures	Books, safety reports, Design and Technology Standards for Schools, information via the Internet, industry
Value issues, e.g. cultural, social and environmental	Keep up with the latest news, events, films, exhibitions, cultural and social issues use the Internet to find out about company values and about environmental issues

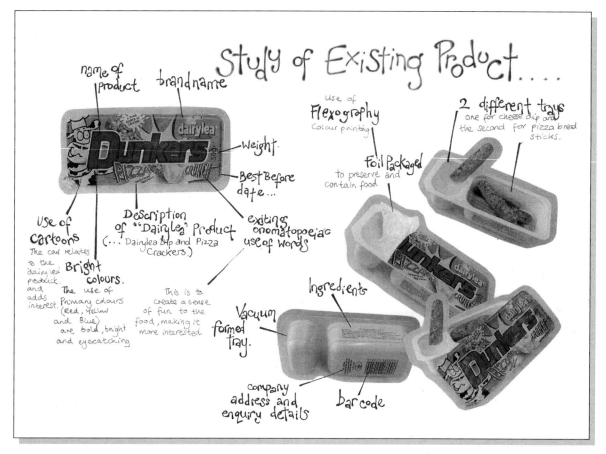

Figure 2.3 *Student research into an existing product*

Writing a bibliography

Reference all your secondary sources of information in a bibliography and include:

- sources of any information found in textbooks, newspapers, magazines
- sources of information from CD-ROMs, the Internet, etc.
- sources of scanned, photocopied or digitised images.

Product analysis

In industry product analysis is often used to obtain information about the design and manufacture of competitors' products. Analysis can include data research, using catalogues or trade literature and analysis of the products themselves, either by eye or using physical analysis of the product's component parts (see Figure 2.3). You can analyse commercial products that are similar to the 2D and 3D elements of the product you intend to design and make. This can provide information about design, style, advertising and marketing trends, corporate image, packaging, materials, components, processes and assembly.

It will also develop your understanding of quality of design and manufacture and of the concept of value for money.

At this stage of your project you may find that some activities need to be carried out simultaneously. For example, research and the analysis of the problem are often bound together – some analysis may be necessary in order to focus some of the research, while analysis of the research is necessary if the gathered information is to be useful when making decisions about what to design.

Analysing research

You will need to analyse your research and select useful information that is relevant to the design of your intended product. Your research analysis should, therefore, provide you with a good understanding of the problem and the desired aesthetic, functional and performance requirements of the product you intend to design and make. This will enable you to move on to the next stage in the development of your product – writing a design specification.

To be successful you will:

- Carry out a wide range of imaginative research, selecting information that will help you make decisions about what to design and make.

4. Develop a design specification, taking into account designing for manufacture

Factfile

A design specification is detailed information that guides a designer's thinking about what is to be designed. It is used to help generate, test and evaluate design ideas and to help develop a **manufacturing specification.**

In the same way that analysis and research are often bound together, the design brief, the research process and the development of the product design specification are all linked and may be developed simultaneously. In order for you to develop design ideas for both of the 2D and 3D outcomes, you will need to develop specifications for each of them. For example, if you are going to 'Design and make a poster to advertise a new perfume', you will have to develop a specification for the poster *and* a specification for the prototype perfume bottle.

Both of these design specifications need to take into account **designing for manufacture (DFM)** – where considerations for design should include the purpose, function, aesthetic and performance requirements of the product, materials, components or systems; market and user requirements; any value issues that may influence your design ideas *and* considerations about manufacturing processes, technology, scale of production; quality and safety issues; time, resource and cost constraints. Figure 2.4 shows part of a design specification for a hotel complementary pack.

Product Design Specification

The **Function** of the hotel complementary pack is to provide a sensible packaging for the customers of the Piccadilly Plaza, which is portable, and promotional for the hotel company, so the customers can carry them around if necessary, and hopefully increase the business of the hotel company even more.

The **Purpose** of the hotel complementary pack is to keep all the accessories of the Piccadilly Plaza's complementary pack in one portable place so the customers can carry around and use it whenever they need it. Also it is there to add some additional promotion for the hotel company.

The **Performance** of the hotel complementary pack must be of the highest standards it should make it easy for the customers of the hotel to access the accessories of the pack quite easily without spillage or ripping the package. It should help the customers to find the accessories easily, without having to look around for things they may have forgotten or can't find.

The **Aesthetics/Characteristics** of the hotel complementary pack should be attractive and eye catching so the customers will notice it when they come into the room. It shouldn't be too colourful or bright, because of children messing about with the hotel complementary packs. The colours most associated with the hotel company are blue and white because it is proven that those colours attract adults the most. The characteristics of the packaging and accessories must include the hotel name and logo, which is promotional.

The British **Quality Standards** of the hotel complementary pack must be met by the ISO 9000 (International), Kitemark Scheme Standards (British) and CE Marking (European). Kitemark is well known and respected for its product quality mark, 80% of the British public recognise the Kitemark as a symbol of quality and safety. 60% are prepared to pay a price premium for products that bear the Kitemark, so suppliers try to achieve Kitemark products. To achieve Kitemark we need a combination of stringent and regular independent product, testing and quality systems assessment. The hotel complementary pack must comply with the British, European and International Standards.

The **Safety** of the hotel complementary pack must also apply to the use of the pack and its production.

<u>USE</u>

As well as adults, children use hotel rooms, and the Piccadilly Plaza have a lot of families who stay in their hotels as well as schools. The pack must not contain any lose, small pieces which can be easily swallowed by children. The pack will contain bits for children but these are to be handled by adults. So the pack will not be brightly coloured because according to research it's proven that blue and white colours tend to be bought by adults.

<u>PRODUCTION</u>

The production I will be using for the hotel complementary pack will be one-off production. This is because I'm making a prototype for what will be made by batch production for a trial period to see if it's successful. The material I will be using is the plastic polypropylene, which will be formed by the vacuum process. This will be done safely by using gloves because it's a process that requires heat. The mould I will use will be made of wood, which I will make using scrap pieces and the band saw and PVA glue. I will wear gloves and have a teacher nearby.

Figure 2.4 *Student design specification for a hotel complementary pack*

You are expected to identify an *appropriate* scale of production, such as a one-off product for a specified client, or a prototype product that could be batch or mass produced for a target market group. The scale of production will impact upon all your design and manufacturing decisions. For example, a one-off architectural scale model would need to meet the requirements of the client. This might be an architect, a local authority planning department or a model-making consultancy. This kind of model would require great accuracy of scale and manufacture and could be simple or complex. On the other hand, a prototype educational toy that could be batch or mass produced would need to meet the requirements of a range of children and their parents, but may require processes to be simplified for ease of manufacture. You would need to take these requirements into account when producing your prototype. In both of these scenarios, you would be working in a similar way to an industrial or architectural model maker.

To support both of these scenarios, of course, you need to develop the 2D element. For the architectural model you could develop presentation drawings that show the building in situ. The presentation drawings may need to be reproduced for display elsewhere (for example in an exhibition or a library) or used in architectural publications, in which case you would need to consider reproduction methods. For the prototype educational toy, you may have decided to produce marketing materials, which would need to be reproduced in high volume.

The design specifications for both of your 2D and 3D outcomes should guide all your design thinking and provide you with a basis for generating design ideas. A design specification, which may start out in outline, can change and develop as research is carried out, until a final specification is reached. This is used as a check when testing and evaluating design ideas and will provide information to help monitor a product's quality of design. The design specification is, therefore, an essential document that sets up the criteria for the design and development of your product. Specification criteria can also be used later on to guide your thinking when developing a manufacturing specification.

B Generating ideas (15 marks)

1. Use a range of design strategies to generate a wide range of imaginative ideas that show evidence of ingenuity and flair

You are expected to generate a wide range of feasible design ideas that cover both the 2D and 3D elements of your coursework project. Your ideas should be based on your design specification criteria. In fact, your design specification should be a great help when designing, as it is much easier to develop ideas from a starting point than starting from nothing!

Imagine starting with a blank sheet of paper and being able to design anything you like. Where do you start? How do you know what is required? No designers work like this because, firstly, there is not enough time to work in this way and, secondly, products are designed for people who have definite likes and dislikes and lifestyles that demand products with specific aesthetic, functional and performance characteristics.

Many designers say that designing becomes easier when there are limitations under which to work because these limitations provide a framework within which to design.

Inspiration for design ideas

Many sources can be used for inspiration – your design specification and research, of course, but you may also decide to base some aspects of your ideas on a theme. You could look at:

- natural forms
- other artists or designers
- an art or design movement, e.g. Art Nouveau
- a theme, e.g. 'minimalist' or 'eco design'
- influences from exhibitions, music or films
- new technology, materials or processes.

Use product analysis to inspire your design work. It is a good way of collecting information about materials, components, assembly, style, colour, quality and price. If you work out the design specification for an existing product, this will help develop a specification for your own product. Ensure that any product you develop in this way is truly original and not just a copy of an existing product.

Generating ideas

It is often a good idea to keep a notebook for quick sketches – these can be pasted into your coursework project folder, rather than be redrawn (your project folder should show evidence of creative thinking rather than stilted copied-out work). Approach your initial design work in a similar way, using hand-drawn sketches, colour ideas and brief notes to explain your ideas, which can be developed and modified later (see Figure 2.5). Use a medium that you feel comfortable with and that is easy to work – pencil, ballpoint pen or fine markers are ideal for quick line drawings and for simple shading. At this stage the examiner is looking for evidence of your design thinking and later on to see how you develop these ideas. It is not always necessary to develop a wide range of totally different ideas, although you should always try to produce variations.

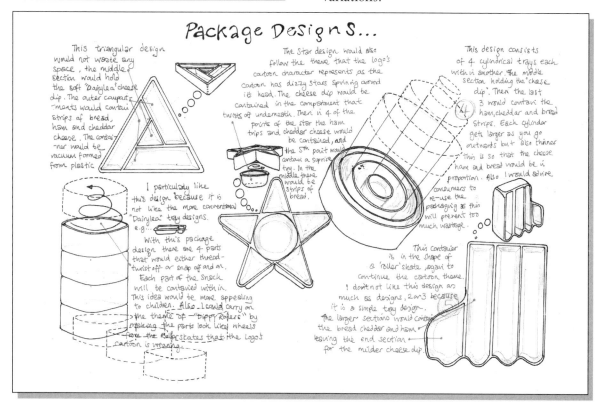

Figure 2.5 *Student generation of ideas for a cheese dip tray*

Factfile

Design strategies

Many designers find it easier and quicker to sketch ideas rather than using CAD, which can sometimes inhibit the development of first ideas. However, CAD may be helpful later on, when you are in a 'What if?' situation or when you need to trial different colour schemes for a design.

To be successful you will:

- Use a broad range of design strategies to generate and refine a wide range of imaginative ideas.

2. Use knowledge/understanding gained through research to develop and refine alternative designs and/or design detail

Showing the influence of research on your design ideas should not be a problem, as this influence should come through naturally from your 2D/3D design specifications, which are based on your research and analysis. Also, you should be using your research and specifications as inspiration for ideas – so be sure to make this explicit when you are developing your initial ideas by adding brief notes to explain your thoughts.

As you experiment with first ideas and gradually refine your thinking, you may find that you start to think about the possible materials or processes you could use, to work out if your ideas are feasible. You may also start to play around with combinations of ideas or work on the fine detail of some of your ideas. Making use of the information you acquired during your research phase should enable you to develop and refine alternative ideas to the stage where you can select one (or possibly two) as the most promising. Figure 2.6 shows how one student evaluated his ideas against the design specification.

The type of product being designed will influence the research you need to do and the kind of ideas that you produce. For example, ideas for a prototype board game may be concerned with its aesthetic and functional aspects – you may have to consider the production of playing pieces as well as storage of component parts. However, ideas for the design of point-of-sale advertising may be concerned with colour, style and font size in relation to the image you are trying to promote about a prototype product.

To be successful you will:

- Demonstrate effective use of appropriate research.

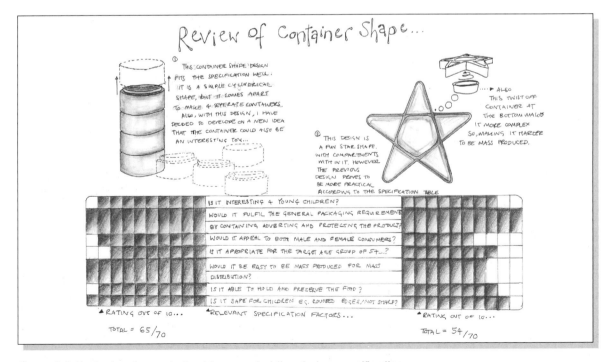

Figure 2.6 *Evaluate all your design ideas against the design specification*

3. Evaluate and test the feasibility of ideas against specification criteria

As your ideas develop you should evaluate and check back to your design specification to see if you are going in the right direction, to see if there is any aspect that you have missed or to look for more inspiration. It is always helpful to talk to others at this stage, as talking over your ideas often provides the step forward you need. There are a number of options here:

- discussing your ideas with a teacher or tutor
- presenting your ideas to a small group of fellow students
- talking to someone acting as a client
- talking to potential users of your product.

In industry this would be a normal part of product design and development, in order to determine the feasibility of ideas, ease of manufacture and **market potential**. You should evaluate your ideas against the design specification so that you are able to justify, using written notes, why your chosen design is worth developing. This can be done using **force field analysis** (see Figure 2.7)

During the course of your design development you may find that there are some aspects that you need to find out more information about. You

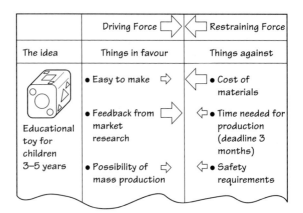

Figure 2.7 *Force field analysis*

may need to modify your design specification in some way, if, for example, you find after evaluating ideas against the design specification, there are some aspects that are not feasible.

> *To be successful you will:*
> - Objectively evaluate and test ideas against the specification criteria.

C Developing and communicating design proposals (15 marks)

1. Develop, model and refine design proposals, using feedback to help make decisions

Your aim should be to develop your chosen idea until you produce the best possible solution to your problem. This can be done by **modelling**, prototyping and testing. Modelling your design proposals in 2D and 3D will provide you with helpful information about the feasibility of your proposals. You can also get feedback by finding out what potential users or a client think about your ideas – show them your proposals and ask them a range of questions. Base any questions you ask on the requirements of the design brief and specification. Using feedback should help you make decisions about the aesthetic and functional characteristics of your design proposals. You can then refine your ideas if necessary, until you find the best possible solution.

> *To be successful you will:*
> - Develop, model and refine the design proposal, with effective use of feedback.

2. Demonstrate a wide variety of communication skills, including ICT for designing, modelling and communicating

You should use a variety of 2D and 3D communication skills to develop, model and refine your design proposals. Communication skills can include writing, drawing or using information and communications technology (ICT) for word processing. Modelling will enable you to experiment with ideas before making any final decisions (see Figure 2.8). It also allows you

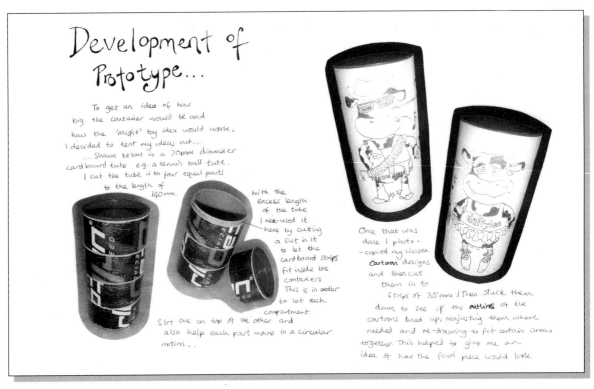

Figure 2.8 *Student development of a prototype*

to visualise your proposals, using a range of hand and computer methods. The method you use will depend on the type of product you are designing and the aspect of its design which is to be modelled. For 3D modelling you should use materials which are quick and easy to cut, shape and join. The use of computers for 2D and 3D modelling is an essential part of industrial product design.

Factfile
Modelling in industry
In industry modelling is a key process because it enables manufacturers to test and modify products and processes. Even though modelling may involve making a number of development models, in the long run it saves time and reduces manufacturing costs because products and processes are tried and tested before manufacture. This avoids costly mistakes, such as the product not functioning properly or the product not meeting customer requirements.

2D modelling techniques can include, for example, drawing **layouts**, using **crating, pictorial, perspective** drawings, **sections, exploded** views, using cut-and-paste, or using

computer-aided design (CAD) software. For example, you can scan in your image and try out different variations, or use **parametric design** software to trial 'What if?' scenarios. However, you should only use ICT if it is appropriate. You will not be penalised for its non-use. Although most graphic designers use CAD for their work, most will have been trained in drawing techniques. Sometimes it is more appropriate to use hand-drawn techniques, for example when thinking through the development of an idea. You should, however, try to develop both hand and computer skills, where appropriate.

Factfile
Prototyping in industry
As a product develops it becomes more accurate until it becomes a prototype product. This is a detailed 3D model made from inexpensive materials to test a product before manufacture. Prototyping is a key industrial process because it enables the product, its assembly and the manufacturing process to be designed – this is called concurrent manufacturing. It can be used to trial a design proposal, to see how materials behave, to test construction processes and to work out costs.

3D modelling and prototyping techniques can be used to test your design proposals before manufacturing your final product. If you are making a prototype that could be manufactured in batch or high volume, you may need to work out assembly processes, materials requirements and the order of assembly of the different component parts. Modelling by hand can involve the use of different materials and techniques, such as:

- using laminated card to model curved structures
- moulding clay or plasticene to build 3D models
- using block models made from polystyrene foam – either by carving a solid block or gluing together pre-shaped blocks
- developing frame models using wire, strips of wood, laminated card, drinking straws or even spaghetti!

As your design proposals develop, your ideas will become more refined until you reach the stage of finding the best possible solution – your final design proposal. If you are working on any kind of scale model (for theatre, film or a building, for example), or on the development of a prototype product, you may need to make further **sketch models** to explore different aspects of the design. These can help you plan:

- the most appropriate assembly processes
- how long different processes might take
- the materials, components, equipment and tools you need
- the order of assembly
- how easy the product will be to manufacture in the time available – do you need to simplify anything?
- any forward ordering of materials, so they are ready when you need them
- estimated costs of materials and manufacturing time
- where and how you will check the quality of your product

To be successful you will:
- Use high-level communication skills with appropriate use of ICT.

3. Demonstrate understanding of a range of materials/components/ systems, equipment, processes and commercial manufacturing requirements

You are expected to demonstrate an understanding of the materials, components and/or systems that are appropriate to the

manufacture of your product. Modelling and prototyping should enable you to do this because they enable you to test and trial materials and components. Selecting the most suitable materials is vital to the success of your project. They should be appropriate to your chosen scale of production. For example, special building materials may be required for the design of an interior. What are the properties of these materials and how can you replicate them in a scale model? What kind of materials are appropriate for a child's toy that is to be produced in high volume?

Selecting materials
Technologists have difficult decisions to make when selecting the right material for the job, especially when new materials are appearing all the time! Deciding which materials to use in a product design is not easy. Making good choices requires understanding of a wide range of materials. One way to make decisions about material properties is to use an appropriate property test. When choosing materials you need to take into account the following:

1. Materials availability
   - How rare or common is the material?
   - Is the material natural or manufactured? How simple, easy or safe is its manufacturing process?
   - Does the material come in standard forms and standard preferred sizes?

2. Moral and social issues
   - Is the material harmful to work with or use, even though it has some very useful properties?
   - Should you take the risk and use the material or not?
   - Does the production of the material exploit a labour market in any country?

3. Cost
   - How high are the processing and delivery costs of the material?
   - What about storage costs? Is it possible to use just-in-time ordering of materials to reduce storage costs?
   - How high are costs of waste materials? Can these be recycled?

4. Method of production
   - Are the properties of the material appropriate to the chosen method of production?
   - How many products need to be made?
   - Will the combination of materials and the production method produce a high-quality product?

5. Influences on the product styling
   - What is the product's purpose? What is it for? What should be its durability and life span?

- What is the product's function? How is it required to work? What does it need to do?
- What are the aesthetic requirements? How is the product required to look? Consider:
  - visual properties, such as colour, reflection, transparency and surface pattern
  - textural properties, how rough or smooth, surface detail and the finish used
  - form, such as the 3D shape and size.

The final choice of materials is often a compromise, which means that it partly satisfies all the product requirements, but does not completely satisfy them all! In some cases some requirements are more important than others such as cost being more important than function. It is only when all the information is collected that a decision can be made. There is rarely one correct answer or absolute best material. Usually, there are several materials that will do the job.

Task

Selecting materials

Copy out the chart below and use it to help choose the most appropriate materials for your product.

a) Model and prototype your product to trial possible materials and components.
b) Choose and use appropriate materials properties tests.
c) Check with your teacher or tutor that you are using the correct sample materials and that all testing is carried out safely.
d) Fill in the selector chart with actual or approximate values and as much detail as possible.

Materials properties selector chart	Material 1	Material 2
Hardness		
Strength		
Stiffness		
Toughness		
Ductility		
Electrical conductivity		
Thermal conductivity		
Durability		
Weight/mass		
Aesthetic properties		
Functional properties		
Environmental		
Moral/social		
Availability		
Cost		
Appropriateness for method of production		

Illustrating the final design proposal

You should clearly illustrate your final design proposal to show what your 2D and 3D outcomes will look like. You should also explain how the 2D outcome will be achieved and how the 3D outcome will be manufactured.

You should use an appropriate graphic style, which could be hand- or computer-generated. You will not lose marks if you choose to use hand techniques as opposed to using ICT. However, you could enhance your design and technology capability by using a range of computer techniques, such as word processing or CAD software (see Figure 2.9). You are not required to know how to use a specific language, computer or software package, but should use what is appropriate and available.

Factfile

Illustrating the final design proposal

Professional product designers aim to illustrate the final design proposal in the most convincing way possible. To do this they use a variety of techniques such as:

- 2D and 3D presentation drawings
- schematics
- detailed drawings of the design
- small-scale and large-scale models
- written reports.

The technique chosen to illustrate the final proposal needs to meet the client's budget, be suitable for the type of information to be communicated and for the character of the product. For example, the design for a pop-up book or its point-of-sale advertising could be presented in a simple, colourful style. A prototype perfume bottle, on the other hand, would require a more sophisticated presentation because it is aimed at a more exclusive target market group.

You should annotate your design proposal, using appropriate technical language. Identify the materials, components, systems and construction processes required to manufacture your prototype product. This may require you to produce front or back views of your product or to produce exploded views to explain any design detail. Your annotation should demonstrate an understanding of the working characteristics of your chosen materials, components, systems, equipment, processes and technology. This information should be related to your chosen scale of production. For example, if you design a

Figure 2.9 Front cover development using CAD – after drawing a final front cover, the student scanned the image and developed it further

one-off product for a client, it may require the use of special materials, tools, equipment or processes, whereas if you design a manufacturing prototype that could be mass produced, it may require the use of easily available materials, of processes that reduce waste or that need to be simplified for ease of manufacture.

To be successful you will:
• Demonstrate a clear understanding of a wide range of resources, equipment, processes and commercial manufacturing requirements.

4. Evaluate design proposals against specification criteria, testing for accuracy, quality, ease of manufacture and market potential

You should evaluate and test your final design proposals against your 2D and 3D design specifications. This should enable you to judge the **quality of design**, accuracy, ease of manufacture and how your proposals meet user requirements. This evaluation may involve getting feedback about market potential by consulting with a client or with potential users in your target market group. Your evaluation should enable you to justify why this is the best solution to the problem, by explaining, for example:

• how your design proposal meets the specification
• how it will meet quality requirements of users
• how easy it will be to manufacture in the time available
• its market potential in terms of costs, potential price range, aesthetics and function.

Think about this!
Understanding the impact of different scales of production
Your choice of scale of production will impact on every design and manufacturing decision you make. The information in Table 2.4 shows some of the different impacts of high- and low-volume production. Use this information to help you justify your choice of production.

Table 2.4 *Impacts of high- and low-volume production*

High-volume production	Low-volume production
Efficient production methods, so lower labour costs	Less efficient production methods, so higher labour costs
Easily repeated stages, so lower skill levels required	Higher levels of skill required for high-quality products
Efficient production through narrow range of process tasks performed by more people	Wide range of process tasks performed by fewer people
Use of standardised components to simplify production	Use of non-standard components for more individual products
Use of more specialised tools and equipment designed for specific processes	Specialised tools and equipment, used flexibly by skilled people for specific processes

To be successful you will:
- Objectively evaluate and test your design proposals against the specification criteria.

D Planning manufacture (10 marks)

1. Produce a clear production plan that details the manufacturing specification, quality and safety guidelines and realistic deadlines

When it comes to the planning stage, you are expected to produce and use a **production plan** for each of the 2D and 3D outcomes. You should aim to manufacture both outcomes within realistic deadlines. This means providing manufacturing details, taking into account quality and safety requirements. Realistic deadlines are those that are achievable. They should match the making of the product to the time available.

The work that you have already done on modelling, prototyping and testing should help you to plan your manufacture. Prototyping enables you to work out assembly processes, plan materials requirements and work out the order of assembly of the different component parts. Other planning decisions may include:

- considering how easy the product will be to manufacture in the time available, whether it is necessary to simplify anything or if any special materials or tools are required
- estimating costs of materials, production costs and a possible selling price.

Producing a production plan

Your production plan should provide clear and detailed instructions for making your prototype product. This should include:

- A manufacturing specification which includes, where appropriate, accurate working drawings with clear construction details, dimensions, sizes, **tolerances** and finishing details, colour tolerances in printing/reproduction processes, quantities and costs of materials and components. It is sometimes appropriate to use CAD software to produce accurate working drawings.

- A **work order** or **work schedule** which can be produced as a flow diagram or in table form and should identify:
  - the order of assembly of the different components
  - tools and equipment to be used
  - assembly processes
  - the estimated time each stage of manufacture will take
  - key stages of manufacture where quality is checked
  - safety requirements and/or procedures.

Figure 2.10 shows a time plan and a flow chart for a clock.

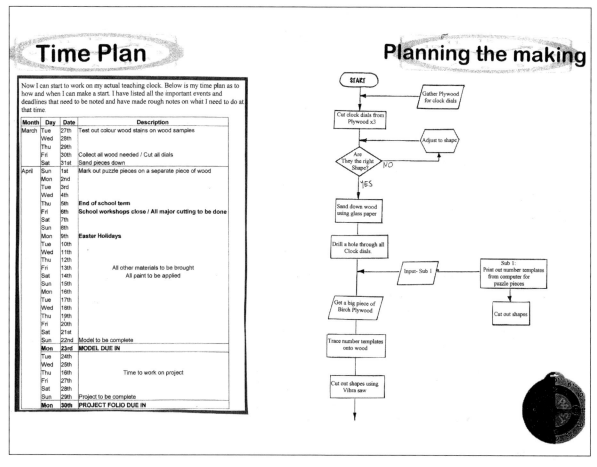

Figure 2.10 *Planning the production of a clock*

- Estimating product costs means producing an accurate price for the product, which would make it sellable *and* produce a profit. In industry, cost levels depend on the method of production, which must be simple and fast so that labour costs are as low as possible. You can cost your product in set stages:
  - work out **direct costs**, i.e. materials and labour costs (how long your product takes to manufacture at a set rate per hour)
  - work out **overhead costs**, i.e. rent, heat and electricity (these are often worked out as a set percentage of labour costs)
  - add together the direct and overhead costs, to give the total manufacturing cost
  - work out your manufacturing profit (a set percentage of the total manufacturing cost)
  - add together the total manufacturing cost and the manufacturing profit, to give the selling price
- Realistic deadlines are those that are achievable. Following a production plan will help you meet your production deadlines, but be prepared to

modify your planning if you make any changes to your product or assembly processes during manufacture. Any modifications should be noted in your production plan, so make sure that you leave enough space to do this.

Factfile
Planning manufacture in industry
Planning manufacture means choosing the best, safest and most cost-effective method of production, the best layout for equipment and people, the best materials and the best way to control product quality. A production plan is a key part of a company's quality control system because it documents each stage of manufacture. This enables checks to be made for quality, so each product is made to the same standard. A production plan also enables faults to be identified and provides feedback so that changes can be made to the production plan if necessary.

To be successful you will:
- Produce a clear and detailed production plan with achievable deadlines.

Think about this!
Planning manufacture
Use the following to help you plan manufacture:

- Ensure your materials are easy to work and handle.
- Check that the performance characteristics of your chosen materials meet the design specification.
- Forward order materials and components if necessary.
- Produce a manufacturing specification and a work schedule.
- Specify any safety requirements and procedures.
- Specify where and how you will check for quality.
- Make sure that you follow your production plan and adapt it if necessary, so you meet your deadlines.

2. Take account of time, resource management and scale of production when planning manufacture

Planning is an important part of any project and many of the activities that you undertake will overlap, because designing and manufacturing is a complex and interrelated activity. In addition, the number of weeks that you have available will depend on your timetable, for example you may have a set number of weeks where you concentrate totally on your coursework project or you may have less time during the week but more weeks overall. In other words, your coursework project time may be short and fat or long and thin!

Planning materials

One of the most important reasons for planning is to make sure that materials, components, tools and equipment are available when required and in the appropriate quantity and quality. Check out your requirements well in advance. As far as quality of materials is concerned, you should ensure that they meet your specification requirements. For example, if you are planning the manufacture of a one-off product for a client you may require the use of special materials or more time-consuming complex processes, so take this into account. However, if you are planning a manufacturing prototype you may need to consider:

- how the use of standard sizes or standard components might reduce manufacturing costs
- that simplifying assembly processes could ensure fast, cost-effective manufacture.

Keep your plan and review your progress as you go. You may need to modify your plan if delays occur. **Gantt charts** can be made using different coloured card and adhesive stickers, so they are adjustable. Alternatively, you could produce a Gantt chart using computer software (see Figure 2.11).

Tasks
1 Planning
Use the following questions to help you plan your production:

- Will my materials, components, tools and equipment be available when I need them?
- How will my scale of production affect my manufacturing processes?
- Can I use standard components, parts or materials to simplify my task?
- Will the quality and quantity of my materials and components match my manufacturing specification?

2 Production planning
Using a Gantt chart is a good way of planning a project because it gives you a picture of the whole project at a glance. It can be used to plan:

- a whole project over the total number of weeks
- for more detailed production planning
- for planning the manufacturing processes in a work schedule.

Produce a Gantt chart for your project:

a) Draw up a chart similar to the one in Figure 2.11, with the weekly dates across the top.
b) In the left-hand column, put in order a list of tasks to be done.
c) Note any tasks that can be done at the same time.
d) Plot the tasks against the time you have available.

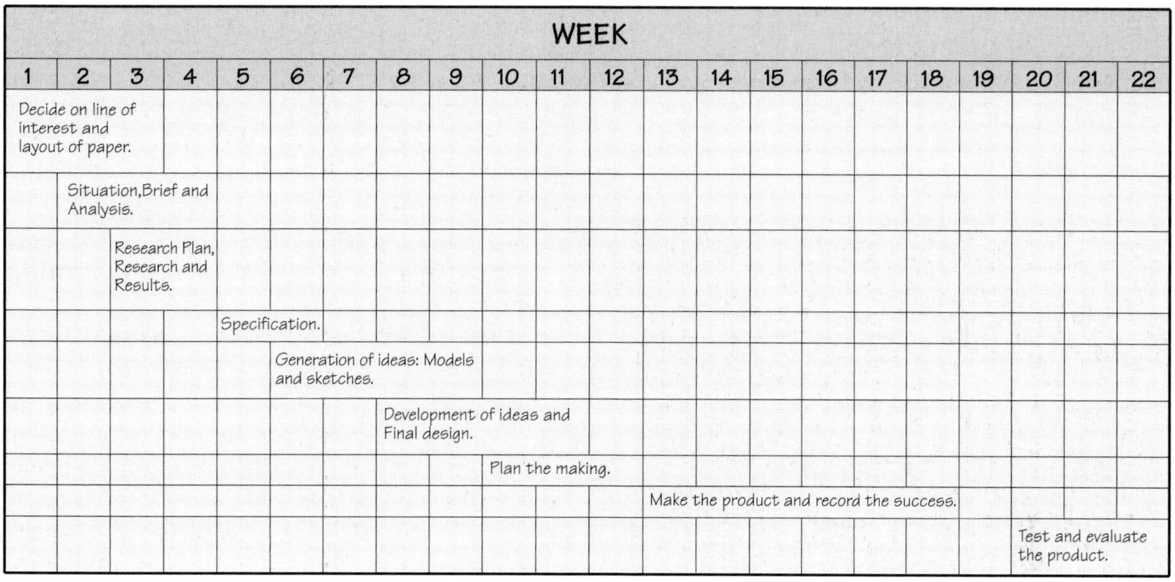

WEEK																					
1	2	3	4	5	6	7	8	9	10	11	12	13	14	15	16	17	18	19	20	21	22
Decide on line of interest and layout of paper.																					
	Situation, Brief and Analysis.																				
		Research Plan, Research and Results.																			
				Specification.																	
					Generation of ideas: Models and sketches.																
							Development of ideas and Final design.														
									Plan the making.												
												Make the product and record the success.									
																			Test and evaluate the product.		

Figure 2.11 *Part of a Gantt chart used to help plan a project*

Help! What if my production plan goes wrong?

The purpose of your production plan is to guide you through your product manufacture so that you use your time well.

- Sometimes you may come across delays, such as having to wait for materials to arrive or for the use of a piece of equipment.
- The first thing you must do is not to panic, but to modify your production plan slightly, so you take into account any lost time.
- If you use a Gantt chart for planning you will easily see your progress, so you can monitor things as you go along.
- If you are held up for any reason, make a note of it in your production plan and get on with something else!

3. Use ICT appropriately for planning and data handling

The aim of using information and communications technology (ICT) for planning and data handling is to enhance your design and technology capability. You will not be penalised for non-use of ICT, although you should use it where appropriate and available.

When planning and data handling, good use of ICT might include using word processing, databases, spreadsheets or CAD software for a range of activities that may include:

- organising and managing data
- production planning using colour-coded Gantt charts, diagrams or flowcharts
- producing manufacturing specifications
- producing accurate **working drawings**
- working out quantities and costs of resources and products.

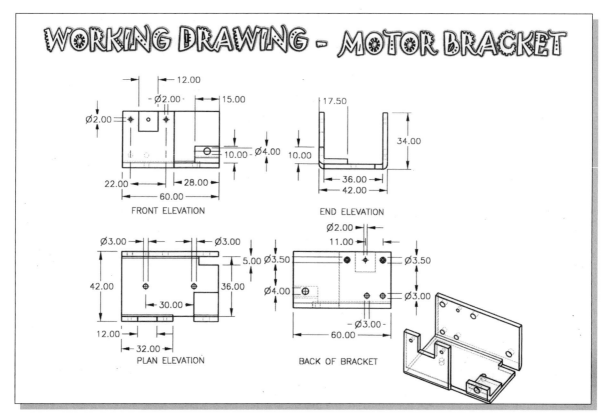

Figure 2.12 *A student working drawing for a motor bracket*

E Product manufacture (40 marks)

1. Demonstrate understanding of a range of materials, components and processes appropriate to the specification and scale of production

You are expected to demonstrate an understanding of the working characteristics of a range of materials and components that are required to manufacture your 2D and 3D outcomes. The 2D outcome should be developed from traditional and modern graphics media. The 3D prototype product should be constructed from modelling materials *and* resistant materials. These could include some wood, and/or metal and/or plastic.

Your chosen materials should achieve the performance requirements specified in your production plan. Your 3D prototype product must be of a high quality for you to attain high marks.

To a certain extent you have already demonstrated an understanding of the materials,

components and processes needed to manufacture your product as you:

- modelled, prototyped and tested materials, components and processes
- annotated your design proposals and explained the working characteristics of suitable materials, components and processes
- specified the materials, components and processes required to make your product.

You are now ready to further demonstrate your knowledge and understanding of materials and components through the actual manufacture of your product.

Scale of production

Your 2D and 3D outcomes may be a combination of one-off and high-volume levels of production. For example, if you made a 3D scale model of a robot for use in a film, you would be designing and making a one-off product to meet the needs of a client. The model may require working to a defined scale or the use of electronic components

Table 2.5 *Comparing the key stages of manufacture for prototype and mass-produced products*

Stages of manufacture	Prototype product	Mass-produced products
Materials preparation	Standard sizes and sections, standard components Calculate materials and components requirements	Standard sizes and bought-in components Calculate materials and components requirements to minimise waste
Processing – cutting and shaping the parts	Cut and shape by hand or by simple machine tools Cast metal components in sand mould Vacuum form plastic shapes	Cut and shape by automatic and CNC machine tools Injection mould plastic shapes
Assembly	Reduce the number of parts to be joined Nails, wood screws, nuts and bolts Woodwork joints, hand welding cold adhesives	Reduce the number of parts for fast assembly Self-tapping screws, 'snap' fitting parts Butt joints, automatic welding Hot glues which bond as they cool
Finishing	Paint/varnish applied by hand with a brush, spray can or spray gun	Paint/varnish applied by dipping or with spray gun by industrial robot Plastic or vitreous enamel coating

to create 'special effects'. Any printed material produced to support the marketing of the film may need to be designed for reproduction in a large batch.

Alternatively, you could design and make a 3D prototype robot toy that could be manufactured in high volume. You would still make this 3D prototype as a one-off, but it would require a different approach throughout its development. Its target market may be children (and adults who buy for children) – and its manufacture would require different materials or processes to those of the one-off film robot (see Table 2.5). Any point-of-sale advertising that you might produce to support the robot toy may also have to be printed in a large batch.

Although most of your practical work will involve making a one-off product, there may be times when two or more identical component parts will have to be made. Whatever scale of production you work to, your prototype will still need to be made and finished to the highest quality. There will also be other complications such as constraints related to the materials, tools and equipment available to you. If problems occur with the availability of resources, for example, you may have to change your original choice of materials or adapt the processes you use. If this happens, do not forget to record any changes in your production plan and justify any new choices of materials, components or processes.

Task

Approaches to manufacturing

A different approach is required when manufacturing a one-off product for a single client or manufacturing a one-off prototype product for high volume.

For one of the following products draw up a table to show the key stages of manufacture and the materials, components and processes required to manufacture the product as a one-off for a single client and as a one-off manufacturing prototype:

- toy with moving parts
- torch.

Think about different materials and process requirements, related to aesthetics, function, ergonomics, user and performance requirements, time constraints, cost.

To be successful you will:

- Demonstrate clear understanding of a wide range of materials, components and processes.

2. Demonstrate imagination and flair in the use of materials, components and processes

Think about this!

Practise the techniques and processes you aim to use during manufacture, so you can demonstrate your ability.

- experiment with working, shaping and joining materials
- use mock-ups to trial structures
- experiment with processes to improve aesthetic qualities such as materials finish.

Be prepared to modify your manufacturing processes if necessary or to adapt details of the design, for example changing the method of joining. Keep all your experimental work in a small box, so you can refer to it if necessary. You can evidence your experimental work by including a photograph in your folder.

One clear way that you can demonstrate imagination and flair is in the way that you handle materials and processes. An understanding of how materials behave and how processes work will enable you to show your skills and ability. This will result in the production of a quality product – one that:

- is attractive to the market or client
- is well made from suitable materials
- is enjoyable or fun to use
- would sell at an attractive price
- is manufactured for safe use and disposal, without harm to the environment.

This is quite a long list of considerations, but if you check you should find that you have taken most of them into account in your design and manufacturing specifications. At this stage, prior to manufacture, you have the opportunity to hone your skills – to experiment with materials, components, techniques and processes, so you can demonstrate your ability through the manufacture of your product.

To be successful you will:

- Demonstrate imagination and flair.

3. Demonstrate high-level making skills, precision and attention to detail in the manufacture of high-quality products

Demonstrating high-level skills involves making the best use of available materials and components, in relation to your design proposals. It also involves using tools and equipment with accuracy, confidence and skill. If you practise your existing skills before manufacturing your product, you should gain an understanding of your ability in relation to your expectations for your product. If your ability falls below your requirement, you have two options, either improve your skills or adapt the process. Improving your skills will result in improving the quality of your work, so you produce a high-quality product.

The making of high-quality products also depends on planning quality into your design and manufacturing process. Refer to your design and manufacturing specifications and to your production plan, where you should find references to quality. Your work order should identify the key stages of manufacture, where you can monitor the accuracy of your work as it progresses, checking against the dimensions and tolerances you detailed in your working drawings.

Figure 2.13 shows how one student demonstrated high-level making skills.

Bike Design

Painting

All the parts were given two coats of Halfords white primer and left for three days to harden. The front triangle was masked out for the 'swoosh'. This was then given three coats of Halfords Nissan Morello pearlescent red. Once dry, the masking tape was removed and the red was masked off ready for the blue coat. The blue used is Halfords Ford Java metallic blue. Once again, the masking tape was removed when the paint was dry and the lining tape was then put on. This was hard to do, as the curves are tight. When applying it, I used a hair dryer to make it bend easily. The frame was then given two coats of clear lacquer. The rear triangle was sprayed with Halfords Aluminium spray paint and when dry given two coats of lacquer. All the decals were then applied. Most of the decals were hand-made using carbon tape and a white vinyl tape. The web address was scanned into the compter and then printed onto acetate and glued to the model.

Figure 2.13 Student work demonstrating high-level making skills

4. Use ICT appropriately for communicating, modelling, control and manufacture

You can use ICT to help your product manufacture, where it is appropriate and available, but you will not be penalised for its non-use. You are not expected to know how to use specific equipment or programs, but you should understand the benefits of using ICT for manufacture. Different uses of ICT include:

- communicating information between CAD software and computer numerically controlled (CNC) equipment
- using software to model 'virtual' products on screen before manufacture, saving time and costs because it reduces the need to make expensive manufacturing prototypes
- using CNC machines to control processes accurately and quickly
- communicating manufacturing information between the design office in one location and the manufacturing site in another.

If you do not have easy access to computer-aided manufacturing (CAM) equipment, you could use a printer or a plotter to print out technical drawings, or to cut out a component parts drawing and use it as a template for making identical parts.

If you do have access to specialised CAM equipment you can use CAD to produce design ideas and then export the digital information to CNC equipment for producing accurate component parts for your product.

Factfile
Using ICT in manufacture
The increasing use of ICT through the use of CAD/CAM systems has had an enormous impact on manufacture. CAD/CAM enables the efficient design and manufacture of products and the control of manufacturing equipment. CAM automates production, repeats processes easily and precisely and enables the production of cost-effective products.

5. Demonstrate a high level of safety awareness in the working environment and beyond

Safety in manufacturing means the safe design, manufacture, use and disposal of products. Manufacturers must follow safety procedures and check standards, regulations and legislation related to product design.

This following of safety procedures ensures that products are safe for the consumer, the producer and the environment. Legal requirements, such as the Health and Safety at Work Act 1974 and Reporting of Accidents 1986, ensure that safe production processes are followed to prevent industrial accidents.

Safe production means identifying all possible risk and documenting safety procedures to manage and monitor the risk.

Ask the following questions at key stages of your design and manufacture:

- What could go wrong?
- What could cause things to go wrong?
- What effect would this have?
- How can I prevent things from going wrong?

You need to demonstrate a similar awareness of safety at all stages of design and manufacture, by making safety a priority in your work.

At the research and design stage, you should take account of designing with safety in mind, both for you, the maker, for your intended user(s) and for the environment. This may involve researching safety regulations related to your product. Safety features should be identified in your design specification.

Your work order should identify specific safety features related to manufacture, including safety guidelines for your chosen materials and the tools, equipment and processes you may use. Modelling and prototyping before production will enable you to test for safety against the criteria that you have identified.

During the manufacturing process you should follow safety guidelines related to safety with people, with materials, with equipment and machinery.

F Testing and evaluating (10 marks)

1. Monitor the effectiveness of the work plan in achieving a quality outcome

Production planning is a key tool in monitoring your manufacture because it enables you to monitor the quality of your outcome. You should record any changes you make to your 2D or 3D outcomes, or to any processes you use during manufacture (see Table 2.6).

You may not need to make any changes, but if you do, however minor, they should still be recorded because they could have an impact on the quality of your product. Recording any changes will also enable you to make an identical product to the same standard.

Sometimes completely unforeseen problems can arise through, for example, using a process or technique that is new to you. Other reasons for making changes could be through not having the right materials, components, tools or equipment available when you want them, or because you are running out of time.

If you do have to make any changes, make sure that you explain what you have done and why – write it down straight away before you forget, or do a quick sketch to explain a change in design or in the construction. Recording any changes to your product will make it easier to evaluate its quality of design and manufacture.

> *To be successful you will:*
> * Make effective use of your work plan to achieve a high-quality outcome.

Manufacturing process	Changes to process	Changes to product	Quality checks
			Check tolerances
			Check dimensions
			Check against specification
			Check finish

Table 2.6 *Record any changes you make during manufacture*

2. Devise quality assurance procedures to monitor development and production

Think about this!
The meaning of quality
In industry, quality means:

* conforming to the specification
* ensuring fitness-for-purpose
* making products with zero defects
* making products right first time, every time
* ensuring customer satisfaction
* exceeding customer expectations.

Use the following questions to help your quality planning:

* Do you aim for fault-free work?
* Do you know what standards are expected?
* Do you check the quality of your work against the specifications?
* Does your work meet the specifications?
* Are you pleased with your work?
* Could you do it any better?

Quality planning is a key process during product development and manufacture and you should devise your own quality control procedures to monitor quality. There is a saying that 'you can't manufacture quality into a product'. In other words, quality must be designed into it.

Check that you meet the quality criteria outlined in your design specification. Use flowcharts or ICT to detail quality control checks against both your 2D and 3D outcomes at the critical stages of manufacture, as detailed in your work order. Quality checks could include using sensory tests of vision and touch, examining the outcome by sight and hand for accuracy and consistency, and checking your working tolerances against those in the manufacturing specification.

> *To be successful you will:*
> * Devise clear quality assurance procedures.

3. Use testing to ensure fitness-for-purpose

Testing to ensure fitness-for-purpose means testing the performance of the outcome to make sure that it meets the requirements of the specification and the user(s). Any testing and evaluation should be holistic in relation to your 2D and 3D outcomes – after all, the two elements are linked, with one supporting the other. For example:

- A prototype product, supported by point-of-sale advertising, would both be linked to the context for design, which may be corporate image. An evaluation of the one element without the other would be incomplete.
- Testing a board game and its packaging during manufacture may involve testing size tolerances, the clarity of reproduction techniques, the construction of the playing pieces, the quality of finish and the stability of the outcome.

You should record the results of any testing that you do during and after manufacture:

- Test the performance of the product against the design and manufacturing specifications.
- Test that the quality of the product is suitable for users – check against user requirements in the specifications.

> *To be successful you will:*
> - Make effective use of testing to ensure fitness-for-purpose.

4. Objectively evaluate the outcome against specifications and suggest appropriate improvements

You should objectively evaluate and justify the success of your 2D and 3D outcomes in relation to the design brief and specifications. Being objective means taking an unbiased view of the outcome. This can sometimes be difficult, as you will have been closely involved in the design, development and manufacture. It is easier to be objective if you use standards against which to judge your product – these standards are the design brief, the design and manufacturing specifications and the quality criteria identified in the work order.

You should produce a written evaluation regarding the success of your outcome against user/client needs, relating to how well the product looks and meets requirements. This should include:

- comments on the performance against specifications, accuracy, design and production quality
- views of intended user(s) through questionnaires, surveys or user trials.

An objective evaluation should provide you with feedback on the success of your product, which will help you decide how and if it can be improved (see Figure 2.14). Your suggestions for improvement should be based around the product's aesthetic and technical success, its quality of design and manufacture and its fitness-for-purpose.

> *To be successful you will:*
> - Objectively evaluate the outcome and suggest appropriate improvements.

Bike frame design – meeting the specification

Size: The model is an exact 1:1 scale model.

Function: When left at 'The Bike Place', a few people expressed interest in buying the frame. The promotion requirements of the company were met with the brochure.

Safety: Paint used was lead free and non-toxic.

Cost: The project cost more than the £50 set out in the specification, but the end result was well worth it.

Material: The model was constructed out of 12 mm and 6 mm MDF as set out in the specification.

Appearance: The model of the frame included the name of the company, the model name and logo.

Shape: The model was an exact 1:1 scale model.

Reliability: The model has a tough lacquered finish, so it's easy to clean and looks good for a long time. It is a good advert for the company.

Finish: The frame was hand sprayed to match its exclusive image.

Possible improvements: I could have had the brochure printed on photographic paper, instead of having it laminated. I could have made the point of sale display out of sheet metal riveted together for more strength, to last longer and to stand out from the other displays. The model could have had more sanding time and been sprayed with a gun and compressor to get a better finish.

Figure 2.14 It is important to evaluate your design against the specifications

Student checklist

1. Project management

- Take responsibility for planning, organising, managing and evaluating your own project.
- Ensure that you have photographic evidence of technical details that evidence your product's quality of manufacture – ideally you need to show the processes you use at each stage of manufacture.
- Ensure that your coursework project folder contains only the work related to the assessment requirements.

2. How to make your AS coursework project a success

- Identify a realistic need and solve a problem for your specified user(s).
- Include folder content that represents a conclusion to your research information.
- Show the influence of research on your design decisions.
- Ensure your project is a manageable size so you can develop your design and manufacturing skills.
- Make good use of your coursework project folder to clearly show how your ideas unfold.
- Demonstrate a variety of communication skills, including appropriate use of ICT.
- Show understanding of industrial practices in your designing and manufacturing activities.
- Make good use of modelling, prototyping and testing.
- Demonstrate high-level making skills, using a variety of materials and processes.
- Allow time to evaluate your work as it progresses and modify it if necessary.
- Plan your time effectively so that you can meet deadlines.

3. Evidencing industrial practices in coursework

- Use **industrial terminology** and technical terms.
- Include a range of design activities that are similar to those used in industry, i.e. developing a design brief, using market research, modelling and prototyping, etc.
- Include a range of manufacturing activities that are similar to those used in industry, i.e. using a production plan, planning quality control, testing against specifications, etc.

4. Using ICT in coursework

- Use ICT, where appropriate, to enhance your design and technology capability.

- Develop the use of ICT for research, design, modelling, communicating and testing.
- Develop the use of ICT when planning, data handling, controlling and manufacturing.

5. Producing a bibliography

- Reference all secondary sources of information in a bibliography. Include all references for information found in textbooks, newspapers, magazines, electronic media, CD-ROMs, the Internet, etc.
- Include references for scanned, photocopied or digitised images. Do not expect to use clip art at this level.
- Do not expect to be given credit for any work copied directly from textbooks or other media, including the Internet.

6. Submitting your coursework project folder

- Have your coursework ready for submission by mid-May in the year of your examination.
- Include a title page with the Specification name and number, module number, candidate name and number, centre name and number, title of project and date.
- Include a contents page and numbering system to help organise your coursework folder.
- Ensure that your work is clear and easy to understand, with titles for each section.

7. Using the Coursework Assessment Booklet (CAB)

- Complete the student summary in the CAB. This should include your design brief and a short description of your coursework project.
- Ensure that the CAB contains a minimum of three clear photographs of the whole product, with alternative views and details of processes where appropriate.
- Your candidate name and number, centre name and number and the module number must be written by the product photographs and appear on back of the photographs.

Help! What if my project goes wrong?

- If your Unit 2 coursework project doesn't meet your expectations, don't worry! You can retake the unit and the better result will count towards your final grade.
- If you find yourself in this situation your teacher or tutor will be able to advise you on the best way forward.

3A Materials, components and systems (G301)

Summary of expectations

1. What to expect

Unit 3 of the exam paper is divided into two sections:

- Section A: Materials, components and systems.
- Section B: consists of three options, of which you will study only one.

2. How will it be assessed?

The work that you do in this unit will be externally assessed through Section A of the Unit 3 paper.

Section A consists of general knowledge questions from within the overall unit content. There are six compulsory questions, each worth 5 marks. You are advised to spend 45 minutes on this section of the paper.

3. What will be assessed?

You must be able to demonstrate clearly that you have an understanding of the topic or process in each question. The examiner will be looking for the following:

- short, concise answers
- often a brief description
- an explanation or a sketch.

4. Content

The following list summarises the topics covered in Section A. You are expected to be familiar with all the concepts and work covered under these headings.

- Classification of materials and components:
  - paper, card, softwoods, hardwoods and manufactured boards
  - metals and alloys
  - thermoplastic and thermosetting polymers
  - composites, synthetics, laminates and manufactured materials
  - ceramics, glass and rubber.
- Working properties of materials and components related to preparing, processing, manipulating and combining.
- Hand and commercial processes.
- Testing materials.

You should apply your knowledge and understanding of materials, components and systems to your project work in Unit 2.

5. How much is it worth?

This unit, with the option, is worth a total of 30 per cent of your AS qualification. If you go on to complete the whole course, then this unit accounts for 15 per cent of the full Advanced GCE.

Unit 3 + option	Weighting
AS level	30%
A2 level (full GCE)	15%

1. Classification of materials and components

Paper and card

Paper and card

Paper and card are produced primarily from hard and soft woods. Softwood fibres are longer, offering greater strength. Hardwood fibres are shorter, offering a smoother, opaque finish. The wood is processed by one of three basic methods to form a pulp – either mechanical, chemical or waste.

Mechanical pulp

The logs of coniferous wood are saturated with water and debarked. They are then ground down, with the resulting pulp screened to accept 1–2 mm pieces with the larger pieces being recirculated. The resulting pulp can only be used for low-grade paper or packaging material; so the pulp is bleached with peroxide or sodium hydroxide. This method offers a high yield but contains greater impurities.

Chemical pulp

After debarking, the hardwood and softwood logs are cut into 2 cm chips along the grain. These are pounded into fragments and screened. The resulting pulp is stored and treated with acid or alkaline. The fibre yield is lower but the fibres are longer, stronger and contain fewer impurities.

In semi-chemical pulp, the wood is processed as above, but the chips are treated with steam which results in longer fibres. This offers a high yield and when the fibres are bleached they become shorter and similar to chemical pulp.

Waste pulp

Recycled paper and card used for waste pulp is often used for lower grades of paper as its strength, durability and colour are not as good as virgin fibres. Waste pulp can be mixed with virgin fibres to produce better quality paper.

Manufacturers blend a variety of pulps and process them with bonding agents and pigments to produce paper with different qualities. These processes help to achieve a consistent colour and bind fibres to create a better surface. A sizing agent can also be added to improve water resistance and prevent ink from feathering on the surface.

The manufacture of paper

Paper is produced on a **Fourdrinier machine**. The pulp is diluted to 99.5 per cent water and held in a head box. A continuous stream of pulp is sent through an adjustable slit on to a moving, woven mesh belt which is vibrated to drain off the water and allow the fibres to interweave. The pulp is then pressed with a mesh on top to drain the water quickly and is passed under a dandy roll to smooth out the fibres, before it travels over suction boxes and rollers which draw out more water. At this stage, the paper contains around 80 per cent water, allowing it to be unsupported by the mesh belt. The paper is then passed through a series of rollers and heated cylinders to produce a paper with 4–6 per cent moisture content. This is the desired content to allow a balance with the relative humidity of the atmosphere.

During the drying process, sizing agents such as starch, resin and alum can be added by spray or press. After this, highly polished rollers called calenders can be used to give a smooth, gloss finish.

The paper is wound on to a roll from which it can be placed into a precision cutting machine to produce the desired size, or directly loaded on to the printing press.

Paper is defined in weight as gsm (grams per square metre), with 80 gsm being the weight of average printing paper. Cartridge paper weighs 90–220 gsm; boards weigh 220 gsm and over. Paper is sized in A sizes, metric, imperial as well as custom measurements. It can also be defined by its:

- durability
- colour and brightness
- texture
- **opacity**.

Paper can be classified into four basic types:

- uncoated paper – newspaper, copier paper
- paperboard (weighing over 220 gsm) – ivory and display boards
- coated paper – billblade, matt and gloss art paper
- specialist – embossed, wrapping, and marbled paper.

The production of card and board follows the same process. It can be formed using three layers of pulp with the centre layer being lower-grade pulp. Card can also be formed by laminating paper to produce the required properties.

Task

Collect a variety of different types of paper.
For each:

a) define what pulp may have been used to make it

b) classify it into one of the four main areas

c) compare its durability, texture and opacity to that of standard 80 gsm photocopier paper.

Hardwoods, softwoods and manufactured boards

Characteristics and faults of wood

The many different species of tree provide an enormous wealth of materials, which are put to many uses. However, any naturally occurring material inevitably produces variable quality, and wood is no exception to this. As a material it can be cut and shaped in numerous ways and this is one of its major advantages. Regrettably, the irregularity of grain, knots, warping and twisting are all disadvantages. Wood is also prone to biological attack from insects and fungi.

To a certain extent some of these irregularities have been minimised by rapid growth in the area of manufactured boards, which have reduced the demand for prime quality timber, but they should not be regarded as a cheap substitute. They present in many ways their own problems, and cost and strength comparisons are sometimes misleading. Table 3.1.1 considers the advantages and disadvantages of manufactured boards.

Table 3.1.1 Advantages and disadvantages of manufactured boards

Advantages	Disadvantages
Large standard-sized sheets	Difficult to join
Uniform thickness	Exposed edges often need to be treated
Stable in most atmospheric conditions	Thin sheets become easily distorted unless held by a frame
Grained boards have good strength to weight ratio	Adhesives can blunt cutting tools quickly
Thin sheets can be bent easily	

Faults and defects in wood affect the overall strength and durability as well as mar the visual appearance. Defects can be caused by a variety of factors:

- Shrinkage. After **conversion** and **seasoning**, shrinkage affects the shape of the cut wood. Movement cannot be entirely eliminated since any change in temperature or humidity will result in some change. Movement exists in three main forms:
  - warping: a **cupping** across the width of the board
  - bowing: along the length of the timber
  - twisting: a twist from side to side along the length of the timber.

- Splits. Logs will develop a radial split if they are allowed to dry out before seasoning. Splits in seasoned timber occur because it has been dried out too quickly through the end grain. In a timber mill the end grain of trees and sawn timber is often painted to stop this rapid drying-out process.
- Knots. These are natural irregularities formed between the junctions of branches. They inevitably weaken the structure.
- Irregular grain. Knots also contribute to variations in the grain direction which can lead to serious weaknesses especially where short grain occurs. It also makes for difficult working of that section.
- Shakes. Separations in adjoining layers of wood are known as shakes. They include:
  - heart and star: caused by shrinkage
  - cut and ring: strains of the wind, felling or seasoning
  - thunder: thinner hair-line cracks formed perpendicular to the grain, commonly found in African timbers.
- Fungal attack. This type of attack causes the wood to decay which results in total loss of strength and weight.
- Dry rot. This thrives in damp conditions where there is no air circulation. The wood is attacked by fine strands, which reduce the wood to a dry powder, and leave a musty smell.
- Wet rot. External timbers subjected to both wet and dry conditions decompose and become alternately spongy when wet and brittle when dry.

Characteristics and working properties of timber

The characteristics and working properties of timbers can be broken down into categories. Knowledge in each of these areas will help you to identify timbers in your work. They are:

- weight
- texture
- durability
- colour
- odour
- ease of working.

A tree essentially consists of two major parts, the inner or 'heartwood' which gives rise to strength and rigidity, and the outer layers or 'sapwood' which is the region of growth where food is stored and transmitted. Growth is a seasonal process where layers are seen as concentric growth rings, known more commonly as annual rings. As the tree grows the wood tissue grows in the form of long tube-like cells which vary in shape and size. These are known as

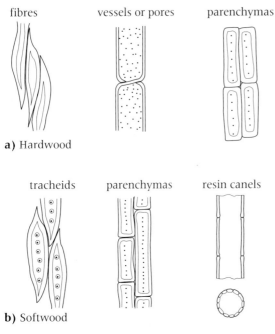

a) Hardwood

b) Softwood

Figure 3.1.1 *Hardwood and softwood cells*

fibres and are arranged roughly parallel along the length of the trunk and give rise to the general grain direction. This variation in cell size, shape and function leads to the botanical distinction of hardwoods and softwoods (see Figure 3.1.1):

- Softwood structure, e.g. Scots pine, red cedar, spruce:
  - Tracheids: these are elongated tubes which become spliced together in the direction of growth and make up the grain. Sap and food pass through smaller openings known as pits that harden as the tree grows older.
  - Parenchymas: smaller than tracheids, these make up the remaining cells.
  - Resin canals: evident in the majority of conifers. They carry away waste products in the form of resin and gum.

- Hardwood structure, e.g. oak, mahogany, beech:
  - Fibres: these constitute the bulk of hardwoods. They are not in any regular pattern or formation and they are much smaller and more needle-like than tracheids in softwoods.
  - Vessels/pores: they form long tubes within a tree which carry food. They are used to distinguish one type of hardwood from another.
  - Parenchymas: these are more prominent in hardwoods and in oak they become quite thick, up to 30 cells thick, and are seen as the familiar 'silver' flashes within the grain.

Sources of different types of wood

The prime source of the world's supply of commercially grown softwood is the northern hemisphere, but particularly the colder regions of North America, Scandinavia, Siberia and parts of Europe. Conifers are relatively fast growing and produce straight trunks which make for economic cultivation with little wastage. With careful management of forests it is possible to control the supply and demand of softwoods. Being relatively cheaper than hardwoods they are used extensively for building construction and joinery. What waste that is produced is used in the manufacture of fibre boards and paper.

The UK imports almost 90 per cent of its timber needs since it is one of the least wooded countries in Europe. The imported boards are usually supplied debarked or square edged ready for further processing at the saw mills.

There are thousands of species of hardwoods grown across the world and many are harvested for commercial use. Most broad-leafed trees grown in temperate climates such as Europe, Japan and New Zealand are deciduous and lose their leaves in winter with the exception of a few like holly and laurel. Those grown in tropical and sub-tropical regions like Central and South America, Africa and Asia are mainly evergreen which means they grow all year round and reach maturity quicker. Hardwoods generally are more durable than softwoods and offer much more variety in terms of colour, texture and **figure**. Since they take a relatively longer time to grow they tend to be more expensive than softwoods with the really exotic timbers being converted into veneers which allows for much greater use of a limited supply.

Conversion

Once a tree has reached full maturity it is felled (cut down). This process is normally carried out during the winter months when growth is dormant and less sap and moisture are in the tree. All branches are removed which makes transportation and storage easier. With timber grown in the UK the bark is normally left on until the logs reach the saw mill as this stops rapid drying out occurring and prevents some of the defects as a result. Conversely, imported timber is often stripped of bark to avoid any risk of insect contamination. Conversion is the term given to the process of sawing logs into commercially viable timber.

Ultimately, the figure and stability of the sawn timber is determined by the plane of the saw in relation to the annual rings. However, this is reflected in the price paid for the timber since some methods of conversion are more wasteful than others. There are basically two methods of

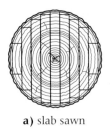 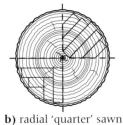

a) slab sawn   **b)** radial 'quarter' sawn

Figure 3.1.2 *Slab and radial sawn logs*

cutting used in the conversion process (see Figure 3.1.2):

- slab, plain or through and through
- quarter (radial) sawn.

Slab, plain or through and through conversion is the simplest, quickest and cheapest of the two methods. The process makes a series of parallel cuts through the length of the log resulting in parallel slices or slabs. The thickness of the slabs can be varied as the log is cut and this type of cutting is frequently used on softwoods where the logs tend not to be that large in diameter.

Quarter (radial) sawn is a much more time-consuming process and involves much more manual handling. It is also a much more wasteful process. However, the timber produced tends to be better in quality and is much more stable in that it is less likely to move, warp, bow or twist.

Essentially, quarter sawing tries to make the annual rings as short as possible and at 90 degrees to the cut surface. This type of cutting results in the grain's figure being exposed and this is quite noticeable in oak where the silver grain is exposed.

Seasoning

Drying or seasoning wood is the process of removing the excess water and much of the bound moisture from the cell walls. As the wood dries water is lost from the cavities until only the cell walls contain moisture. This is known as the fibre saturation point and it occurs at about 30 per cent moisture content. On further drying moisture is lost from the cell walls and shrinkage starts. At some point the loss of water stops and the wood is in balance with the relative humidity of its surroundings and this point is called the equilibrium moisture content (EMC). It is important, however, to reduce the moisture content to less than 20 per cent since this has a number of implications:

- It makes the timber immune to rot and decay.
- It makes the timber less corrosive to metals.
- It increases overall strength and dimensional stability.

Not all moisture should be removed from the timber. Depending on the use to which the timber is to be put there are recommended levels of moisture content as shown in Table 3.1.2.

Table 3.1.2 *Recommended levels of moisture content*

Moisture content	Uses
Below 18%	General outdoor use such as fences and sheds
Below 10%	Indoor use in centrally heated homes: stairs, door frames and skirting boards

Seasoning can be carried out in two different ways: natural-air seasoning and kiln seasoning. Natural-air seasoning, as its name suggests, is where slabs of timber are stacked and natural air is allowed to flow around them. It takes an average of one year to season 25 mm of thickness of wood in this way. Kiln seasoning uses steam to heat and remove the excess moisture in a controlled environment. Very precise levels of moisture content can be achieved using this method.

Manufacture of strong and lightweight forms

In order to manufacture plywood and blockboard sheet veneers (thin sheets of wood) are produced by two methods:

- Rotary cutting – a log softened by steam is mounted between two centres and rotated slowly against a blade. This produces a continuous sheet of veneer.
- Slice cutting – a steam-softened log is mounted on to a flat, moveable machine bed. Moving it against a blade produces sheets of veneer.

Plywood

Layers (or plies) of soft or quick growing hardwood are glued together so that the grain is at right angles to each other. Plywood always consists of an odd number of plies so that the grain on the two outer layers runs in the same direction – this ensures that stresses are balanced. These combined layers are then put into a press to dry.

Plywood can be graded for interior and exterior use. This is dependent on the water resistance of the adhesive used. A range of thicknesses are produced, with the thinner ply being flexible enough to bend into a curved shape. The amount of plies used to make a sheet will determine the strength.

Blockboard

Blockboard is produced by **laminating** blocks of softwood between two plies. The core blocks are arranged in odd numbers, with the heart side of the wood alternating from top to bottom to prevent warping. The outer plies run in the same direction and at 90 degrees to the grain of the core blocks.

Both plywood and blockboard are commonly produced in large sheets of 2440 x 1220 mm. They are manufactured to produce a sheet material with great strength in all directions. Their strength derives from the crossing of grains, in layers, during the production process. Both materials can be faced with a decorative hardwood veneer or plastic laminate.

> ## Task
> What type of products could be manufactured from:
> **a)** plywood
> **b)** blockboard?

Metals and alloys

The major proportion of all naturally occurring elements are metals and they form about one quarter of the Earth's crust by weight. Aluminium is the most common (8 per cent), followed closely by iron (5 per cent). With the exception of gold, all metals are found in the form of oxides and sulphates. The ores have no pattern of distribution around the world but some countries have larger deposits than others. Metals are divided into three basic categories:

1. Ferrous – the group which contains mainly ferrite or iron. It also includes those with small additions of other substances – mild steel, cast iron. Almost all are magnetic.
2. Non-ferrous – the group which contains no iron – copper, aluminium and lead.
3. Alloys – metals that are formed by mixing two or more metals and, on occasions, other elements to improve properties. They are grouped into ferrous and non-ferrous alloys.

Metals usually have one or two loose electrons in their outer electron shell and therefore they are quite likely to become easily detached. The **metal crystals** have a regular arrangement held together by electrostatic attraction. It is this movement of electrons which accounts for metals' high electrical and thermal **conductivity**. This mobility also leads to a degree of plasticity in metals in the form of ductility and malleability. Once a bond is broken another is formed.

With the exception of mercury, all metals are solid at room temperature. In their molten form they are held together only by weak forces of attraction which mean they lack cohesion and will flow. As the metal solidifies, the energy is reduced within each atom, giving out heat, and the atoms arrange themselves according to a regular pattern or **lattice structure**. Their overall properties are affected by this lattice structure. Most metals crystallise into one of three basic types of lattice, as shown in Figure 3.1.3:

- close-packed hexagonal (CPH)
- face-centred cubic (FCC)
- body-centred cubic (BCC).

Iron is a very important metal since it changes from BCC to FCC at 910°C. Above 1400°C it changes back to BCC again. In the FCC form it absorbs carbon which is essential in the process of steel making. When cooling the changes occur in reverse.

A pure metal solidifies at a fixed known temperature with the formation of crystals, in either a cube or hexagonal structure. On further

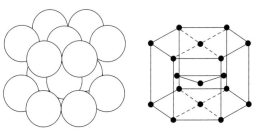

a) close-packed hexagonal (CPH)
zinc, magnesium
(weak, poor strength to weight ratio)

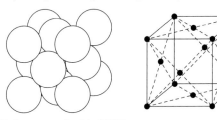

b) face-centred cubic (FCC)
aluminium, copper, gold, silver, lead
(very ductile, good electrical conductors)

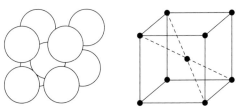

c) body-centred cube (BCC)
chromium, tungsten
(hard, tough)

***Figure 3.1.3** Different metallic structures*

cooling, the crystals continue to grow as **dendrites** until each one touches its neighbour. At this point grains are formed and boundaries become visible when viewed under a microscope.

Table 3.1.3 on page 72 summarises the working properties and uses of common metals.

In order to obtain the metals in any useful form, they have to be extracted from the ore before processing can take place. Mining or quarrying removes the ore from the ground, whereupon it is crushed to remove much of the unwanted earth, clay or rocks. The metal, now in a concentrated form, is roasted which causes the ore to change chemically into an oxide of the metal. The remaining stages of reduction break the chemical bond between the metal and the oxygen in the ore to leave a pure metal ready for further processing.

The production of ferrous metals

During the reduction process of iron, the ore coke, in the form of carbon and limestone, is fed into the blast furnace and heated to 1600°C. The limestone is used to extract the impurities from the ore to form a molten slag which floats on the iron and is tapped off separately.

The iron at this stage is called pig iron and it is still too impure for general use and needs further refining. In its molten state the pig iron, which contains about 3–4 per cent of carbon, is transferred into a further converter furnace which is more able to control the carbon content and any further impurities.

To make steel, the carbon content needs to be reduced significantly and this is normally done in the basic oxygen furnace where large volumes can be handled. Essentially, oxygen is blown into the liquid to combine with the carbon and other impurities. Lime is added which causes the impurities to float on the surface as slag.

After careful checking and analysis of the composition of the melt, it is either poured into ingots, poured directly into castings or directly into a continuous casting machine. From larger ingots or billets, the steel is removed to the mill and reheated before being rolled to form rods, flats, square tubes or channel sections.

The result of hot rolling gives a black oxide film and this type of steel is called black mild steel. Bright drawn mild steel (BDMS) is oiled and re-rolled cold before being drawn through dies to create the accurate sizes.

Alloy steels are formed to create metals with enhanced properties. Mild steel is generally alloyed with such metals as chromium, tungsten and nickel. Whereas basic carbon steels lose their hardness at higher temperatures, high speed steels retain their hardness and cutting edge even at red heat. Resistance to corrosion can also be increased, as in stainless steel, which contains 12 per cent chromium and 8 per cent nickel.

The production of non-ferrous metals

Aluminium is only available commercially in the form of bauxite, a hydrated form. It is very difficult to break down and therefore a process of electrolysis is needed. This is very expensive in terms of electrical energy. Firstly, the bauxite is crushed, mixed with caustic soda and then heated under pressure, whereupon it melts. Once filtered and washed it is roasted to produce alumina before it passes through the electrolysis stage. The proportion of aluminium is alloyed to improve its strength and hardness. Copper ores contain about only 4 per cent copper and they undergo similar processing to that of the production of aluminium to extract the pure metal. Crushing, floating and the addition of lime all help in the removal of impurities before the final stages of electrolytic refining.

Task

Define the materials that are used in the manufacture of the blade for a carving knife, and a cold chisel. Explain how the properties and composition of each material contribute to the function of the product.

Thermoplastic and thermosetting polymers

Thermoplastic and thermosetting plastics cover a wide and diverse range of substances that exist in both a natural and synthetic form. Natural resources such as cellulose from plants, latex from trees and shellac, a type of polish extracted from insects, play only a small part in the plastics industry. Synthetic resources, especially crude oil, supply the majority of the raw material for the production of plastics. This single resource of hydrogen and carbon accounts for the majority of plastics.

The refining of crude oil in a fractioning tower is the process that gives rise to the product hydro-carbon naphtha which is subsequently cracked into fragments using heat and pressure to form ethylene and propylene. In naturally occurring compounds, the molecules, consisting of only a few atoms, are short and compact. In plastics, the molecules do not stay as single units but link up with other molecules to form large chains of giant molecules. This process is called **polymerisation**.

Table 3.1.3 The working properties and uses of common metals and alloys

Material	Melting point °C	Composition	Properties	Uses
Cast iron	1000–1200	Pig iron, scrap steel, various additions dependent upon use	White cast iron – very hard and brittle Grey cast iron – easily casts and good corrosion resistance	Heavy crushing machinery Bench vices
Steels	1400	Alloys of carbon and iron	Dependent upon carbon content and other elements	
Low carbon steel		Less than 0.15% carbon	Soft, ductile, malleable	Wire, rivets and cold pressings
Mild steel		0.15–0.3% carbon	Ductile and tough Cannot be hardened and tempered	General construction steel, car bodies, nuts and bolts
Medium carbon steel		0.3–0.7% carbon	Harder than mild steel, less ductile	Springs, axles and shafts
High carbon steel		0.7–1.4% carbon	Hardness can be improved by heat treatment	Hammers, cutting tools and files
Alloy steels Stainless steel		Medium carbon steel + 12% chromium + 8% nickel	Corrosion resistant	Kitchen sinks, cutlery
High speed steel		Medium carbon steel + tungsten, chromium and vanadium	Retains hardness at high temperatures Brittle but can be hardened and tempered	Lathe tools, drills and milling cutters
Aluminium	660	Pure metal	Malleable and ductile Very conductive of heat and electricity	Aircraft, boats, window frames and castings
Duralumin		Aluminium + 4% copper + 1% magnesium	Work hardens, ductile and machines well	Aircraft parts
LM4 casting alloy		Aluminium + 3% copper + 5% silicon	Increased fluidity and improved hardness Good corrosion resistance	General purpose casting alloy
Copper	1083	Pure metal	Malleable and ductile Excellent conductor of heat and electricity	Wire, central heating pipes and car radiators
Brass	927	65% copper + 35% zinc	Corrosion resistant Casts well Good conductor of heat/electricity	Casting, ornaments and marine fittings
Bronze	900–1000	90–95% copper + 5–10% tin Sometimes includes phosphor	Harder and tougher than brass Hard wearing Corrosion resistant	Castings, statues and bearings
Lead	327	Pure metal	Soft and malleable Corrosion resistant Easy to work Immune to attack from chemicals	Protection against radiation from X-rays Roof coverings and flashing
Tin	232	Pure metal	Soft Corrosion resistant	Tinplate and soft solders
Zinc	420	Pure metal	Ductile and easily worked A layer of oxide prevents it from further corrosion	Coating for steel (galvanising), rust-proof paints, die casting

Plastics are subdivided into two main groups and one minor group, with the formation of the chains the key feature that separates them:

1. *Thermoplastics.* These plastics are made up from long chains of molecules with very few cross-linkages. The smaller cross-links are known as **monomers** and the polymer chains are held together by a mutual attraction known as Van der Waals forces (see Figure 3.1.4a). This physical attraction is weakened by the introduction of heat. As the molecules move, they become untangled and the material becomes pliable and easier to mould and form. When the heat is removed the chains reposition and the material becomes stiff once again. Thermoplastics have a plastic memory, which means they have the ability to return to their former state after heating provided that no damage or chemical decomposition has happened during the heating process. Polythene, polystyrene and polypropylene are all examples of thermoplastics. Polythene is extensively used in the production of toys and carrier bags; polypropylene is used for containers with built-in hinges and chair shells where its good resistance to work fatigue is exploited.

2. *Thermosetting plastics.* Thermosets set with heat and thereafter they have little plasticity. During the polymerisation process the molecules link both side to side and end to end. This cross-linking process, known as covalent bonding, makes for a very rigid material, and once the structure has formed it cannot be reheated and changed (see Figure 3.1.4b). Polyester resin is used for paper weights. Urea formaldehyde is a stiff, hard, strong plastic and it is used for electrical fittings.

3. *Elastomers.* This third group of plastics falls between the two basic groups. A limited number of cross-links allows some movement between chains. Rubber is a type of elastomer and it is used to make tyres for cars.

Glass reinforced plastic (GRP), often referred to as fibreglass, consists of strands of glass that are set in a rigid polyester resin. The strands are woven into matting which is available commercially in different weights. The polyester resin, albeit a thermosetting plastic, exists in a liquid form that has a catalyst or hardener added to it along with a coloured pigment for decoration purposes. The glass fibre strands provide the basic strength while the resin with its additives bonds the fibres together and provides a very smooth surface finish.

In order to achieve a high standard of finish from any GRP work, a high-quality mould must first be produced. The external or 'finished' side of the work must be finished to a very high standard since any defect or imperfection in the mould will be replicated in the finished piece.

The mould can be made from virtually any material but medium density fibreboard (MDF) and hardboard are often used with the additional use of wire and plaster of Paris for complicated shapes. If a porous surface has been used it is essential to seal the surface prior to use and a proprietary mould sealer should be used. It is common to make a full-size model of the finished piece which is then used to produce the GRP mould. In order to be able to remove the work from the mould it should be made with tapered sides and it should have no undercuts. A release agent is also essential and the mould should be coated several times before lay up proceeds. Good mould design should see no sharp corners and large flat areas should be avoided.

The stages in laminating follow a structured process. These are the basic stages:

- Polish mould with the releasing agent.
- Prepare matting into appropriate sizes.
- Mix gel coat with pigment, hardener and catalyst.
- Apply gel coat to an even thickness of 1 mm.
- Wait about 30 minutes before stippling matting over mould making sure it is wet through.
- Build up layers to the appropriate thickness.
- Full curing takes approximately 24 hours before separation can take place.

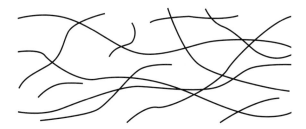

Figure 3.1.4a *Van der Waals bonding – low density polythene*

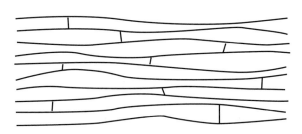

Figure 3.1.4b *Covalent bonding – polyester resin*

GRP is used in a vast range of products where its great strength to weight ratio can be fully utilised. This means that much stronger shapes and products can be built that weigh much less than when being produced by other means. It is also very resistant to corrosion. GRP is used to make sailing boats and canoes. Some high-speed train front nose cones are also made from GRP.

Injection moulding is the most widespread and versatile process used for producing moulded plastic products, from bowls to television casings. Although the cost of a typical mould is high, the unit cost per component produced becomes very small for high volumes making it an ideal process for mass-produced components. The process is best suited to thermoplastics, but a few thermosetting plastics are used depending on the conditions the product is to be used in.

The process is simple. The material is heated to a plastic state and injected into an enclosed mould under pressure. The mould is opened and the product is removed with the use of ejector pins. Injection moulding is a highly automated process that produces high-quality products that require no further finishing other than to remove any sprue pins, gates and runners which are chopped off and reused. The machine itself consists of a hopper into which are fed the plastic granules, a heater and a rotating screw mechanism. The screw mechanism acts as a ram that injects the plasticised material into the mould before it is allowed to cool (see Figure 3.1.5).

Blow moulding is the process used to form hollow products and components. A hollow length of plastic called a parison is formed by extrusion and is lowered down between an open split mould. The mould is then closed to seal up

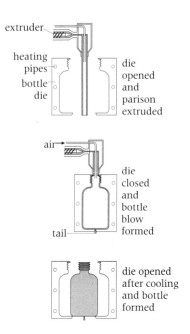

Figure 3.1.6 *Blow moulding*

the free end of the parison and compressed air is blown into the mould forcing the plastic to the sides of the mould cavity where it is chilled and sets (see Figure 3.1.6). Blow moulding is a highly automated process that produces little waste and requires only the flashing to be trimmed. An estimated 1.5 billion PET (polyethelene teraphthalate) plastic bottles are made and thrown away each year in the UK. Like injection moulding, the initial mould cost is high, as is the machinery, but with components being produced in the volume of PET bottles, it is easy to see how the unit cost is very low.

Task
State the impact on the environment that the production of PET bottles has – from production of plastic to disposal of bottles.

Vacuum forming is used to produce simple shapes from thermoplastic sheets. It is possible to vacuum form acrylic. However, the ideal degree of plasticity in acrylic is not reached until a temperature of 180°C. At 195°C acrylic starts to degenerate, therefore making it difficult to achieve a uniform heat across the whole sheet. However, 'Perspex TX' is an extruded form of acrylic and this becomes plastic at 150°C making it more suitable for use in industry when products as large as baths can be formed. More commonly used materials include high-density polystyrene, ABS (acrylonitrile-butadiene-styrene) and a flexible grade of PVC (Polyvinyl chloride).

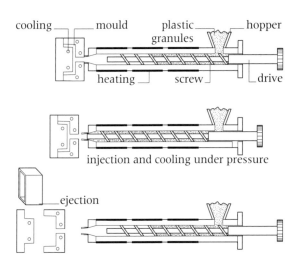

Figure 3.1.5 *Injection moulding*

Vacuum forming requires a mould of the finished component to be produced first and this must be to a high standard with the sides tapered slightly (to between five and ten degrees) to ease the removal of the formed component once completed. The process works by removing the air trapped between the mould and the sheet, thus reducing pressure below the trapped material. Atmospheric pressure pushes the heated plasticised sheet on to the mould. The plastic sheet to be vacuum formed should be clamped around its edges in an airtight plate with a rubber seal. Heat is applied by elements normally housed in a hood that is held above the material.

When a material has reached its plasticised state the heaters are removed and the pump is used to expel the air below. There is, however, one basic problem with vacuum forming and that is that on deforming the material can become quite thin. One technique used in industry to overcome this problem is that when the material reaches its plasticised state, it is blown to uniformly stretch the whole surface before the mould is raised on the table and the air expelled.

The whole process can be summarised into four basic stages:

1. The plastic is heated using radiant heaters.
2. The plasticised material is then blown to stretch it.
3. The table or platen is raised into the dome area.
4. The air is expelled causing atmospheric pressure to force the plastic material down on to the mould (see Figure 3.1.7.).

Vacuum-formed products range from acrylic baths to the plastic packaging found around Easter eggs.

Fusion

Plastics can be fused together to form strong joints. Sheet plastics can be fusion welded using ultrasound, where tiny vibrations cause a rise in temperature at a localised point, and with the application of pressure the two separate pieces become fused. It is also possible to fusion weld rigid sections together by heating them at the joint with a hot-air torch and adding a filler rod of the same material. The process can be used where adhesives cannot because of the risk of chemical attack which would render the joint useless. Recently, many novelty products such as inflatable post-cards and clocks have been manufactured using this fusion welding process.

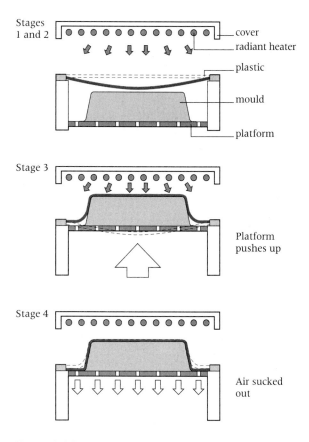

Figure 3.1.7 Vacuum forming

Task

Assess critically the methods used in a variety of plastics-forming processes, using the questions below to guide you. If possible, watch a video that shows industrial and commercial processes. Also try handling some products made by these forming processes.

- How is the force applied?
- How is the heat applied?
- How easily is the process controlled?
- How complex is the mould
- Is any finishing necessary?

Composites, synthetics, laminates and manufactured materials

When two or more materials are combined by bonding, a composite material is formed. The resulting material has improved mechanical and other properties and, as with most composites, it will have excellent strength:weight ratios.

Composites consist of a reinforcing material that provides the strength and a bonding agent, termed the 'matrix', in the form of glues or resins.

Wood is a natural composite in that its fibres, which provide the strength, are held in a matrix of lignin, the glue.

Initially, glass fibre was the most widespread form of reinforcing material and was available in a variety of forms: a string of fibre woven together, loose strands, or, most commonly, in a non-woven matting of short strands. Glass fibre is held in a matrix of polyester resin and it is best suited to large structural items such as boat hulls, septic tanks and pond liners.

Think about this!

Consider the following environmental issues:

- Polyester resins contain styrene and can cause skin irritation.
- Catalysts (hardeners) are usually peroxides and these too can cause skin irritation and they will cause damage if they come into contact with the eyes.
- Toxic gases in the form of 'styrene' are given off and can cause respiratory problems in the throat and nose.
- Fibres themselves can cause irritation to the skin.
- When cutting or finishing glass fibre the dust generated can cause respiratory problems in the nose and throat.
- Styrene vapours present fire hazards.

'Formica' is a layered composite that uses melamine formaldehyde resin to bond layers of paper together, the outside one being a decorative pattern or colour. A clear hard heat protective resin is added over the top layer and the whole sandwich is heated under pressure where the resin cures and strong cross-links are formed.

'Tufnol' is another type of layered composite that consists of woven linen impregnated with a phenolic resin. It is extensively used for gears, bearings and slides in machines.

More recently, carbon fibre has been developed in a similar form to that of glass fibre. Carbon fibres are, however, much stronger and are used in structural components for aircraft, propellers, protective clothing, body armour and sports equipment such as golf clubs, skis and tennis and squash rackets. The extensive replacement of fibre glass with carbon fibre in the aircraft industry has led to major weight reductions of between 15 per cent and 30 per cent which has resulted in better fuel economy. Carbon fibres are available in various forms but most frequently they are laid up using resin to produce strong lightweight structures.

One of the most widespread and commonly used composite materials is medium density fibreboard (MDF). The fibres are made from wood waste that has been reduced to its basic fibrous element and reformed to produce a homogeneous material. Fibres are bonded together with a synthetic resin adhesive to produce the uniform structure and fine textured surface. There are various types of MDF board, which have a less dense central core but still retain the fine surface.

This type of fibreboard can be worked like wood and with a veneered surface it makes an excellent substitute. It finishes well with a variety of surface treatments and it is available from 3 mm to 32 mm thick and in sheets 2440 mm by 1220 mm wide.

As is the case with all composites there are some dangers involved in the use of them. As a result of the very fine fibres and synthetic resin adhesives great care must be taken when undertaking any form of cutting, drilling or sanding. Respiratory equipment should be used since the dust can cause irritation of the skin, throat and nasal passage and in school/college a dust extractor must also be used.

Tasks

1 Find out what COSHH stands for.
2 Assess the risks involved in the use of MDF and GRP.
3 How are those risks controlled?
4 List the safety precautions that should be taken when using MDF and GRP.

Ceramics, glass and rubber

Ceramics

Most ceramics are brittle and hard since they are chemically bonded by either covalent or ionic bonding. They are far less ductile than most metals and easily fractured at the slightest attempt to deform them. As a general group of materials, the following are all considered to be ceramics:

- glass products
- bricks
- roof tiles
- sanitary ware (toilets and basins).

Ceramics are also able to withstand high temperatures, pressures and have good resistance to chemical corrosion. Modern ceramics are capable of operating at much higher temperatures and are now being used as replacements for alloy steels as components in combustion engines. They are also used on the nose cones of US space shuttles to insulate and protect the crew and their instruments on re-entry to the Earth's atmosphere where surface temperatures on the nose cone can reach 1500°C.

Ceramics are manufactured from three main materials: clay, sand and feldspar (aluminium potassium silicate). To this combination fluxing agents are added to lower the temperature needed in their manufacture and refractory compounds are added to increase resistance to temperature. Once all the components have been finely ground to a powder they are mixed into shape and fired to a temperature between 700°C and 2000°C.

Glass

Glass is made from silica sand, lime and sodium carbonate. Other materials can be added in order to produce various types of glass:

- Coloured glass – oxides of transition metals such as iron, copper and nickel are added.
- Lead glass – lead oxide is added to produce higher values of refractive index, used for cut glass and neon signs.
- Borosilicate glass – contains a high proportion of boron oxide and this type of glass is sold under the trade name of 'Pyrex'.

The raw materials for glass are fed into the furnace where they are heated to 1500°C. Once molten, they are tapped out in a continuous flow and floated on to a bath of molten tin that has a perfectly smooth surface. As the glass comes from the production line it is cut to size and this type of glass is known as plate glass. Glass has many uses because of its transparency and resistance to chemical attack. It is brittle and is much stronger in compression than in tension. In the construction industry, buildings can be made with glass walls that allow light in but reflect light so that people cannot see through. The other type of glass used in the construction industry is safety glass and there are two types:

- Laminated glass consists of two thin sheets of plate glass with a sheet of non-brittle plastic material trapped between them. They are bonded with an adhesive and the plastic centre holds fragments together if the glass is broken.

- Tempered glass has been **annealed** to give a low stress uniform structure. It is heated to 400°C and it becomes very tough and much stronger in compression. It is used in doors and vehicle windows.

Rubber

Rubbers are defined as materials which show 'elastic' properties. The materials possess long chain molecules known as 'polymers', hence the widely used definition of elastomer. Rubber and elastomers have the ability to be stretched repeatedly and return to their original length immediately.

Natural rubber is produced by the Hereabrasiliensis tree as an emulsion of cis-poly-isoprene and water (latex). It is extracted by cutting the tree and is then dried to form a clear crepe rubber.

Synthetic rubber is produced by reacting suitable monomers to form polymers. They are obtained as a water emulsion or a suspension in water or solvents. These raw polymers are then developed by compounding them with other materials to enhance their elastomeric properties:

- Fillers such as carbon black and ground silica provide reinforcement.
- Oils, waxes and fatty acids improve processability.
- Pigments are used for colouration.

Permutations and combinations allow elastomers to be produced for a specific function. Rubber compounding takes place in either open rubber mills or internal mixers. By the use of a shearing action and the varying of temperature the raw rubber is broken down to allow the compound to be mixed in. The final mixture is then removed in the form of a sheet.

After compounding most rubbers need to be vulcanised or cured. These processes create cross-links in the molecular structure and produce the physical properties required so giving the finished rubber chemical and thermal stability.

To form a product the raw material is moulded. There are three main ways to mould rubber:

- Compression – the rubber is shaped in a mould using heat and pressure.
- Transfer – the rubber is forced under pressure through a series of feed gates into the cavity.
- Direct injection – the rubber is fed into a screw extruder where it is pre-plasticised before being forced into the cavity.

Classification of components

Pencils

Graphite pencils

Wood-encased graphite pencils and retractable graphite pencils (also known as clutch pencils) are available in a variety of grades, ranging from 8B which is very soft to 9H which is very hard. The letter B denotes the blackness of the lead while the letter H denotes the hardness. It is essential that the correct pencil is used for the task in hand in conjunction with the correct paper.

Softer pencils, either used alone or in various combinations, are more suited to rough or grainy paper to create different effects when sketching or illustrating. Harder pencils are commonly used on smoother surfaces such as layout paper for technical drawing. HB or 2B pencils are the most suitable choice for rapid freehand sketching.

Coloured pencils

Coloured pencils are commonly made from a mixture of chemical pigment and kaolin, which is a type of clay. Professional quality, coloured pencils are available in an extensive range of colours which vary in hue, tone and intensity. They are not usually graded in the same way as graphite pencils but are often quite hard so that they can be sharpened to a fine point for use in detailed work. However, different types of coloured pencils can be used to create a variety of effects. Water-soluble pencils used in conjunction with a watercolour brush create colour blends and washes while chalk-based, pastel pencils create the same effects as pastels and chalks but minimise the mess.

Paints

Paints commonly used in paper-based graphic work are listed below and are all water-based media.

Water colour

These paints are supplied in paste or tablet form and when diluted have a transparent quality. When applied they should be worked from light to dark on good quality, watercolour paper to prevent distortion and wrinkling. This type of paper is passed through water, left to dry and then stretched to prepare it for use with watercolour paint.

Gouache

Gouache (also known as designer's gouache) is made by adding coloured pigment to a precipitated chalk which is bound together with gum arabic to create a paste form. Individual colours can be mixed together to create different shades and the consistency of gouache can be altered with the addition of water. Most gouaches are opaque, but some range from almost opaque to transparent. The properties of gouache also vary in opacity, permanence and staining. These variations are marked on the tube using individual manufacturer's codes to denote the properties of their product. When dry, gouache can leave a relatively unstable, powder surface which remains water soluble.

Acrylic

Acrylic paints are made from synthetic resin similar to that used in the manufacture of perspex and polythene. Acrylics come in paste form and most can be thinned using water to change the consistency. Some acrylic paints are unsuitable to mix with water and require a special dilutant; these paints will be marked accordingly on the container. When thinly diluted, acrylic paint can produce a similar transparency of colour to watercolour paints. The benefit of using an acrylic paint is that it produces a rich intense colour that stands up well to wear, producing a tough, long lasting finish. It dries quickly forming a plastic coating that is water resistant.

Inks, pens and marker pens

Inks

There are two main classifications of inks:

- *Waterproof* – these inks are permanent and are oil- or spirit-based.
- *Non-waterproof* – these inks are soluble and water-based.

Inks are available in a range of colours but they are not as extensive as paints. However, the basic colours can be mixed in a similar way to paints to produce a wider range of colours.

Pens

An extensive range of pens is available on the market, which are designed to fulfil a variety of functions. The majority of permanent or soluble ink pens are gravity fed. The design of the nib and the method of ink flow are the main features that distinguish these pens from one another.

Technical pens are predominantly used for precision and accuracy in technical drawing as they provide a consistent ink flow for detailed line work. The nib of the pen is made from steel tubing and ink is fed down it from a refillable cartridge. The size of the tube determines the width of the line and the interchangeable nibs range in size from 0.13 mm to 2.0 mm.

Ballpoint pens distribute ink via a steel or tungsten carbide ball that rolls within a plastic or metal skirt. The quality of the line will fluctuate depending on the type of ink-flow system used, for example some roller ball pens use hundreds of tiny beads behind the nib to maintain an even flow of ink

Plastic tip pens are used to produce accurate line widths, including very fine lines. A network of fine channels in the plastic nib is used to draw ink to the surface of the tip using a capillary action.

Fibre tip pens have a nylon or vinyl nib, which can vary from a firm tip for precise or finely detailed work to a suppler tip for sketching. The nib is usually constructed from synthetic fibres bonded in resin to make them more durable. The quality of fibre tip pens ranges from a standard fibre tip, colouring marker to examples similar to technical pens that use a tubular nib and are available in a similar range of nib sizes.

Fountain pens are traditional writing implements that have been used for many years; the design is derived from the quill. Fountain pens are predominantly used for writing, calligraphy and sketch work. The nib works by using a capillary action to draw the ink between a thin channel that looks like a hairline split in the middle of the nib. A reservoir of ink behind the nib maintains an even and constant flow. However, fountain pens are not always the most practical pen as they can release large deposits of ink if agitated.

Marker pens

The range of marker pens available is extensive. They can be divided into three categories:

1 Ink – the ink used in marker pens can be divided into two groups:
   - water-based ink which is non-permanent and soluble
   - spirit-based ink which is permanent and waterproof.

   The drying times for these inks will vary but certain markers, usually spirit-based, will state if they are quick drying.

2 Composition of the nib – the most common materials used for the nib are:
   - felt
   - fibre
   - nylon
   - foam.

   Felt nibs are easily damaged with heavy use whereas the other materials are designed to be more resilient.

3 Size and shape of the nib. These vary from fine to broad nibs and can create a number of effects and line thicknesses:
   - round
   - square
   - bullet
   - chisel
   - fine
   - brush.

In most marker pens the ink is soaked into a fibre core, which is in contact with the nib allowing the ink to soak up from the core to the nib. The most widely used marker pens for visual media and presentation work are known as studio or graphic markers. These may be referred to by their brand names such as Magic or Pantone markers.

Unlike alternative media such as watercolours or gouache, markers can be used instantly. They require no mixing or preparation and a wide array of colours is available. In addition to providing a comprehensive selection of hues and tones, another reason for this extensive variety of shades is to create matches with other materials like paper, overlays and printing inks. This allows consistency in colour at every design stage from initial sketches to the final product. Ranges of cold and warm greys are also produced to enable monotone visuals to be produced.

Studio markers are produced in spirit-based and water-based forms. Spirit-based markers are colour-fast but tend to bleed and cockle paper if not used on marker paper. They give a flat, even colour allowing the colour to be applied in layers one on top of another. Water-based markers do not tend to bleed as much but in comparison to spirit-based markers they do not offer such flat and even coverage. In addition, water-based markers do not tolerate being overworked with layers of colour as this can result in a patchy appearance. As a result, spirit-based marker pens are the most commonly used.

Nib shapes for studio marker pens vary but the chisel tip is the most widely used as it offers both a broad line and a fine line when used on its tip. Many manufacturers offer fine tip markers for detailed work in the same colours as chisel tip markers, which has resulted in twin tip markers being produced with a fine tip at one end and a chisel nib at the other.

Task

What are the advantages and disadvantages of:
a) marker pens
b) fountain pens
c) acrylic paint?

Acetate film, corriflute and papier-mâché

Acetate, the collective name for cellulose triacetate and diacetate, was produced as a non-flammable material for the production of motion picture film. It is manufactured by reacting cellulose with acetic anhydride using a catalyst of sulphuric acid. The low flammability of acetate allows it to be used for a variety of purposes. It is still frequently used for motion picture films and overlay cells in animation. It is also commonly used for overhead transparencies and the preparation of images for printing.

Corriflute is a sheet material made from polypropylene thermoplastic. It is a strong, lightweight material derived from sandwiching corrugated flutes between two sheets. It is widely used in sign making as it can be screen printed on to. It is an inexpensive material that is available in a variety of colours.

Papier-mâché is a quick and inexpensive material for producing 3D shapes. It is produced by saturating paper or tissue with a water-based adhesive, such as wallpaper paste. Depending on the effect required, papier-mâché can be used in a pulp or in strips. When forming shapes a mould should be used. Papier-mâché has no real structural properties when it is wet so a support will be needed until the water has evaporated. The resulting adhesive-bonded paper form will be susceptible to moisture. Therefore, the surface will have to be treated with a protective finish such as PVA glue.

Projected images using light

Overhead transparencies

Overhead transparencies (OHTs) can be prepared and used in a number of ways. The two most commonly used methods to prepare OHTs are photocopying and printing via a computer printer. This can be done in black or colour but before doing so you need to check that:

- the photocopier or printer is suitable for use with acetate sheets
- the correct acetate is used for the task in hand.

Permanent and water-based markers designed specifically for use with OHTs can be used to mark the acetates. This enables the user to add information to the OHT during a presentation to build up information or assist in explaining processes.

Overhead and screen projectors

Primarily, overhead projectors (OHPs) are used as a presentation device and have become a standard way to display visual information to a large group of people. OHPs are used in conjunction with acetates. The transparencies are placed on the horizontal glass surface of the projector and the light from a bulb underneath projects the image on to a vertical screen using a magnifying lens and a mirror to enlarge the image. Adjusting the height of the magnifying glass and the distance from the projector to the screen will alter the size and sharpness of the image or text.

Artists and designers have also used the OHP as a tool to enlarge images for their artwork. The projected image becomes a guide or template for painting or sketching. This is a useful technique to scale up images.

Slide projectors use colour or black and white 35 mm photographic transparencies. These are individually framed in card or plastic and can be prepared by a photographic retailer directly from film; 35 mm film negatives can be used with specially designed clip frames but they tend to be inferior in quality.

The projector passes light through the slide and through a lens that can be adjusted to focus the image on a vertical screen. The distance from the screen to the projector also determines the size and clarity of the projected image.

Slides can be placed in the projector one at a time but most projectors usually work on a carousel allowing a number of slides to be placed in sequence and changed when needed during a presentation. The carousel can usually be detached, allowing another carousel with slides to be added in sequence. In practical terms the carousel projector is compact and portable. Once loaded in sequence the person presenting the work can switch from one slide to the next manually or in some cases with remote control. The projector can also be pre-set on a time-lapse system programming an allotted time between each slide being shown. Many designers and artists use 35 mm slides as a reference system or a database of their 2D and 3D work. This offers two advantages:

- Slides can be used as a basis to produce photographs or for use in publications.
- Slides can be easily and cost effectively sent to clients.

Rubber stamps

Rubber stamps are primarily used for symbols, instructions, certification and messages that are used repeatedly but not printed or marked on an object or surface. These stamps allow a permanent mark to be made quickly and easily, thus saving time on repeated tasks.

Rubber stamps can be made to order directly from a specialist stamp maker or through a print shop or stationer. Any artwork and text can be produced to any size required. However, rubber stamps can become clogged and over-inked, affecting the clarity of the text and image when it is stamped on to the required surface. Ready-made rubber stamps bearing commonly used instructions or adjustable stamps that feature the date are widely available.

Self-adhesive letters

Self-adhesive letters are made from plasticised PVC, which is commonly referred to as vinyl. These letters are available in a wide range of colours and sizes and in both matt and gloss finishes. The nature of the material gives the letters a soft and flexible quality that allows them to be adhered to a variety of contours.

The adhesive used for vinyl lettering is contact or impact adhesive which has been lightly coated on the back of the letters and covered with a treated paper to protect them and make them easy to peel off. Self-adhesive letters are primarily produced for signs and information graphics but they are also used in the production of models. Therefore, the material has to be durable and the adhesive must be suitable to stick to a number of surfaces.

In addition to vinyl lettering, self-adhesive images and shapes can also be produced. The CAD software used in their production enables a variety of font styles and scanned images to be cut from sheets of vinyl on a machine similar to a plotter with a cutting blade fitted (see Figure 3.1.8). Using this technique, complete words and sentences can be produced together, thus avoiding mistakes and the time-consuming task of placing letters individually.

Dry transfer lettering

The major manufacturers of dry transfer lettering such as Letraset and Mecanorma produce over 500 typefaces in 25 point sizes and in numerous colours. They also produce lettering in a variety of alphabets such as Arabic and Greek as well as borders and technical symbols, etc.

The lettering is on sheets with an entire alphabet featuring capital and lower-case letter forms and a selection of numbers, punctuation marks and related symbols such as the percentage symbol (%) and the ampersand (&). Each letter is repeated several times depending upon the frequency with which it is used, i.e. there are more vowels than less commonly used consonants such as x and z.

The lettering is produced by printing an ink image on to a transparent plastic film that is then coated with an adhesive. When the letter is placed in the required position on the work, pressure is applied to the upper surface of the sheet using a spatula, pencil or burnisher until the letter has stuck firmly to the surface. It appears grey rather than black when it has been released from the sheet. Corrections can be made by removing the unwanted letters with masking tape. As each letter is applied separately, using this product can be time consuming. It also requires the user to space the letters accurately so some sheets incorporate a spacing system. Dry transfer lettering can be used on models as well as in presentation work.

Model-making kit parts

Kit parts are manufactured for use in scaled architectural and interior models. Therefore, they are manufactured to the common scales used in the production of models (1:100, 1:75, 1:50, etc.).

The kit parts save time in the model-making process by allowing the model maker to purchase standardised parts or intricate models that may be required in multiples. They also give the visual scale of the model and can be divided into:

- internal – staircases; furniture
- external – foliage; street furniture; vehicles
- figures – in various ages, positions and occupations.

They can be manufactured from either plastic or metal and finished in full colour or single colour.

Photographic mounts

Mounting photographs and artwork is done primarily to improve its visual appearance and enhance the overall presentation of the work. It also protects the work, particularly if it is likely to be handled frequently.

There are a variety of ways in which photographs and artwork can be mounted. In fact, the final stages of presentation for visual work are an integral part of the design process as

Figure 3.1.8 *A plotter-cutter*

they contribute to the overall aesthetic of the work. The type of material used and the techniques employed for mounting the work may detract from the visual impression of the work as well as enhance it.

Artwork can be mounted individually or in a group. A variety of card and paper can be used but the most commonly used material is mounting board, a rigid card covered in facing paper available in a variety of colours, textures and finishes. Foam core board can also be used to provide a very sturdy mount but edges and corners are prone to damage and can look untidy. To prevent this it may need to be covered in paper.

The following are the most commonly used methods of mounting:

- *Flush mounting.* The artwork is fixed directly on to mounting board which is exactly the same size leaving no visible border. This method is used only to provide strength and support for the artwork.
- *Top mounting.* The artwork is fixed directly on to the board but this time with a border. This is a quick and easy method of presenting single or multiple items of artwork, allowing work to be overlapped. Coloured paper can be put behind the work to frame it and maximise the visual impact.
- *Sink mounting.* This is a particularly good way of presenting photographs as it prevents the corners from being lifted through constant handling and gives a neat and professional appearance. This method requires the top layer of mounting card to be removed where the artwork is to be placed. The artwork is then placed in the space so that both surfaces are level.
- *Window mounting.* The most commonly used method for presenting finished work, window mounting has become a standard method of presenting work within picture frames. A bevelled-edge cut (45-degree cut) is commonly used as it looks neat and draws in the eye. The artwork is held in position on the back of the frame often using adhesive tape, which allows the artwork to be removed, replaced or repositioned. Strong paper or card can be added to the back of the work to provide support. Window mounting can be used in conjunction with top mounting. Frames can be placed around the artwork using a thinner paper or card instead of mounting board. Window mounting can also be used to display transparencies if they are positioned between two frames; this can be done individually or in multiple batches if film negatives are being used.

Photography

Cameras can be classified into four main groups:

- viewfinder/rangefinder (compact camera)
- twin lens reflex (TLR)
- single lens reflex (SLR)
- studio, view and technical cameras.

Viewfinder (compact)

The user views the subject differently from the lens. The image recorded on film is not in the same position as the image in the viewfinder. If the picture is taken at a distance the **parallax error** is hardly noticeable; however, in taking close-ups the error can become visible.

Twin lens reflex

TLRs comprise two cameras. The bottom camera is fitted with all the elements of a conventional camera. The top camera is fitted with a lens and projects the image via a mirror on to a glass screen. The image will appear the same size as it does on the negative but it can suffer from parallax error in close-ups. In the viewfinder the subject is reversed from right to left making it difficult for the user to follow moving subjects.

Single lens reflex

The subject is viewed directly through the lens helping in a composition with no parallax error. The orientation of the subject is in its normal state allowing the camera to better follow moving subjects. The SLR camera allows an easy interchange of a wide variety of lens. Whatever is seen in the viewfinder will be on the film; therefore, effects such as filters can be experimented with. The SLR, it must be noted, can be quite noisy when taking pictures. The SLR viewfinder will black out at the time the picture is taken and causes a time delay in the region of 1/30 of a second from shutter button being pressed to the shutter opening.

Studio, view and technical cameras

Mounted on to a tripod, these cameras connect a panel with the lens and shutter to a ground glass screen by bellows. The lens panel is moved backwards and forwards to achieve size, focus and perspective control. Large film sizes can be used to produce high quality photographs, but the set-up time is usually a lot longer.

Lenses

Lenses can be adjusted manually and changed for varying effect. Lenses on compact cameras tend to be non-adjustable but cameras with automatic lenses are becoming more common. Each of these lenses is often grouped into the following categories:

- normal lens – produces a view similar to the naked eye
- telephoto lens – produces longer focus lengths for a telescopic view
- wide-angle lens – produces a wider field of view (distortion of the image will occur)
- zoom lens – produces a range of views from wide angle to telephoto and is, therefore, extremely versatile.

Film

Most compact and SLR cameras use 35mm film in 12, 20, 24 or 36 exposures in black and white and full colour. Although available for SLR cameras, 120 and 220 film is usually used by TLRs and studio cameras.

Colour and black and white film are categorised by speed. The speed of the film is described by a number using the ISO system. In general terms, slow-speed films range from ISO 12/12 to ISO 64/19, medium ISO 100/21 to ISO 200/24, and fast ISO 320/26 and over.

The film speed refers to the sensitivity to light – the faster the speed, the more sensitive it is. Therefore, fast film is used when the light levels are low, or for instance, in action shots. This will result in a grainy image. In contrast, slower films produce far less grainy images, but also need a lot of light. They tend to be used for big enlargements and studio work. For general purpose work, medium-speed films can be adequately used.

Additional equipment

Filters are placed on to the lens of a camera to change the light passing through, thus altering the image produced on the film. A variety of filters is produced to enhance tone and colour as well as adding effects.

Artificial lighting falls into two categories:

- *studio lighting* – continuous light for use in producing studio work
- *flash lighting* – flash units produce an intense burst of light when the photograph is taken. The flash has to be synchronised with the shutter speed (usually marked on the dial) to ensure correct use. A guide table will also accompany the flash unit. The table links the film speed, aperture and distance from the subject allowing for the correct use of the unit.

Adjustable tripods are used to hold the camera in a steady, fixed position. This allows slow shutter speeds to be used under 1/60 to avoid the image blurring because of any camera shake/movement.

Task

Describe what photographic film and equipment would be needed to take a picture of:

a) a fast moving vehicle at night
b) a football game
c) a close up portrait.

Screw thread forms

There are several different types of screw thread, each used for very specific purposes but essentially they have three main uses:

- converting rotary to linear motion
- obtaining a mechanical advantage
- fastenings.

Screw threads have two basic forms: the V-thread and the square thread. The most common type of screw thread is the 'V' type and it exists in the form of an isosceles triangle with a crest and a root either rounded or flattened (see Figure 3.1.9a).

The square thread takes its name from the profile shape of the thread (see Figure 3.1.9b). It is not as strong as the V-thread but it allows a large force to be applied and it is therefore used in vices and cramps. Buttress threads are used where a force has only to be applied in one direction (see Figure 3.1.9c). They are commonly used in woodwork vices that are fitted with a quick release mechanism and a half nut for rapid opening and closing. Acme threads are used extensively for the transmission of motion with an engaging nut as on the centre lathe (see Figure 3.1.9d). Here the tool post moves automatically along the bed until the tool reaches the end of the cut and the nut is disengaged.

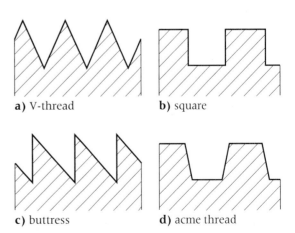

a) V-thread   **b)** square

c) buttress   **d)** acme thread

Figure 3.1.9 *V-, square, buttress and acme thread*

Nuts and bolts

Made from low or medium carbon steel, bolts are forged or machined and the threads cut or rolled. Sometimes they are also made from high tensile steel, alloy steel or stainless steel, and in some cases they are protected from corrosion by galvanising.

Coach bolts and studs are two other forms of bolts. Coach bolts are mainly used for fastening metal parts to wood and have a domed head with a square underneath it that acts as a locking device. A stud is a headless bolt with a thread at each end and a plain middle.

A nut is a collar and must fit the bolt with the same thread form and it should be of the same thread diameter. Nuts can be hexagonal or square in shape and are normally forged with a chamfer cut on one or both faces. In a special type of nut a ring of fibre or nylon is inserted in a groove inside the nut and provides extra frictional forces when the nut is fitted.

Rivets

Riveting is a simple way to make a permanent joint between two or more pieces of metal, either in the form of a hinge pin or in a rigid form as in a ship's plating. For general engineering purposes, rivets are normally made from soft iron and are therefore ductile and easy to work by hammering or pressing. Rivets exist in many different forms but the three most common types are countersunk headed, flat headed and snap or round headed. The choice of rivet to be used depends solely on the materials being joined and the location of the joint. In general, access is required on both sides when riveting but this is not always possible. In these circumstances a different process has to be used called pop riveting. This process was developed for the aircraft industry where extensive use of thin sheet material is joined with access generally only available from the one side. However, since the joint is hollow it is relatively weaker than when using conventional rivets but much lighter in weight as pop rivets are normally made from aluminium alloys.

Gears

A gear is a toothed wheel with a specially shaped profile (see Figure 3.1.10). This allows it to mesh with other gears, thereby transmitting forces and motion when fixed on rotating shafts within machines. Two gears connected together form a gear train, where each gear turns in the opposite direction when fixed to a parallel shaft. If the gears are different sizes, then they will turn at different speeds and this is termed the velocity ratio (VR). Where several gears are introduced

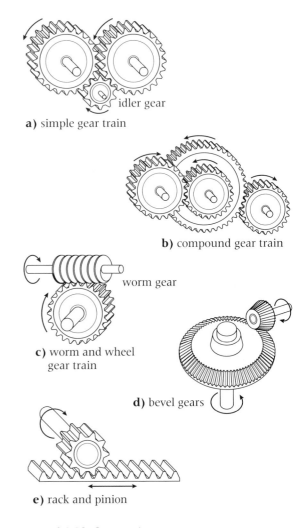

a) simple gear train

b) compound gear train

c) worm and wheel gear train

d) bevel gears

e) rack and pinion

Figure 3.1.10 Gear systems

on to identical shafts and meshed with other gears a **compound gear train** is formed and large speed changes can be achieved. Worm gears and bevel gears are special gears that transmit rotary motion through 90 degrees. Bevel gears need different sized gears to achieve any speed change, as is the case with the hand drill.

A rack and pinion allows rotary motion to be converted into linear motion as can be seen on the pillar drill. When the hand wheel is rotated the chuck moves down in a linear fashion.

Cams

A cam is a mechanism that is normally used to convert rotary motion into a reciprocating motion (although the cam itself may have an oscillating motion). The cam is fixed to, or is part of, a rotating shaft and a follower is held against it either under its own weight or by a spring (see

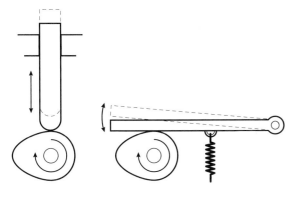

Figure 3.1.11 *Cams and followers*

Figure 3.1.11). As the cam rotates the follower moves depending upon the shape of the cam. Followers vary in shape and their use depends on the type of cam they are being used with. The four main types are:

- knife-edge
- flat
- roller
- roller and rocker.

Bushes and bearings

All rotating shafts need to be supported and bearings and bushes provide the engineering solution to this problem. Two different load types present two different problems to be contained: radial load and axial load. A radial load is simply caused by the rotating shaft whereas an axial load is a combination of a radial load with an extra load being pushed along the shaft into the bearing itself.

Bearings come in many forms, each having a specific use and application, but the three most common types are plain, roller and ball bearings. Friction is a problem in bearings and it is overcome with lubrication. In situations where contamination could be a problem from lubricants, (e.g. in the food industry), nylon or other plastic materials are used.

A bush is the simplest form of bearing and it is basically a cylindrical sleeve that fits into a hole and acts as the bearing surface.

> **Task**
>
> Look around the workshop and try to identify different forms of screw thread. Identify what type they are and work out why they have been used in that situation.
>
> Carry out the same exercise, only this time look for different types of gears. Identify what type of gear it is and why it has been used in that situation.

---

2. Working properties of materials and components related to preparing, processing, manipulating and combining

Aesthetic properties

The aesthetic value of a material relies upon how it is perceived in its appearance. An aesthetic judgement is made based on how pleasing the material is to the viewer. The materials used in the design and manufacture process may be evaluated on their own aesthetic appearance as well as the aesthetic contribution they will make to the product as a whole. The aesthetic properties of materials can be defined by their:
- colour
- style
- texture.

Colour

Most materials have an innate colour or colours. This can be enhanced or changed using various techniques (including heat, chemicals, pigments, dyes or natural finishes) to alter its appearance for functional or aesthetic purposes.

Style

The choice of materials used can also help the designer/maker to achieve a distinctive style for the product. A connection is made between the materials used and the aesthetic style that is being created; therefore, certain materials are more effective in conveying a style than others. For example, the manufacturer of traditional dining furniture would choose a material such as oak to achieve a traditional style rather than a contemporary or industrial material such as fabricated stainless steel. The use of stainless steel would communicate a very contemporary style to the viewer. However, a mixture of styles that would not normally be seen together or a traditional design that used contemporary materials would create a post-modern style for the product.

Texture

Like colour, most materials have an innate texture. The textual quality of a material can be achieved:

- visually – the look of a surface
- physically – the feel of the surface.

These textual qualities can be changed or enhanced for a desired effect.

Functional properties

Durability

The durability of a material relates to the type of object that is being produced. The choice of the material is dependent upon:

- the life span
- frequency of use
- demands placed upon the object (i.e. weather, corrosion).

A variety of materials can perform the same task. Therefore, when selecting materials, it is important to examine each material individually for performance and attributes.

Flammability

All materials can be changed and altered by the effects of heat. Materials that are susceptible to low heat or a naked flame are described as flammable. Paper, card and wood, due to their fibrous nature and low moisture content, are obvious candidates.

The shape and form of the material is also a determining factor in how quickly the material will burn. This is also accompanied by how well air can circulate in and around the material.

Regulations are stringent regarding flammable material in domestic and working environments. Materials which are flammable and release toxic gases when burnt need to be treated with fire retardant chemicals. Manufactured foams used in upholstery and soft furnishings are a good example of this type of material.

Mechanical properties

A mechanical property is associated with how a material reacts when a force is applied to it. The material will deform in one of two ways:

- **Elastic deformation** describes the behaviour of the material that returns to its original shape and form once the deforming force has been removed.
- Plastic deformation occurs when the deforming force permanently deforms the material even after the deforming force has been removed. This property allows materials to be pressed into new shapes that they retain once the force has been removed. The extent to which material can undergo permanent deformation in all directions under compression without cracking or rupture is known as malleability. Lead is a very malleable material and it is used to make joints watertight on roofs.

Extensive deformation can in some instances lead to fracture. However, some materials can undergo extensive deformation without fracture and these materials are said to be ductile. Copper, aluminium and silver are all very ductile and can be drawn through a die into thin wires. Permanent reduction in cross-section is achieved without causing any rupture or dislocations within the material. Conversely, those materials that exhibit little or no deformation before fracture are termed brittle.

Toughness is the ability to withstand sudden impact and shock loading. A tough material will also resist cracking when subjected to bending and sheer loads. Hardness is the ability to withstand wear, scratching or indentation. It is an essential property in all cutting tools. Diamond is the hardest of all materials but tungsten carbide is a much cheaper manufactured alternative for use in cutting tools. The strength of a material is defined as the ability of the material to withstand forces without permanently bending or breaking. Strength can be broken down into five main areas:

- tensile strength – the ability to resist stretching or pulling forces
- compressive strength – the ability to withstand pushing forces
- bending strength – the ability to withstand the forces attempting to bend the material
- shear strength – the ability to resist sliding forces acting opposite to each other
- torsional strength – the ability to withstand twisting forces under torsion or torque.

3. Hand and commercial processes

Sketching

Sketching is an informal drawing process that communicates ideas quickly. A range of media and techniques can be used when sketching. HB to 2B pencils sharpened to a conical point are best for this purpose as they enable the user to employ a range of lines from faint to heavy that can be erased easily. The sketch and its constructional lines can be drawn lightly first to ensure they are correct. Pressure can be applied to produce bolder lines to finish off the sketch.

Layout paper is commonly used for sketching; it allows images to be traced, which is an advantage in the quick development of ideas. Graph paper helps to achieve accurate proportioning and sizing when sketching ideas in **orthographic** views. **Isometric paper** may assist in transferring these orthographic views to isometric while maintaining proportion. These types of paper can be used as guides when they are placed underneath the layout paper or can be drawn on to directly.

The sketched image can be enhanced (see below) to offer a more realistic view or assist the viewer in understanding it. The sketched image can be drawn in a variety of views. A combination of views may be needed to produce a more detailed illustration, particularly with 3D objects.

2D views
- Orthographic views – plan, front and side.
- Layout – the arrangement of images and text in relation to each other.

3D views
- Perspective – 1, 2 and 3 point.
- Oblique.
- Isometric.
- Planometric.

With both 2D and 3D sketching additional techniques may be required to assist in communication to the viewer. These include:

- cut-aways
- exploded views
- hidden detailing
- sectional views.

Constructional lines

Faint constructional lines are used in sketch work to ensure good proportion and positioning of elements is achieved. For example, constructing a box for the drawing can help to calculate key points within the image. This box can also be divided with constructional lines to help determine good proportions for the sketch. Constructional lines function as a framework for the image, which is particularly useful in 3D views as shapes like circles will become ellipses.

> **Task**
> Sketch a selected object using constructional lines. Sketch this in 2D and 3D.

Enhancement

Enhancement techniques are used in both sketch work and formal pictorial drawings to produce a more realistic image and assist the viewer in understanding the ideas being communicated. It can be divided into five key areas:

- line
- tone
- texture
- colour
- background.

Line

Thick and thin line technique is a quick and effective way of enhancing a drawing. The outside edges or the lines that meet the background of the drawing are marked with a thick line. In contrast, the internal lines are thinner. The drawing is done in one line width (thin line), then a pen with a noticeable difference in nib width is used to trace over the outline of the drawing. This effect can also be achieved with a pencil if enough pressure is added.

Tone

All 3D objects are subject to a light source that causes the object to have a tonal range from light to dark. The surface closest to the light source will be the lightest, with its opposite side or the surface farthest away from it appearing the darkest. The 'tonal rendering' technique mirrors this, making the drawing appear more solid and realistic. When tonally rendering a drawing a decision must be made regarding where the light source is coming from. This is usually from over your left shoulder but it can be changed to vary the effect. Objects with flat planes (i.e. a cube) can be rendered with three separate tones: light, medium and dark. Surfaces that are curved (i.e. a cylinder) will show a more distinct tonal range that graduates from light to dark. The drawing can then be rendered in one of two ways:

- Graduation – a smooth and subtle transition from light to dark where the intensity of tone can be varied depending on the medium used (pencils, watercolour, acrylic, gouache, airbrush, marker pens, etc.). Pencils create a gradual blend through the varying amounts of pressure applied during application and paints can be altered through mixing and dilution. Marker pens can build tone in layers of application or by using a range of tones within the same colour.

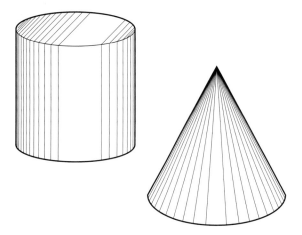

Figure 3.1.12 *Solid lines can be used to create tone*

- Solid – this method is used for media that do not provide subtle gradations in tone (pen, ink, dry transfer, etc). It relies upon the use of application techniques to represent light and shade in alternative ways. For example, solid lines or dots can create a gradation depending on how closely they are spaced together or apart (see Figure 3.1.12).

Texture

Incorporating the look or feel of a surface into a drawing can help to identify the material being portrayed. This can be achieved in monochrome, but the use of colour will add visual impact and along with tone will assist in building a realistic image. Various techniques and types of media are used in combination to render drawings in the likeness of materials such as wood, metal, plastic, glass, concrete and fabric. Some types of media are more effective in representing materials – for example, the reflective surface of chrome can be represented using an airbrush. Textures can also be applied using dry transfer sheets and computer-generated textures. These techniques are best suited to cover large areas.

Colour

Colour is used to enhance drawings aesthetically but it is also a useful tool to convey information. This can be done in a number of ways:

- Colour can be used to highlight sections of an illustration, drawing the viewer's eye to key areas of the visual.
- It can also be used to assist the viewer in understanding the information presented. This can range from the inclusion of simple

directional arrows to the colour coding of a complex system, which distinguishes components or identifies working processes. Colour code systems are effective when used in conjunction with a key.

Background

Backgrounds are used to enhance the overall visual presentation of a drawing. A background may be used to place a product or image in context or to group a number of visuals together. A variety of media can be used for this purpose including marker pens, coloured pencils and crayons, which have the advantage of easy application. Backgrounds can be:

- abstract – this type of background uses colour, shapes and effects to lift the image and draw the viewer's attention to it
- contextual – this type of background depicts a relationship to the image. It might be an illustration of the environment the image/product appears in, a user interacting with a product or a theme relating to it.

Colours used for the background must be chosen carefully as they may detract from the main image. Neutral and complementary colours are good options but should be tested for suitability with samples. The position of the visual should also be experimented with to increase its impact.

The final image can be cut out and placed on the prepared background. This is a quick and efficient method as it ensures the drawing has clean and clearly defined edges without the use of masking.

Task

Sketch a cube and a cylinder:

a) Render them both with solid tone.
b) Try this again, but render them with graduated tone.
c) Evaluate the methods you have used.

Drawing

Drawing can be broken down into pictorial, information and working drawings.

For all of these types of drawing the basic equipment used is a 90-degree set square with measurements and a drawing board with parallel motion or a T-square. These are used to ensure accuracy and consistency throughout the drawing process. Additional equipment may be needed such as a compass, protractor, etc.

Pictorial drawings

Pictorial drawing consists of a number of methods used to visualise objects in a 3D form.

Perspective

Perspective allows an object to be drawn as it is viewed by the human eye. Parallel lines appear to converge at a vanishing point (VP) the further away they are. Lengths, heights and widths will appear to foreshorten as they recede into the distance. There are three types of perspective:

- **1-point perspective** – only one VP is used in relation to length or width
- **2-point perspective** – two VPs are used, one for length and one for width
- **3-point perspective** – three VPs are used for length, width and height (commonly used for drawing buildings).

The position of the VP will be placed on a horizon line or on eye-level. For length and width the VP must be on the same line. In the case of a third VP, it will be placed above or below the horizon line (depending on the view) and in relation to the centre of the object.

Where the object is drawn in relation to the horizon will determine what view is required (see Figure 3.1.13):

- Worm's eye view – the object is above the horizon and the underneath will be seen.
- Street level – the horizon line passes through the object; the top or underneath cannot be seen.
- Bird's eye view – the object is below the horizon line allowing the top to be viewed.

Estimating depth

Two methods are used to estimate depth as shown in Figures 3.1.14 and 3.1.15:

1 Division:
- A box is drawn into which the object will fit.

- The front vertical is divided equally into the number of units the object is to be divided into.
- For each of these divisions draw a line to the VP. Then draw a diagonal line from corner to corner across the side.
- Where the diagonal cuts across the division lines a vertical line can be drawn.
- Each unit should be smaller than the previous one giving the appearance of foreshortening.

2 Addition:
- A single cube is drawn to which cubes will be added.
- A diagonal line is drawn across one face from corner to corner.
- Another diagonal is drawn parallel to the first from the corner of the cube.
- Where the diagonal crosses the line leading to the vanishing point a vertical line is drawn.
- This is continued until the required number of cubes are drawn.

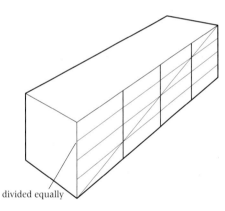

Figure 3.1.14 *Estimating depth in perspective – division*

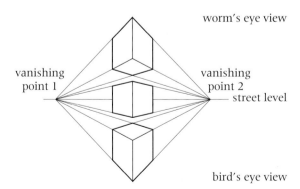

Figure 3.1.13 *An example of 2-point perspective in three views*

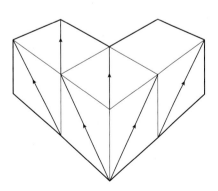

Figure 3.1.15 *Estimating depth in perspective – addition*

Circles

Circles in perspective will appear as an ellipse because of foreshortening. They can be drawn freehand or in conjunction with a flexi-curve or French curve:

- A square is drawn that is the correct size for the circle to fit into.
- Two diagonal lines are drawn, corner to corner.
- A cross is drawn where the diagonal lines intersect to find the four mid-points of the square
- The circle can be drawn using these guides. Opposing curves should always be the same.

Isometric

Isometric drawing is an accurate method of showing the three faces of an object. All lines are true or scaled lengths, therefore, foreshortening does not occur. All vertical lines are drawn 90 degrees to the parallel rule and horizontal lines on the object are drawn at 30 degrees to the vertical using a 60/30-degree set square. Isometric paper can be used to assist the process.

In addition, circles can be drawn using an isometric ellipse template which offers a range of sizes. Compasses can be used for larger circles to strike arcs (using the correct method). Irregular curves can be achieved with a French curve or flexi-curves.

Planometric

Planometric, also known as **axonometric**, technique is often used for interiors and buildings as it allows a bird's eye view of the object.

The plan view is drawn 45 degrees to the horizontal using a 45-degree set square. All dimensions are true or in scale, thus circles and curves are drawn as they appear in view.

Vertical lines can also be drawn in the same way, but they can appear distorted and too high. Therefore, it is acceptable to reduce the vertical scale by three-quarters, two-thirds or even one half in relation to the plan.

Oblique

This method is not often used and has largely been superseded by the methods outlined above. Oblique projection is similar to planometric in that the face (front or side views) is true. The receding lines can be drawn at 45, 30, 60-degree set square angle from the face.

The face is drawn from a horizontal line with the vertical lines being 90 degrees to this line. The receding lines are drawn at half size to prevent the distortion of the drawing.

Exploded drawings

Exploded views of an object help to explain how the object is assembled. Parts of the object are drawn separately and at a distance from each other but on the same axis. Arrows or lines can be used to help explain the relationship of the parts to one another.

All of the drawing methods outlined above can be used to create an exploded drawing.

Task

Take a simple object, such as a pen, and draw it in:

a) perspective
b) isometric
c) planometric
d) oblique
e) exploded isometric.

What are the strengths and weaknesses for each method?

Information drawings

Information drawing is a process that is used to display data visually. The pictorial representation of data helps to simplify information, making it more immediate and comprehensible to the viewer. Information drawings can be displayed in a number of forms:

- charts
- tables
- diagrams.

The data has to be classified into groups or sections, and they need to be quantified by a number, scale or percentage. This allows the data to be displayed so that the sections are visibly comparable.

The use of icons, colour, texture and tone can assist in distinguishing sections, along with 2D and 3D forms. It should also be appropriate to the type of data being displayed.

The use of ICT software, such as spreadsheet packages, can produce a variety of pictorial information quickly and easily. Once the information has been entered, the user can cross reference it with other criteria. This allows a wider analysis and a range of varying results. The outcome can be displayed as bar, pie and line charts.

Working drawings

Working drawings are produced to display all of the required information of a product in 2D form so that it can be produced or manufactured. The range of drawings includes:

- orthographic projection
- sectional views
- assembly drawings
- parts drawings.

A standardised method of producing working drawings has been defined by the British Standards Institution (BSI). All working drawings are laid out with a margin and title block. The title block should include:

- name
- date
- title
- projection used
- scale
- drawing number.

Letters and numbers are written in block capitals to avoid confusion; letter stencils are used to ensure consistency. Drafting film is the best paper to use as it allows repeated corrections to be made in both pencil and ink. A hard pencil (2H) should be used to draw the object and construction lines. When the drawing is correct the object should be lined in with a technical ink pen. Photographic reproduction (blueprint/dye line) of the drawings can be made. This allows the original to be kept neat and untouched.

Orthographic

Orthographic drawings show a 3D object in a series of 2D views – front, end and plan. The purpose of orthographic drawings is to allow the manufacturer or producer a detailed drawing from which they can produce the product. It is easier to draw 2D in detail and easier to read the information this way.

The scale needs to be calculated along with the spacing between each view to ensure all the views fit on to the paper. There are two methods of orthographic projection:

- 3rd angle – the plan view is placed at the top (see Figure 3.1.16). The front view is placed underneath. This is what is seen on the bottom line of the plan view. The end view is placed to the right of the front view. This is what can be seen on the right of the front view.
- 1st angle – the front view is placed at the top (see Figure 3.1.17). The end view is placed to the right of the front view. This is what is seen on the left side of the front view. The plan is placed underneath the front view. The front view is the bottom line of the plan.

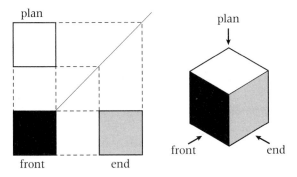

Figure 3.1.16 *3rd angle orthographic projection*

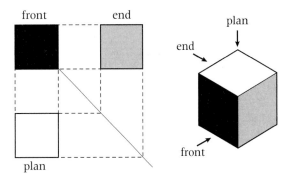

Figure 3.1.17 *1st angle orthographic projection*

When producing an orthographic drawing, three views to be placed should be laid out in the correct order for the respective angles. The dimensions are transferred from one view to another using drawing instruments. This will ensure accuracy and consistency between the three views. A 45-degree angled line from the front elevation should be drawn in between the plan and end elevation. This will assist in transferring dimensions from one view to another. Details of the object can then be built up in all three views.

Sectional views

Sectional views allow a cross-section of the object to be viewed. This is shown by 45-degree hatched lines. The sectional views must relate to the orthographic projection. A curve should be drawn to mark the cutting plane and labelled.

Assembly and parts drawings

An assembly drawing shows the position of each separate part of a product and how the parts are fitted together. Each item is identified by a number or letter which is referenced to a parts list that describes the materials and components.

The parts drawings are detailed views of each separate part to be manufactured. These drawings are referenced to a cutting list which specifies the quantity, material and dimensions of the parts.

Tasks

1 Draw a simple object in both 1st angle and 3rd angle orthographic projection.
2 Evaluate the advantages and disadvantages of each drawing.

Computer graphics

Computer graphics are pictorial images that are produced on a computer. These can be produced on a PC or Apple Macintosh based software. More complex graphics will need to be produced on more powerful machines using software specific to the requirement. An example of this would be SGI machines for animation.

Graphic formats

The graphic format used to save the file will be dependent upon what the image is used for. The quality of the image colour can be specified in each format with reference to 'bits per pixel'. Graphic file formats can be divided into:

- index, definition and preview – BMP, DIB, RLE
- format originating from specific applications – WPG, MAC, etc.
- formats designed for high compression – JPG, JAS, TIFF
- miscellaneous – EPS, RAW, etc.

Computer graphics can be divided into two groups as outlined below.

3D

The 3D graphic applications software packages construct objects to be viewed at any angle and synthesise realistic images. The object can be constructed in a combination of orthographic and pictorial images. Once the desired shape is achieved the object can be tonally rendered to create a 3D solid. At this point, the following techniques can be applied to enhance the object.

- shading – transparency and textures
- colour
- ray tracing – reflection and transparency
- radiosity – light reflection from diffuse surfaces and quality of light
- light source
- viewpoint/camera angle – this can be static or moveable to allow the viewer to be taken in and around the object
- animation – (not applicable to all software).

There are many software packages that can produce 3D graphics. The most commonly used are ProDesktop, 3D studio and Lightwave.

2D

In contrast to 3D, 2D graphics only allow the image to be viewed at one angle. This can be as basic as drawing geometric shapes or as complex as the manipulation of digitised images. In addition, 2D images can be animated to appear 3D. This is a similar technique to that used in traditional animation where 2D images are shown in quick succession. This should not be confused with 3D graphics.

The 2D software applications packages offer a range of tools to construct and manipulate images. The basic tools allow shapes and lines to be made, to which colour and texture can then be added. Effects and filters can be applied to vary the look such as blur or mosaic. Digitised images, such as scanned pictures and existing graphic files, can be combined and manipulated in the same way.

There are a number of software packages that can do this. Desktop publishing and advanced word processing packages can offer a basic construction. For greater manipulation of images, software such as Adobe Photoshop and Paintshop Pro are commonly used.

Typography

Typography is the way of designing communication by means of the printed word. This is done predominantly on the computer in desktop publishing applications such as Quark. Before this is done preliminary ideas will have to be produced to ensure an overall visual consistency. The four major elements are as follows.

Layout

The layout of the page will have to be consistent if several pages of text are produced. Therefore, the setting of the page size, margins and number of columns will have to be set as a template for the text to be applied. The position of the page number and footline (the chapter or title) will also have to be positioned and set.

Text

For the main body of the text an easy-to-read font should be chosen and maintained throughout. The size of the font is dependent upon what is being produced. For layouts where there is a bulk of text 10- or 11-point fonts should be used.

Headings and subheadings

The purpose of a heading is to draw the reader's attention to an article or text and give an impression of the theme or content. A subheading will draw the reader's attention further and perhaps give more information regarding the text. Think about the way headings and subheadings are used in newspapers and magazines. Headings and subheadings can also be used to break up the layout of a page or illustrate where a break has occurred in the text.

The typeface used for a heading is often the same as that used in the main body of the text. It can be different but the general rule is to restrict the use of typefaces to two. There are a number of effects or adjustments that can be made to create a contrast between the heading and the text:

- size
- italics
- lower and upper case
- colour
- character shape – serif/sans serif
- character width – expanded or condensed
- position on page
- density – solid/outline/positive/negative.

The subheading is secondary typography (heading is first, subheading is second, text is third) and should be smaller than the heading. There are two types:

- External – these subheadings appear directly beneath the heading to add further detail or explanation. They might also be placed in the margin alongside the text.
- Internal – these subheadings appear in the text at the beginning of a paragraph or section. They can be distinguished in the same manner as the heading. Lines or boxes can also be used to add emphasis.

Captions/breakouts

Captions assist in the explanation of photographs or images which appear within the text. The style of captions should be consistent throughout to give a distinctive and uniform look to the publication. The position of the caption is important as it should be near to the image but separated from the rest of the text to avoid confusion. The caption is usually placed directly beneath the picture and can be set apart with a line or box.

Breakouts or callouts are interesting or relevant sections of the text usually in the form of a short quote that are separated from the main text and contrasted with it. This can be achieved using the methods outlined above or the breakout/callout may simply be surrounded with an empty space.

Task

Using a newspaper or magazine article, identify the body text, headings, subheadings and breakouts.

Modelling and prototyping

The communication of design ideas can be clearly shown in a 3D form. It can be achieved using:

- virtual modelling – realisation of ideas constructed in a computer software modelling package such as 3D studio or ProDesktop.
- material modelling – using a variety of material and media to produce a physical manifestation of design ideas
- computer-aided design/computer-aided manufacture (CAD/CAM) modelling – a combination of the two methods above in which the virtual model or individual parts of it are produced by a computer numerically controlled (CNC) machine from a suitable material.

A variety of different models can be produced which are appropriate for different stages of the design process. Care should be taken to choose materials, processes and techniques that are best suited to the type of model being produced.

As models are used to help shape the design and production processes, different models fulfil different requirements. Some models are constructed to focus on aesthetic considerations, i.e., shape, texture, finish. However, complex or costly processes can be substituted in the modelling stage. Applied finishes such as paints allow the model to appear to be made from another material. This enables the model to be made from materials that can be more easily shaped and manipulated.

The use of scale is an important aspect of producing models. Scaling models down can save the model maker time, materials and space while still giving an accurate representation of the product. This is particularly useful for large-scale projects such as architectural models. Smaller, detailed products may have to be scaled up to allow for precise evaluation and development.

Sketch models

Sketch models are produced in the earliest stages of the design in conjunction with preliminary sketches to explore possibilities and confirm ideas. The materials used to produce a sketch model tend to be ones that are inexpensive and can be worked easily, quickly and where surface finish is not important. Therefore, card, paper, expanded polystyrene, styrofoam and wood are usually used for sketch models.

Block models

Block modelling helps to determine shape, dimensions and surface finishes by constructing an accurate representation of the final product. Its external appearance should show surface detail including screws, joints, surface texture and colour. As the name suggests the block model does not deal with internal details such as moving or working parts.

Working models

Working models progress from block models. Although they include all the surface detail and styling, they also incorporate the working or moving parts of the final product. Depending on the type of product being made, the working model may include hinged or removable sections. Internally the working model must allow the correct space and dimensions for any components or machinery that will enable the product to function. For example, if the product needed a circuit board fitted inside, the working model would have to allocate the correct position and amount of space. However, the amount of internal detail actually applied to the model should be minimal, for example mounts and fixings could be ignored.

Prototypes

Prototypes are the culmination of the design and modelling processes as they offer the most accurate representation of the final product. They are made from the same materials and components as the end product and include all external and internal details. Prototypes are usually made by hand to ensure that aspects of the design and production are correct before investment into tooling and machinery is undertaken.

Architectural models

Scale models of buildings and interiors are produced to allow the detail of architectural drawings to be visualised in a 3D form. Architectural models can vary from simple constructions made from white card to elaborate and highly detailed models that include surface finishes or effects that give the viewer an idea of the materials to be used. Card and its derivatives are most commonly used as they can be shaped easily with basic equipment. With its plain surface, card also has the advantage of having no indication of scale; other materials such as wood have natural surface markings like grain and therefore would be out of scale in a 1:500 model.

Architectural models are often made with removable parts or sections so that the viewer can see the internal layout. Accessories and parts can be bought to assist in the construction of architectural models to save time while still including detail. These include scaled models of people, vehicles, shrubs, etc. or scaled surface effects such as paper printed with brick or tile designs.

Task

Discuss how models contribute to the design process.

Production of nets and surface developments as a form of model making or box construction

A net, also known as a development, is a flat 2D shape that can be cut, bent or folded to produce a 3D shape. It is used when working with sheet material such as card, metal and plastic.

The accuracy of the drawing is of the utmost importance. When drawing nets the equipment used should offer accuracy and consistency. Therefore, complicated nets may need to be drawn on paper, then temporarily fixed to the sheet material.

To produce an accurate net the final 3D shape will need to be drawn. This will enable the shape, size and layout of the net to be drawn more easily. The net will need to show:

- cut lines – a continuous line where the material is to be cut
- fold lines – a broken line where the material is to be folded, scored, bent, heated, etc.

Annotation will assist in labelling edges, sides and features in relation to one another. In joining edges together a flap or tab will have to be included in the net. This will allow for neat edges as well as a larger surface area for fixing. The corners of the tabs should be cut off at an angle for ease of assembly and a neater finish.

Casting

Pouring molten metal into a mould that contains a cavity of the required shape produces a casting. Initially, the mould or pattern needs to be made of the component to be cast and it should be slightly oversized to allow for shrinkage of the casting on cooling. Most patterns are split along the centre line and are located with dowels to form a solid pattern. Patterns should avoid sharp corners and undercuts, and all vertical surfaces should be tapered to allow for easy extraction from the sand. The process of casting varies depending upon the type of pattern being used.

On a large-scale production site producing such items as engine blocks and metalwork vices, casting is carried out on a continual basis. Casting can also be used to produce a single item in the school/college workshop but this makes it labour intensive since the mould will be used only once.

Grinding/abrading

Abrasives can be used on all types of woods, metals and plastics. Disc sanders, belt sanders and orbital sanders are all commonly found in workshops and they work by moving abrasive papers over the work. Certain plastics can also be used on these machines.

Polishing with emery cloth, wire wool and various pastes on a buffing wheel, can abrade metals.

Grinding of metals is carried out by either a disc grinder, offhand grinder or a surface grinder. Each of these methods uses discs that have been made from abrasive powder that have been cemented together. The first two methods are essentially used for cleaning up and finishing, while surface grinding is a precision machining process that grinds very hard metals to a smooth and accurate finish.

Joining

Joining processes can be categorised as follows:

- Permanent – once made they cannot be reversed without causing damage to the work piece.
- Temporary – although not always designed to be taken apart, they can be disassembled if needed without causing damage.
- Adhesives – these fall into two groups, natural and synthetic. The synthetic types tend to be toxic substances and therefore need to be handled with care. It is thought that most adhesive bonding can be classified as a chemical reaction.

Factfile
Adhesives

- Polyvinyl acetate (PVA) – a popular white woodworking glue is easy to use and strong providing the joint is a good fit. It is not waterproof.
- Synthetic resin (cascamite) – much stronger than PVA and is supplied as a powder to be mixed with water. It is a good joint filler and it is waterproof making it ideal for external use.
- Epoxy resin (araldite) – a two-part resin and hardener that needs to be mixed in equal parts. It takes a while to harden fully, and is expensive.
- Acrylic cement (tensol cement) – available in various forms with Tensol 12 being the most common. It is ready for immediate use.

Contact adhesives

Contact adhesives are used for large areas such as sheet material. Two surfaces are coated and left for 15 minutes. On contact with the other surface adhesion is instant.

Rivets

Rivets are used extensively in sheet metal work although they can be used to join acrylic and wood (see Figure 3.1.18). Conventional rivets are available in various forms in addition to pop rivets which are hollow and used widely in the aircraft industry.

Wood screws provide a temporary method of fixing unless, of course, the materials have been glued together. They can be used for joining

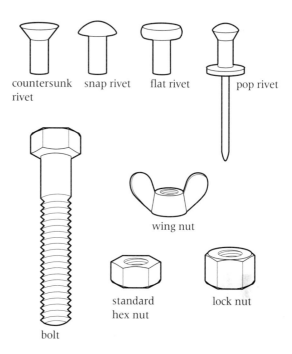

Figure 3.1.18 Nuts, bolts and rivets

wood to wood and metal or plastic to wood. Screws are classified by their length, gauge size, head type and material. They are normally made from steel or brass with the steel type undergoing a number of surface treatments such as chrome plating or galvanising to increase resistance to corrosion.

Nuts and bolts

Nuts and bolts are another method of fixing materials temporarily (see Figure 3.1.18). Bolts are commonly available with hexagonal heads, and together with nuts of matching thread size and form, lock together to form strong mechanical joints.

Task

Using a series of annotated sketches show the main stages of the following processes:

- pop riveting
- making a fruit bowl on a wood lathe
- sticking two pieces of acrylic together.

Airbrushing

The airbrush is designed and constructed around a basic system of compression. Air is forced through a narrow passage to cause a drop in pressure. A tube that is connected to a reservoir of liquid (paint, ink or varnish) is placed where the flow of gas is fastest. The drop in pressure will cause some of the liquid from the reservoir to be drawn up by suction or gravity into the stream of gas. The gas accelerates as it is forced out through the nozzle at such a speed that the air and paint mix together or atomise to form a fine spray.

Some models atomise the liquid externally, which means the air stream and the paint meet at the airbrush tip rather than inside the nozzle. As a result, the gas and paint do not mix completely and produce a less uniform and even spray.

The airbrush can be adjusted to control the amount of liquid that mixes with the air stream. The adjustment depends on the type of airbrush being used. The three main types are single-action, double-action and independent double-action, with the latter being the most versatile for the artist to use.

Most paints such as watercolour, gouache and acrylics are suitable for use with airbrushes as are inks and varnishes; this allows the user a wide scope to create a variety of effects and properties from different media. However, some materials must be diluted to create a form acceptable for use with the airbrush. Airbrushing has three main advantages over other painting techniques:

- Depending on the distance of the airbrush to the surface the user can produce a variety of effects, from a broad area of even colour to a fine line.
- An even gradation of tone or a progression from one colour to another can be achieved without noticeable breaks in the overall texture.
- When using an airbrush there are also no perceptible marks or strokes left from a brush.

Airbrushing allows a drawing to be rendered to a finish that is realistic with an almost photographic quality.

Photography

Cameras use three basic features to capture an image. These features may be adjusted manually or automatically depending upon the type of camera and the effect desired by the user.

Focusing

Adjusting the focus ring on the lens enhances the sharpness of the image. Various visual aids are incorporated into the lens design to assist the user to achieve sharp focus manually.

Shutter speed

The shutter speed and aperture both control the amount of light reaching the film. They are used in combination to affect the overall look of the photograph.

Shutter speed is adjusted by a control on the camera. Markings displayed are in fractions of a second (i.e. 60 is 1/60 of a second). This indicates the length of time the shutter will be open for and thus how long the film is exposed to the image. Other markings on the dial are:

- 'B' – allows shutter to remain open while button is pressed
- 'T' – allows shutter to open when button is pressed, and close when pressed once more
- 'Flash', which is the shutter speed to be used in conjunction with a flash.

The shutter controls the sharpness of the image. Slow speeds (60 and under) may result in blurring or camera shake, unless the camera is in a fixed position. If the image is of a moving object, then quicker speed will need to be used (500) to capture the object as a still image.

Aperture

The aperture also controls the amount of light reaching the lens and affects the appearance of the picture. It is adjusted by a ring on the outside of the lens, marked with a series of numbers. The smaller the number, the wider the aperture, resulting in more light entering into the lens. The aperture also controls the depth of field (i.e. the distance between the subject and the camera). The further away the subject is, the greater the depth of field, and vice versa. The wider the aperture is, the less the depth of field and, conversely, the smaller the aperture, the greater the depth of field. This principle can be utilised to ensure that either the majority of the picture is in sharp focus or only a small section.

Together, the shutter speed and aperture control the amount of light and, therefore, allow the film to be correctly exposed. This is essential in producing a good negative. Most cameras will have a built-in light meter. The type of system used is dependent on the camera, as some automatically set both shutter and aperture. Others may just set one. Priority may have to be taken of one over another. For example, the shutter speed may have to be fixed to avoid camera shake; or the aperture closed for a greater depth of field. The speed of the film used will also determine adjustment.

Photographic processing
Black and white film

There are four stages in processing black and white film:

- development
- wash or stop bath
- fixation
- washing and drying.

First, the film is loaded on to a spiral. This has to be done in total darkness or in a changing bag. The spiral is then loaded into a development tank; this will allow liquid to be poured in without any exposure to light.

The correct amount of development solution for the kind of film and tank size is poured in at a temperature of 20°C. This temperature should be maintained throughout the process. The length of time the film should remain in the tank is dependent upon the processing instructions for the film and the development solution. The development should be no shorter than five minutes. The tank will also have to be agitated for a duration of 15 seconds every minute. This is done to ensure an even development result.

Once the development process is complete, the solution is poured out of the tank. The stop bath is then poured in, agitated and left for 15 seconds. The solution is poured back into the container and then the fixer is added.

The time period for fixation will be outlined in the instructions. Agitation should take place every minute, as described in the development. The fixer can be poured back into the container for reuse. The number of times the solutions can be reused will also be mentioned in the instructions. The temperature of the stop bath and fixer should be within 4–5°C of the development solution.

At this stage, the spiral with film needs to be washed thoroughly with running water for 30–40 minutes. If the water is too cold it will cool the film too quickly resulting in marking. A gradual reduction in the film temperature to the level of the wash water is needed. After washing, the film can be removed from the spiral and hung in a dust-free environment to dry.

Colour film

Colour film development is a longer and more complicated process. It is usually done by machine. The film consists of three layers of black and white emulsions which are blue, green and red sensitive. The film is produced as a black and white negative using silver halide. When this is removed a colour negative remains.

Finishing techniques for function and decoration

The relationship between finishes, properties and quality is one that the designer should consider very early on as one of the major design considerations.

Finishes in their various forms are used to improve the product's functional properties, aesthetic qualities and generally serve to improve quality overall. For example, bathroom fittings and taps are coated with chrome to give them a cleaner, more durable and attractive surface. Motorway barriers are plated with zinc to give them a galvanised finish which will resist corrosion. Tin cans are made from cheaper sheet steel but they are plated with tin to stop them rusting and contaminating the food.

Surface coating

Anodising

Anodising is a surface treatment associated with aluminium and its alloys. The whole product is immersed in a solution of sulphuric acid, sodium sulphate and water. The product itself is used as the anode and lead plates are used as a cathode. When a direct current (DC) is passed through the solution a thin oxide film forms on the component. When finally washed in boiling water, coloured dyes can be added before the surface is finally lacquered. As a process it can be used to finish components to a consistently high-quality finish.

Preparation for finishing

Before any metal surface can be painted, the surface needs to be degreased and cleaned with paraffin. A primer should then be applied and this can be sprayed on or brushed on depending on the circumstances. For a professional finish an undercoat should be applied before the final topcoat. 'Hammerite' is a type of paint that requires no surface preparation other than to remove any old flaking paint prior to application. It is a one-coat application for ferrous metals and is available in a wide range of colours in either a smooth or 'crackle' finish.

Painting

Painting wood also involves the application of a number of coats. Prior to any painting, all knots should be sealed to prevent any resin from escaping. In between coats the surfaces should be rubbed down with fine glass paper. Topcoats are available in various forms:

- Oil based – commonly known as gloss paints, these are available in a wide range of colours. They are durable and waterproof and are excellent for use on products that are outside such as window frames and doors. Certain paints used on boats never need rubbing down for repainting.
- Emulsion – available in vinyl or acrylic resin, these paints are water based but not waterproof.
- Polyurethane – tough and scratch resistant, these paints harden on exposure to air. They are used widely on children's toys.

Painting, other than spray painting, is very time consuming in its application. Great care also needs to be taken to ensure an even application over the entire surface. It does, however, allow you the flexibility to create original pieces by way of choice or mix of colours therefore creating a range of one-off finishes.

Varnishing

Varnishes are a plastic type of finish made from synthetic resins. They provide a tough waterproof and heatproof finish. Polyurethane varnish is available in a range of colours with different finishes; gloss, matt or satin. They are best applied in thin coats with a light rub down with wire wool in between. New varnishes based on acrylic dry more quickly, have less odour and brushes can be cleaned in water. As they do not use solvents, they are environmentally friendly.

Preservation methods

Wooden products that are used outside such as fences and garden sheds need to be protected from the wet, insects and possibly fungal decay. Creosote is an oil-based preservative that is widely used on timbers for external use. This type of product is purchased already treated where it will have been dipped in a bath or pressure treated to absorb the solution. For furniture use, hardwoods and tanalised softwoods are used. Tanalised timber has been treated with preservatives that have been driven into the wood under pressure.

Self-finishing

Plastics generally require a little finishing on the edges and no surface finishing. This is mainly due to the exceptionally high quality of moulds used in the various manufacturing processes. Textures can be introduced to the mould surface and will subsequently be manifest in the final moulded products.

Surface decoration

Engraving

Most materials can be engraved with special tools or by a chemical reaction. Hand-held tools such as chisels are used on the more resistant materials, while etching with acids permanently engraves the surface of metals.

Some techniques are now computer controlled allowing a multitude of fonts and styles to be quickly converted and output. A variety of materials such as woods, metals and even composite materials such as reconstituted stone can all be cut with the appropriate tooling (see Figure 3.1.19).

Figure 3.1.19 Engraved glass goblet

Pyrographic methods

Woods can be decorated by burning patterns into the surface. A tool similar to a soldering iron is fitted with various shaped bits. These are then heated and pushed into the surface allowing total flexibility and creativity over the finished design.

Transfer techniques on plastic

Vinyl is used widely by sign writers to create large and decorative designs and images for signs and vehicle graphics (see Figure 3.1.20) Data prepared on CAD packages is easily output to plotters and cutters before being transferred to a backing material. This process, however, is not a direct transfer process like screen-printing.

Figure 3.1.20 Vehicle graphics using vinyl

This versatile process can be applied to many surfaces and circular items too. Artwork is generated and is then transferred on to stencil film with the use of ultraviolet light. The stencil is then fixed below a mesh and the work piece held below the whole arrangement. Inks are then applied and squeezed through the mesh and artwork on to the work piece. The process can normally be repeated at great speed industrially where such items as T-shirts and ceramics can be painted and over printed using different colours.

Spray paints

Spray painting offers a quick solution to covering large surface areas evenly. Used extensively in the car industry and now controlled by computers and robots, this process produces a high-quality surface finish. Layers of paint are built up to provide a hard surface and one that protects the material below from corrosion and oxidation.

Etching

With the use of acids, designs and patterns can be made on the surface of metals by etching. An acid resistant such as paraffin wax or a mask is applied to the areas not to be exposed. The acid then chemically attacks the metal surface to leave a contrasting surface texture or finish.

Printing

Computer printers

The two most widely used computer printers are inkjet and laser printers.

Inkjet printers deposit extremely small droplets of ink on to paper to form an image. The data to be printed are sent by the computer applications software to the printer driver. The printer driver translates the data and sends them to the printer via the connection interface. The printer receives the data and stores them in a buffer. The buffer can hold data so that the computer can be used for other tasks while the printing process takes place.

The control circuitry in the printer activates the motor to feed the paper, which is carried by rollers, into the printer. Once the paper is in position the motor activates a belt to move the print head across the page. The print head contains a number of nozzles that spray ink on to the page. The ink can be black, cyan (blue), magenta (red), yellow, or precise combinations of these to form any colour. The motor pauses between each spray of ink to form the shape of the characters and create spaces. However, this occurs at such high speed it appears to move in one steady continuous motion.

At the end of each printed line the motor moves the paper a fraction so that the next line can be printed. Finally, when a full page is printed the motor moves the paper through the printer on to an output tray.

Like the photocopier machine, the laser printer (see Figure 3.1.21) uses static electricity, an electrical charge that is built up on an insulated object. The laser system works on the principle that oppositely charged atoms attract one another and will cling together.

The laser printer receives the data from the computer in digital form. The printer's core component is a photoreceptor which typically takes the form of a revolving cylinder or drum. The drum is constructed from photoconductive material that can be discharged by light photons.

The drum is positively charged by a roller or wire (corona wire) with an electrical current running through it. The drum revolves and as it does so a fine laser is beamed across the surface to discharge selected areas. By doing this the laser inscribes the image or text to be printed as an electrostatic image.

When the image is set on the drum, positively charged toner powder coats the surface. The toner having a positive charge is attracted to the discharged areas and clings to them. It is repelled from the positively charged background.

A roller or wire gives the paper a negative charge that is stronger than that of the electrostatic image. As the paper rolls under the drum its stronger charge pulls the toner away from the drum's surface. The drum is then discharged by exposure to light to remove the image allowing the process to begin again.

The paper, which now holds the toner powder, is also discharged and quickly passed through

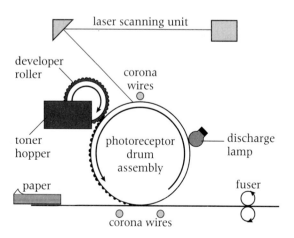

Figure 3.1.21 *The basic components of a laser printer*

heated rollers that melt the powder, making it fuse with the fibres of the paper to fix the image.

To print a multicolour image on a laser printer a similar process takes place. However, it has to be done four times, once for each colour: cyan (blue), magenta (red), yellow and black.

Toner powder in the three primary colours and black is used in various combinations to create the full spectrum of colour. Depending on the printer model this can be achieved in various ways: the printer creates an electrostatic image for each colour and applies toner from separate units; all four colours are added to a plate, or in complex machines a separate printer unit is available for each colour and the paper moves from one to the next picking up each colour.

Screen printing

Screen printing is a widely used and versatile printing process. It is used in graphics for signs, banners and posters and in manufacturing processes to print on fabrics, wall coverings, glass and ceramics. Despite being commonly referred to as silk screen printing, silk is rarely used to make screens. Modern hand and commercial printing screens are made from synthetic or metal gauzes that offer durability and stability.

Screen printing works on a relatively simple method of ink transfer. Ink is forced through holes in the gauze that form an image on the screen; essentially the screen functions as a stencil. Ink is pushed through the apertures in the fine mesh by pressure, applied using a plastic, rubber or steel blade often referred to as a squeegee. Whereas other printing processes (e.g. letterpress) rely on impact to print, screen printing is based upon a non-impact principle of ink transfer. This principle makes screen printing a versatile process because it can be used to:

• print on to a wide range of materials
• print on to surfaces that are curved or uneven
• print thick opaque ink deposits.

Fine gauze material is stretched at high tension over a metal or a wooden frame to make the screen. The three most common screen materials are polyester, polymide and stainless steel.

The image is formed on the screen using a photo-stencil. The screen is completely covered with a light sensitive polymer emulsion. When the screen has dried the opaque positive image is then placed on it and exposed to ultra-violet light. The areas of coating that have been exposed become insoluble in water, thus fixing the image. The screen is sprayed with water after exposure, dissolving the unexposed areas and leaving the stencil image.

Screen printing production methods vary according to the scale of manufacture required. They range from manual hand-bench methods for short runs of printing (for example one to 100 units) to fully automatic methods (rotary or cylinder) which are capable of producing high-volume run lengths (for example 7000 units and above).

To print a multicolour image a separate screen must be produced for each coloured section of the artwork. Precise multicolour images are best produced using manual, part or fully automated equipment such as carousel machines.

Photographic printing

Black and white photographs are printed from the negative on to light-sensitive paper applicable to the task. This is done in a dark room. A low-wattage bulb is used with a filter to prevent the paper being exposed.

The negative is placed into an enlarger which projects light through the negative on to a flat surface. The enlarger can be adjusted to vary the size of the print.

A test strip determines the length of time the negative is exposed to the paper. Sections are exposed five seconds at a time so a range of times can be seen (5–25 seconds). When these are developed the best exposure time is used for the final print.

The print is developed in a similar way to film but uses chemicals specifically designed for prints. The paper is placed in the liquid developer for around one to two minutes until the correct image is produced. It is then placed in a stop bath with tongs and transferred to the fixer.

The paper can then be washed with running water and left to dry. The time taken for each stage depends on the paper and the developer/fixer used. Colour prints are produced using a similar but slightly more complex process.

Offset lithography

Offset lithography (litho) was developed from lithography for commercial use. Both processes are based on the principle that oil and water do not mix and will always separate.

A lithographic plate bearing an image is treated using a greasy medium which will attract oil-based ink. The non-image area is treated with water, which repels the ink. In lithographic printing, paper is rolled across the plate and the inked image is transferred on to it.

Offset litho uses the same principle but the plate does not print directly on to the paper or the required surface. Instead, the lithographic plate is pressed on to dampened rubber rollers which transfers a solution of water and gum

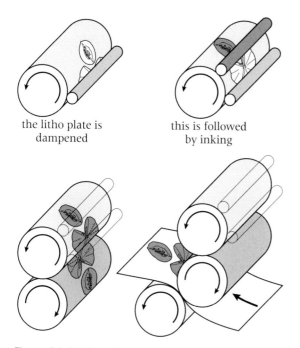

the litho plate is dampened

this is followed by inking

Figure 3.1.22 *The offset litho process*

arabic on to the plate (see Figure 3.1.22). The water solution is accepted on the non-image area, while the image area repels it. Then the ink is transferred on to the plate; it is accepted by the image area and repelled by the non-image area by the water solution. The inked image is then 'offset' on to a rubber blanket cylinder that transfers the image to the paper.

To print a multicoloured image, separate blanket cylinders carrying each colour are required. In multicolour lithographic machines single units are joined together in a row to form multiple unit presses. The sheet is passed through the machine on a system of transfer cylinders which print each colour separately.

There are a variety of printing plates used in offset litho. The most common of these are as follows:

- *Pre-sensitised metal plate.* A negative of the artwork/image is placed on to a thin sheet of metal coated with light sensitive material, before being exposed to ultra-violet light. The image is then fixed using chemical fixer and lacquer.
- *Chemical transfer plate.* A paper negative of the artwork is made and placed on to a paper, plastic or metal foil plate. This is passed through a processor containing developer which transfers the image on to the plate. The image is then fixed with chemicals before the plate is used.

- *Electrostatic plate.* The image of the artwork is reflected on to a special plate made of paper, plastic or metal foil. The zinc oxide surface layer acts as a semi-conductor and an electrostatic charge is created. Grease-receptive particles are drawn to the area which forms the image. The plate is then heated to fix the particles in place.

For small print runs, a basic hand-operated lithography machine can be used. Paper or plastic plates can be produced quickly and cheaply by drawing the image directly on to the surface. High-volume production processes require fully automated machinery which is capable of producing vast quantities. High-quality electrostatic or pre-sensitised metal plates are used for this level of production.

Task
Examine a four-colour printed publication such as a magazine. By looking closely, can you see the colours?

Photocopying
Photocopying is a widely used form of electrostatic printing. Photocopying machines duplicate original images or text. The process is constantly evolving as technology develops and exact processes can vary depending on the machine being used.

- The original of the artwork/text to be reproduced is positioned face down on the glass surface of the machine and held in place with a cover.
- Inside the machine, a metal drum with a photoconductive surface receives an electrostatic charge.
- The original is illuminated and the reflected image is projected through a lens system on to the electrostatically charged drum.
- On the areas where light falls the electrostatic charge disappears. Charged particles remain only in the image area.
- As the photocopying process is dry, a resin-based toner is used instead of liquid ink. This is poured over the surface area of the drum. The powder has an opposite electrical charge to the image area on the drum surface and so settles on it, while the rest falls away.
- The copy paper is electrostatically charged; when it is positioned near the drum surface it draws the powder image on to its surface.

101

- Finally, the powder image is heated so that the resin melts, fixing it to the paper.

Letterpress

Letterpress is one of the oldest methods of printing. It is still in use but its applications are limited as it has been superseded by faster and more efficient processes such as offset litho.

A reverse image of the artwork/text is transferred on to a **plate** photographically. The unwanted areas surrounding the image are treated with acid and etched away. The non-image area is now lower than the image area which stands in relief. Ink is applied to the raised surface area of the plate and this is then pressed on to the paper.

Letterpress was traditionally used as a large-scale manufacturing process for products such as newspapers and books. However, it is now used in high-quality, small-scale production for products such as limited edition books or stationery. Letterpress can be used to print monochrome and coloured images. To print multicoloured images, individual plates have to be produced for each section that is to be coloured.

Typography

Typeface characters sit on an invisible baseline that does not move even when the font is enlarged or reduced (see Figure 3.1.23). Only the ascender and the descender lines shift when the font size is changed.

A huge range of fonts is available. A distinction can be made between fonts which are:

- serif – contain small counter strokes drawn perpendicular to the free ends of the character
- sans serif – characters without the stroke.

Several fonts are derived from a type family; these families consist of a number of typefaces that originate from the same style but have distinct variances like weight and proportion. For example, Latin bold, Latin condensed and Latin wide could be used for headings, subheadings and body text to achieve distinction but maintain visual grouping. Fonts can also be customised or created using software applications like Fontographer.

The size of font is measured in 'points', of which there are two types used:

- pica is equal to 0.35135 mm, which is divided into 12 picas
- computer (used in desktop publishing) is equal to is 0.3527785 mm (1/72 inch).

The choice of font used will depend on its legibility as well as its aesthetic appearance. Legibility is the ease and speed by which the reader can decipher each character and word. For example, lower-case letters are easier to read than upper-case letters (capitals). The spacing between words should be the width of the letter 'i' and there should be more space between individual lines than words. The spacing can be adjusted between letters and lines by the following methods:

- Leading – the total distance from baseline to baseline. The term leading comes from traditional typesetting when strips of lead were inserted between the lines of type to fix the spacing. On computer applications leading is set at 120 per cent of the point size used for the font. In characters of higher points (such as those used in headings) 120 per cent may appear excessive and can be altered to 100 per cent or less.
- Tracking – controls the distance between each letter within a word. Tight tracking is used in large font sizes for headings or headlines, as they can appear too long or take up too much space on the page layout.
- Kerning – pairs of letters such as AV, ay, To, TT, wy, etc., appear to clash when they are placed

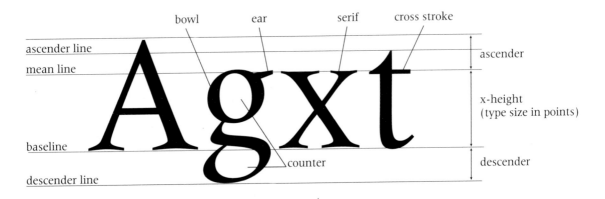

Figure 3.1.23 *Structure of a font*

together. **Kerning** is used to correct the visual disparity between these characters, particularly in large font sizes. Kerning enlarges or reduces spacing between these letters to improve their appearance.

Task

Type a selection of fonts that are serif, sans serif and grouped in a family.

Finishing techniques

Print finishing is a term that refers to a system of operations that prepare the printed product for distribution. There are numerous finishing techniques; the major ones are outlined below.

Imposition

A number of pages are arranged on to a large sheet so that when it is folded the pages will appear in the correct order. This is called a signature. A simple leaflet can be arranged so that the sheet is printed on one side, then turned over and printed on the other. Pages 2 and 3 are printed on the reverse of 1 and 4. This can then be folded to read as a four-page leaflet. Larger, more complex products like books require larger sheets that have to be folded and cut to make signatures.

Folding

Folding can be done by hand or using a machine depending on the amount of material that needs to be folded. For high-volume operations fully automated folding machines are required.

A sheet with a number of pages arranged on it is passed over a flat bed in which there is a long slot. A blunt blade designed to fold but not cut the paper pushes the sheet into the slot. A set of rollers folds the paper at the required point. This process is repeated until all the pages on the sheet have been folded to make a signature. The doubled page edges are cut or slit automatically by the machine.

Scoring

Heavy material such as board is difficult to fold and needs to be scored beforehand. The material is scored on the outside edge of the fold to break the top layer of fibres, making it more flexible. It is important that the scoring is not too heavy as it will crack or cut the material, or too light as this will affect the accuracy of the fold which may feather along the crease. Scoring can be done by hand using a ruler and blade, or on a machine if the material is too heavy.

Cutting

The instrument used for cutting is a guillotine. The size and degree of automation of the guillotine varies between models.

Hand-operated machines are used in classrooms and workshops for occasional and small batch work. A basic guillotine consists of a flat base board with a bar fixed on the edge to carry an encased blade that moves along the length of the edge. Some guillotines have an adjustable edge against which the material is positioned to ensure an accurate cut. Other models will feature a grid or measuring system to fulfil the same function.

Fully automated machines are used to cut large volumes of printed material in one operation. These machines can be programmed to precise specifications and can make a series of cuts using a number of blades. Automated machines use adjustable back and side gauges to position the material correctly for the cut.

Gathering and collation

Gathering can also be referred to as collation. Gathering involves bringing together components of the product into the correct sequence to make up a complete product. This can be done by hand, for example piles of identical pages (i.e. page 1s, 2s, 3s can be laid out in a row and one sheet can be selected from each pile to compile a document). This process is suitable for smaller quantities; for larger quantities machines are used.

Fully automated machines work on a similar principle to the hand process. Items of the same printed material are placed in separate boxes or stations (also known as hoppers) along a belt in sequence. One item from each pile is fed on to the belt sequentially to compile one complete job. This can be done at high speed and can be incorporated into the stitching and binding process to streamline the task.

The term collation also refers to the process of checking the gathered sections to ensure they are in the correct order. This can be done by hand or by a machine using collation marks (also known as back or black step marks) which are placed in a set pattern so that when sections are compiled the marks appear in a sequence or progression on the reverse of the sheets.

Decorative finishing techniques

There are a number of finishing techniques that fulfil an aesthetic function. They include:

- mounting – attaching a sheet on to a strong backing for support or display purposes
- embossing – raising the surface of paper or card in the shape of an image or text so that it stands in relief
- laminating – applying a transparent film of plastic to a surface to protect it and enhance the appearance.

Binding

Binding is a process used to fasten or hold together a number of printed sheets. Many products are bound such as magazines, books, reports and leaflets. Binding can range from the simplest forms, for example stapling, plastic or ring binding, to fully automated processes. There are various methods of binding. In addition to aesthetic considerations, the quantity of paper to be bound and the cost are determining factors as to which process is used. The three main methods are:

- Stitching and sewing. There are two types of wire stitching used in this process. Paper that is folded to make a signature requires saddle stitching down the spine while loose pages are attached at the spine by side stitching. The advantage of saddle stitching is that the book can be laid flat when read whereas the side stitched book cannot. Sewing is a more expensive option but it is neater and more durable – it is often used in hardback binding. Sections or signatures of the book are sewn together and a case of board is covered with material. The case is attached to the sewn sections with strips of adhesive tape that have been sewn into the spine. The strips are covered with endpaper, which are the leaves at the front and back of the book that cover the inner sides of the case.

- Mechanical binding. There are several types of mechanical binding. However, most involve drilling or punching holes through the paper and threading plastic or wire through to keep the pages together. Spiral and plastic comb binding are two of the most frequently used.

- Perfect binding. This is commonly used for paperback books. The sections of the book are cut with a guillotine and the spine is roughened so that it will accept the glue. After the sheets have been gummed along the spine the cover is creased and glued in to position.

4. Testing materials

British and International Standards

Products will be marked with the following testing standards:

- BS – British Standard
- BS EN – British adoptions of European Standards
- ISO – International Standards.

The **Kite Mark** is a visual motif displayed on products and indicates stringent and regular independent testing, and quality testing to British, European and international standards (see Figure 3.1.24).

The CE mark is a form of marking designating the manufacturer's claim that the product meets the requirements of all relevant European Directories. It is a legal requirement that a product covered by a European Directive must have a CE if it is to be sold in Europe (see Figure 3.1.24).

The purpose of these standards is to allow the customer to see if the product conforms to a level of safety, quality and fitness for the product's purpose.

Figure 3.1.24

Think about this!

Your coursework project in Unit 2 should involve testing of some description. You should, however, as with any design work, consider the standards to which you have to design to. The relevant standards should be consulted in the design stages.

Safety testing under controlled conditions

Any testing carried out by the BSI, or on your own work, should be carried out under controlled conditions. It is essential that the tests are identical, so that when testing a range of products, they will all be subjected to, for example, being dropped from the same height or being hit with the same force.

Fire testing obviously needs to be carried out under controlled conditions. Fire regulations are an important aspect that have to be considered when using many modern plastics and textiles since some very toxic gases can be given off when burning. Many of these materials have now been banned and special grades of upholstery foam have to be used for furniture and bedding, and special finishes are used to improve resistance to fire.

The use of information and communications technology in testing

As information and communications technology (ICT) has developed, testing is increasingly being carried out using computer modelling. Traditionally, car safety was tested using a crash dummy. Now, complicated computer programs are being used to simulate how the car will react when hit in a number of ways and this means that the cars no longer have to be destroyed during testing.

Virtual reality modelling

Virtual reality modelling provides a very useful tool in the visualisation of architectural design. Buildings can be modelled on screen to show the internal and external organisation and layout. This technique can also show how a building may look in different materials and at different times of the day and year as the weather and position of the Sun changes.

Test programmes before manufacturing

ICT also allows manufacturers to run test programmes to assess the viability of making and manufacturing in the long term. They are able to assess tooling costs and times and how they might need to hold the work during machining.

Testing of products sometimes involves testing them to destruction and although this is appropriate for smaller products, it is not possible on aircraft or ships. In this case, computer models are used with results obtained from testing smaller scaled versions. Testing is carried out in wind tunnels or water tanks and the results gathered are used to build the computer models for the full-sized product.

Ultrasound testing

Ultrasound testing is used to inspect items such as alloy wheels or welded underground pipes that cannot be tested to destruction. A frequency between 500kHz and 20MHz is passed through the component, and where a flaw exists a reflection is caused on the ultrasound vibration.

Standard destructive tests and comparative testing in the workshop

Standard destructive tests and comparative testing can be carried out in your school/college workshop (Table 3.1.4). A tensometer can be used to carry out tensile testing of materials. In order that a fair test is undertaken, the specimen pieces have to conform to British Standard sizes. The force applied and extension of the material is recorded and can then be plotted on a force-extension graph (see Figure 3.1.25). The tensometer can also be used to test the strength of joints, the bending of materials and their compressive strength. With special tooling they can also be used to test the hardness and shear strength of materials.

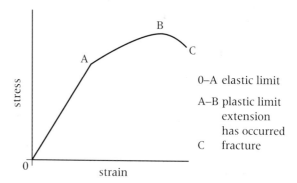

Figure 3.1.25 *Force/extension graph*

Table 3.1.4 *Standard destructive tests and comparative testing in the workshop*

Test	Method	Reason
Tensile	Tensometer	If a material is not strong enough for its purpose it would fail in use. The tensometer measures tensile strength of materials to determine what limits they can be put to.
Hardness	Brinell Vickers Rockwell	Materials must be capable of withstanding wear and indentation. If the incorrect material is chosen the product may fail. These three tests measure a material's ability to withstand scratching or resistance to indentation.
Toughness	120D impact test	The measurement of resistance to shock loading. The material may break and fracture if not tough enough, again leading to component/product failure.
Ductility	Bend tests	If a material is too ductile it may stretch or deform, resulting in mechanical failure of the component or product.

Exam preparation

Practice exam questions

Below are some specimen examination questions. They are similar to ones you will get in your examination.

1 Choose two models from the list below and compare the working properties of the materials used. You should comment specifically on making joints when modeling these materials.

    a) a mock up in Styrofoam for a new mobile phone
    b) an architectural model in PVC sheet and rods of a new railway station building
    c) a model in balsa wood of a supermarket shelf unit.
    d) a model in 700gm card to demonstrate the operation of a piston. (5)

2 a) Give a practical application where each of the following graphical techniques would be used. Justify your answers.
    i)   self adhesive lettering (2)
    ii) dry transfer lettering. (2)

    b) State why both of these techniques are used in preference to more traditional lettering methods. (1)

3 ICT is increasingly used in the production of display posters.
    a) Give FIVE advantages ICT has compared with traditional methods for this application. (5)
    b) State what methods that could be used to reproduce the poster in colour for a print run of:
    i)   10 000 (1)
    ii) 500 (1)

4 A teddy bear, which measures 300mm from head to toe and 200mm from paw to paw, is to be sold in an open front box to enable the potential buyer to squeak it's tummy. Sketch a well proportioned net for a suitable box to hold the teddy. Indicate any folds with dotted or broken lines. (5)

5 A plastic bottle has a feint line running down each side and across the bottom. The bottle top has a indented nipple on the top of it. Using annotated sketches, describe the process used to make the:
    i) bottle (3)
    ii) top (3)

6 a) Name the classes of plastic to which the following structures belong:
    i)   polythene (entangled chain) (1)
    ii) polyester resin (covalent cross links) (1)

    b) Explain how the structure of each plastic determines its characteristics. (3)

UNIT 3B1
Design and technology in society (G302)

Summary of expectations

1. What to expect
Design and technology in society is one of the two options offered in Section B of Unit 3, exam Paper 2. In this option you will consider the consequences of design and technological activities on society, the way in which professional designers work and how anthropometrics and ergonomics are used in designing and making.

2. How will it be assessed?
The work that you do in this option will be externally assessed through Section B of Unit 3, Paper 2. There will be two compulsory, long-answer questions, each worth 15 marks. These questions will ask you to apply your knowledge of the subject matter. You will normally be asked to relate your answer to a specific product or group of products, or to give examples of the wide range of influences on design and the consequences of design and technology activities for individuals and for society at large.

You will be expected to show a clear understanding of the technology associated with this option and to be able to apply it to an open-ended question. Your answers should be from the viewpoint of the specialist materials which you have chosen to use.

You are advised to spend 45 minutes on Exam Paper 2.

3. What will be assessed?
The following list summarises the topics covered in this option.

- The physical and social consequences of design and technology on society:
  - the effects of design and technological change on society
  - influences on the development of products.

- Professional designers at work:
  - the relationship between designers of one-off, batch-produced and high-volume products and clients, manufacturers, users and society
  - professional practice relating to design management, technology, marketing, business and ICT
  - the work of professional designers and professional bodies.

- Anthropometrics and ergonomics:
  - the basic principles and applications of anthropometrics and ergonomics
  - British and International Standards.

4. How to be successful in this unit
To be successful you will need to:

- have a clear understanding of the subject knowledge covered in this unit
- be able to apply your knowledge to a given situation
- organise relevant material clearly and coherently, using specialist words and terms where appropriate
- use clear sketches and drawings where appropriate to illustrate your answers
- be able to write a clear and logical answer to an examination question, using accurate spelling, grammar and punctuation.

5. How much is it worth?
This option, with Section A, is worth a total of 30 per cent of your AS qualification. If you go on to complete the whole course, then this unit accounts for 15 per cent of the full Advanced GCE.

Unit 3 + option	Weighting
AS level	30%
A2 level (full GCE)	15%

1. The physical and social consequences of design and technology on society

The effects of design and technological changes on society

Developments in design and technology do more than just make life easier, quicker or more satisfying. They change the way we live our lives. An innovation can shift patterns of employment, give or deny access to goods and services, bring wealth and freedom, or poverty and servility to millions. This can be seen in even the simplest, earliest use of tools by human beings when the shaping of crude knives and spears and the daubing of cave walls with images of animals led to fundamental changes in the way people worked together to hunt and gather food.

Until very recently, the rate of change on society has been evolutionary, i.e. successive generations only had to contemplate very minor change throughout the whole of their lives. Things were generally stable and predictable. The size of the population that needed to be fed and housed grew slowly, and largely insignificantly.

Only in the last 250 years, since the Industrial Revolution, has the rate of change in the western world become noticeable within the time-span of a generation. During that time we have come to expect increasingly frequent changes to the way we live our lives, and are starting to accept and plan for the fact that the way we do things today will not be the same as tomorrow. At the same time, there has been a rapid rise in population, creating much greater demands for food, housing, education, health care and mobility.

The latest developments in genetic engineering, materials technologies and information and communication technologies suggest that this rate of change, and its consequent impact on society, is progressing at an even greater rate. If anything we are moving from an age of continuous change into an age of discontinuous change, where changes happen suddenly and unexpectedly, and are almost impossible to predict and plan for.

Design and technology has improved the lives of millions of people across the world, and will continue to do so. It has also brought hardship and dissatisfaction. At the same time, it has brought about change that has resulted in far-reaching physical and social consequences.

Mass production and the consumer society

Just a few hundred years ago, the objects we used in our daily lives were all made by hand. The only way of reproducing words or images was to copy them, again by hand. Monks painstakingly copied medieval religious manuscripts.

Printing, the reproduction of marks on a surface, is probably the first process that enabled copies of something to be made in quantity – long before the Industrial Revolution. The history of the reproduction of images is older than that of type. Thousands of years ago the Chinese and Indians started printing textiles using wooden blocks with a design carved out of them. The Egyptians used relief stamps to transfer decorations on to pottery (see Figure 3.2.1). From the eighth century the Chinese used similar methods to print designs representing words, and printing from movable type originated around 1040 CE in China. In Europe, printed textiles were first used around the sixth century. The first woodcut prints on to paper were not made until the late fourteenth century, but the process was slow, and the wooden blocks soon wore out.

The engraving of decorative designs into metals is also an ancient idea, and its application as a printing process also dates back to fifteenth-century Europe and the production of playing cards and religious images. The simpler process of etching also developed around the same time, but did not provide a durable enough plate to provide for longer print runs. Shop signs, containing graphic representations of the products being sold, were essential in the Middle Ages as few people could read.

Figure 3.2.1 *Amratian vase, dated before 3100 BC*

Figure 3.2.2 *Johann Gutenberg, a goldsmith by trade, invented printing from movable pieces of cast metal type. His invention revolutionised the printing industry*

The use of movable metal type, and the first printing press, was developed by Johann Gutenberg in Germany in 1455, in response to the need for books from a growing community of scholars and traders (see Figure 3.2.2). This coincided with the development of large-scale paper manufacture in Europe and in new printing inks. For the first time, it became possible to reproduce texts and simple decorative motifs in quantity. There were extensive job losses among the scribes and copyists of the time as the demand changed to skilled workers able to produce the metal type. The first book in English was printed by William Caxton in 1473 in Belgium, and the first book to be printed in England was in 1476.

The earliest typefaces imitated handwritten styles, but gradually the more familiar less ornate typestyles of today were developed. These were easier to produce, print from and read. The emphasis was on the labour-intensive production of high-quality type which would be used to print on to high-quality paper. During the fifteenth to seventeenth centuries printed products remained expensive, therefore, and only accessible by the wealthy, and those who had been taught to read.

Engraving remained the main method of reproducing visual images until the development of the process of lithography in 1798, in which nitric acid was used to burn away non-printing surfaces, providing a much sharper image. The printing and distribution of the musical compositions of Mozart were one of the first applications of this process.

The Industrial Revolution

Before the Industrial Revolution, 3D products were made by hand. The craft process brought about complex, elegant and highly efficient artefacts, illustrated manuscripts and structures, in which form and function interact with a degree of perfection that in many ways are unsurpassed today. The craft processes and designs evolved slowly over many centuries, and were passed down to successive generations through word of mouth, years of apprenticeship, and constantly and subtly refined through processes of trial and error. Thinking, planning and making were closely related.

The invention of the steam engine in the mid-1700s led to the mechanisation of many craft processes, and to the speed and ease of transport over increasing distances. This meant that goods could be made at much lower cost in increasing numbers and distributed and sold more cheaply to a much wider market. It also brought about radical changes in the way that products were created. The shapes and forms and dimensions of products needed to be drawn up in some way so that many people could make identical copies. This also meant that different people could make different pieces of a product that could then be assembled together, which in turn meant that larger and more complex products could be manufactured. The person who drew up the plans increasingly ceased to be the person who made the product.

The process of design changed radically too. There was no time for evolutionary trial and error. Ideas had to be explored rapidly on paper and through models and prototypes. The classic drawing systems (such as orthographic and perspective views and symbols, which had been in use for centuries) were adapted and developed to create conventions that enabled designer and maker to communicate effectively.

Unfortunately, in the drive for profit by the

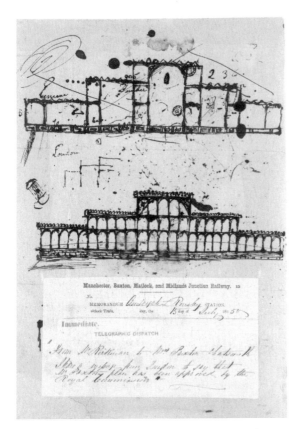

Figure 3.2.3 Paxton's first design – on an envelope – of Crystal Palace

Figure 3.2.4 An ornate Victorian sideboard

new industrialists, both the quality of the products of the Industrial Revolution and the quality of life for the workforce were often sacrificed. In Victorian times, industrially produced products were not generally well received, and often excessively ornate and highly decorated to cover up poor manufacture, as typified in the products displayed in Crystal Palace at the Great Exhibition of 1851 (see Figures 3.2.3 and 3.2.4).

The division of labour and the introduction of the production- or assembly-line approach to manufacture developed most quickly in the USA during the mid nineteenth century in industries such as slaughterhouses and garment production. In the early part of the twentieth century Henry Ford pioneered the mass production of motor cars.

Meanwhile the Industrial Revolution brought commercial demands for novelty and speed, and a better informed public and workforce. Many industrial process were mechanised, and printing, papermaking and typesetting became increasingly automated, often improving speed at the expense of quality.

Manufacturers needed to promote their new products to a wider market and required a wider range of printed promotional materials such as posters, leaflets and advertisements. The packaging industry also became established, along with the newspaper industry. It was during this time that the bigger, bolder more distinctive 'display' and more mechanical looking sans serif typefaces were developed.

The first steam-operated mechanical printing press was invented in the early nineteenth century. It enabled 1200 sheets to be printed per hour, instead of 300. Two machines were initially purchased by The Times and the newspaper was printed on them from 29 November 1814, to be followed by all the major European newspapers. The machine was eventually superseded in 1863 by the much faster rotary press.

Figure 3.2.5 A mechanical printing press

Figure 3.2.6 *A Linotype typecasting machine*

Figure 3.2.7 *A transport poster from the 1930s*

By 1886, the first mechanical typesetting machine had been developed. The 'Linotype' cast lines of type rather than individual letters, entered by means of a keyboard (see Figure 3.2.6). The speed and economics of this machine revolutionised the newspaper industry. The similar 'Monotype' was used for the reproduction of books. Developments in lithographic production processes during the last quarter of the nineteenth century produced the first colour advertising posters.

The reproducible photographic print was also developed during the mid-nineteenth century, but photographs did not come to be mass reproduced in books, magazines and newspapers until the early twentieth century with the development of the half-tone screen.

The development of the consumer society

As international commerce and transport systems developed during the late nineteenth century, new opportunities for 3D and graphic products emerged such as luxury ships, urban transport systems, hotels and theatres (see Figure 3.2.7). Technological development has often been led by the need for military advantage in conflict, and World War I prompted the development of new goods and services. The introduction of the National Grid in the late 1920s provided the catalyst for the widespread growth for electrical 'labour-saving' household products such as radios, cookers, toasters and vacuum cleaners.

As the standard of living improved, the demand for new products increased. Advertising became an increasingly important industry, using market research, packaging and product styling to sell the new products. Mass-produced products were designed with the efficiency of their production processes in mind.

The ideas behind quality control systems and ergonomics were pioneered during World War II to help ensure the reliability of planes and the efficiency and survival of pilots and weapon systems. The development of early computer and **transistor** technology was accelerated during this time.

Early in the 1950s, the first 'phototypesetters' appeared, but they were not particularly successful. It was not until the 1960s that new machines became available that were faster and began to utilise emerging electronic capabilities in terms of control and computer memory. These machines made features such as kerning possible, and generally offered the designer much greater freedom and flexibility. Letraset (dry-transfer lettering) was introduced in 1961.

The Whole Page Facsimile Transmitter, better known today as the fax machine, was commercially developed in the UK at this time, though it took over 12 minutes for an image of a page to be transmitted and reproduced. The idea of the photocopier (or electrophotography as it was called) had been patented in New York in 1937 and eventually demonstrated in 1948, but was not successfully commercially developed until the 1960s and 70s.

During the 1950s and 1960s, the consumer market grew rapidly with an increase in the population. Cheaper printing and the advent of television provided increased opportunities for advertising. Established products were regularly given new 'features' to stimulate demand, and designed to have a much shorter usable life, to be thrown away when the fashion changed. People no longer bought products just for their function, but to make a lifestyle statement that expressed their affluence, taste and aspirations (see Figure 3.2.8).

Task

Identify a 3D product and a graphic product that you own. Discuss the following:

a) What influenced you to choose to buy it (price, features, appearance, etc.)?
b) What information did you have about the product before purchasing it (e.g. through ads, product reviews, packaging)?
c) How long do you expect to be using it?
d) What will you do with it when you stop using it?

Figure 3.2.8 In the 1970s, products were purchased to make a lifestyle statement

The 'new' industrial age of high-technology production

Since the mid-1980s, there has been a growth in 'high-technology' manufacturing. Not only has there been growth in industries such as electronics, but also in the use of high-technology production methods, new types of materials and new processes. These new developments present designers with opportunities to create new and exciting products.

Miniaturisation

The most important technological development in recent years has been in the field of microelectronics. Not only have products become smaller through advances in microchip technology, but previously unimaginable products have been developed (see Figure 3.2.9).

The first transistor radios and televisions appeared in the late 1950s and the colour video recorder for home use was developed in 1966. By far the most influential product in the 1980s was the 'Walkman', which had an unprecedented impact on people's lifestyles. Later developments include products such as wristwatch-style TVs and portable CD players.

In the past, it was easy to recognise the function of a product, such as a typewriter or watch. Nowadays, the impact of miniaturisation may mean that products effectively 'disappear', in that the product no longer visually represents its function, or the function is not clear. The designer then has to convey this function by other means. If a calculator is built into a wristwatch, is the product designed as a calculator or a watch? How is the purpose of such a product made clear?

Figure 3.2.9 The Casio calculator (left) takes advantage of miniaturisation (possible with electronics) to include a wide range of calculator functions in a small case. With the Zelco calculator (right) high-tech production produces an attractive and ergonomically sound shape

Figure 3.2.10 *Interactive watches, similar in style to this one, will allow the wearer to access the Internet and connect with a PC*

Swatch of many functions

Swatch watches were first marketed in the early 1980s when the Swiss watch industry was fighting for survival against Japanese market expansion. Swatch decided to target the lower end of the market in order to produce affordable watches that were designed around the concept of style and fashion. The technological aspect took a back seat.

Now things have changed and the marketing climate is different. Technology is now a major driving force for design, having become seriously 'trendy'. The most recent developments by Swatch are interactive watches with so-called 'access technology'. This type of watch can act as a ski pass and metro ticket. It has already been in use in European ski resorts.

The next development for Swatch is the 'Internet Swatch', which has the potential for you to email, access data from a PC, book tickets and browse the web from your watch (see Figure 3.2.10). Swatch is also working on 'Swatch Talk', a watch that also acts as a mobile phone. The aim is to make people who wear the Swatch brand feel 'up to the minute', with the opportunity to be permanently connected to the Internet, allowing mobility and interactivity.

Task

Investigate other high-tech products that are interactive. How many functions does the product have? Why is it necessary for one product to have more than one function? Explain how the styling of the product enables the customer to understand the purpose of the product.

The global market place

The need to be increasingly competitive in the world market means that many companies market and sell their products world-wide. To be successful in the **global market place**, a company must have a product that appeals to different cultures (see Figure 3.2.11). A product may have to be remodelled to include different design features depending on where it is to be sold. Remodelling may involve:

- increasing the amount of recyclable materials used in the product
- whether a plug is fitted or not
- the fitting of different visual displays
- using devices to suppress noise levels.

Sometimes a product is sold under another name in different countries, even though they are basically the same design. For example, the Vauxhall Cavalier as it is known in the UK is sold as the Opel Ascona in the rest of Europe.

Global manufacturing

Global manufacturing is closely linked to the growth of **multinational companies**, which operate in more than one country. In the past, multinationals were mainly associated with mineral exploitation or with plantations, such as for cotton or food. Since the 1950s, many multinationals have been involved in **global manufacturing**, especially of cars and electrical goods. Today, global manufacturing is growing at an increasing rate, mainly due to international competition and developments in information and

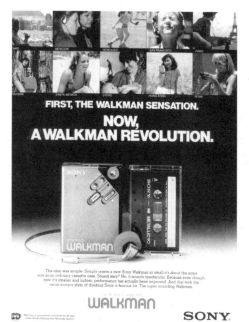

Figure 3.2.11 *In this ad, a designer has tried to show how a product spans different cultures*

communications technology (ICT). Global manufacturing covers a wide range of activities, such as:

- petroleum (BP, Exon)
- motor vehicles (Ford, General Motors)
- electrical goods/electronics (Philips, Sony, Hitachi)
- financial services (Barclays)
- food and hotels (Coca Cola, McDonalds, Trust House Forte)
- textile and garment manufacturing (DuPont, Marks and Spencer).

High-speed revolution

The high-speed information revolution that has come about through developments in ICT will continue to increase international competition. For many companies, competition means reducing labour and material costs. Global manufacturing is a means of doing just this, since moving manufacturing to another country can make use of lower labour costs, thus reducing one of the highest costs of manufacturing. The trend towards global manufacturing often includes designing products like electrical goods or cars in one country and manufacturing in another. This is increasingly cost effective because of improved electronic communications. Large-scale products such as aircraft engines may be designed and engineered by teams working together across the world and assembled using component parts that are sourced from supply chains in different countries (see Figure 3.2.12).

Issues related to global/local production

Issues related to local/global production are concerned with the effects of the global economy and of multinationals on quality of life, employment and the environment. While the head offices of many multinationals are often located in more economically developed countries (MEDCs) (such as western Europe), some multinationals are based in newly industrialised countries (NICs) such as Singapore or Taiwan. Less economically developed countries (LEDCs), such as those in Africa or Asia, have generally welcomed multinationals and the benefits that locating manufacturing there brings. However, there are disadvantages for both NICs and LEDCs countries in global manufacturing.

Advantages of global manufacturing for NICs and LEDCs countries:

- It provides employment and higher living standards.
- It may improve the level of expertise of the local workforce.
- Foreign currency is brought into the country to improve their balance of payments.
- It widens the country's economic base.
- It enables the transfer of technology.

Disadvantages of global manufacturing for NICs and LEDCs countries:

- It can cause environmental damage.
- The jobs provided may only require low-level skills.
- Managerial roles may be filled by employees from MEDCs.
- Most or all of the company profits may be exported back to MEDCs.
- Multinationals may cut corners on health and safety or pollution (which legislation would prevent them from doing in their home country).
- Multinationals can exert political pressure.
- Raw materials are often exported or not processed locally.
- Manufactured goods are for export and not for the local market (where many could not afford them anyway).
- Decisions are made in a foreign country and on a global basis, so the multinational may pull out at any time.
- With increased mechanisation, there is a reduced need for the local workforce.

Influences on the development of products

The design and manufacture of products is a complex affair. Why some products are successful, why they are made as they are and how they are used and disposed of are issues that affect every one of us. How do we choose which products to buy? Do we really need all of them? How do we recognise a well-designed product? Why is design important?

The Design Council recently conducted a survey of 800 UK manufacturers about the contribution

Figure 3.2.12 *The skills of designers, engineers, suppliers and many others were used in the development of engines for the Airbus A330 plane. This would be impossible without the use of electronic communications which enables collaboration with partner companies across the world*

of design to the UK economy. The results were very clear. Ninety two per cent of businesses agreed that design helps to produce a competitive advantage and 87 per cent believed it increases profits and aids diversification into new markets.

Design is clearly important. It is interesting to note that in sixteenth century England the word 'design' meant a 'plan from which something is to be made'. Today, we normally use the term 'design' to mean the drafting and planning of industrial products. One outcome of industrialisation has been the requirement for the profession of the 'designer'.

Tasks

Global manufacturing highlights the many moral and ethical questions, the so-called value issues, that are inherent in product design and manufacture. Investigate further one of the following issues related to global economy manufacturing, to find out who are the winners and losers in terms of jobs, quality of life, and the environment:

1 What are the effects of the global economy on lifestyles in LEDCs? Is it ethical to advertise products that many people cannot afford? Should MEDCs impose their values on traditional cultures?
2 Do multinationals have a responsibility to society and the environment? What are the effects of building new factories and transporting raw materials and products? What are the effects on energy demands in NICs and LEDCs? Should multinationals follow sustainable manufacturing practices?
3 The effects of deforestation and over-use of the world's natural resources are plain to see. How can we avoid the effects of global warming or the loss of bio-diversity in many countries?
4 Moving manufacturing away from MEDCs causes unemployment there. As competition increases, many NICs are also affected by further relocation of manufacturing, such as the move from Hong Kong to China. How can the threat of unemployment be overcome in a global economy?
5 LEDCs have to pay off debts to world financial institutions rather than spend money on food and development in their own countries. This may mean a country exporting food to pay debt during a famine. Is this ethical?

For a company to make a profit, product design must be **market led** and **market driven**. It must take into account the needs of the target market group and the various influences of market trends, including colour and style. For many consumers these days, design has become an important means of self-expression. Consumers choose products not just for what they do, but for what they tell the world about them. Products are no longer simple, functional artefacts. We buy a diver's watch not because we want to spend hours under the sea, but because a high performance product has a sense of glamour. We buy a chair not just as a comfortable product to sit on, but also because it can express a sense of tradition or modernity.

Now that products are mass-produced and sold in millions, the primary purpose of the designer is to inject a sense of personality in a mass-produced object. For example, the first telephone or early typewriters were the result of mechanical solutions. Nobody had any preconceived idea of what they should look like. As new products were invented, it became necessary to define their form.

Task

Compare the design styles of a product such as the radio. What did the first radio look like in comparison with models today? Explain the differences in terms of the shape, form, styling and colour. Is it easy to see how to use the product? What sizes are the controls?

Product reliability and aesthetics

Product reliability is no longer a major issue. Most products carry guarantees and there is consumer legislation to support product quality and safety. Most brands within a given product category perform equally well. The main reason for buying a specific product is often how it looks and feels – its aesthetic qualities. The product's function, relating to its performance is often taken for granted. The job of the designer then, is increasingly to give a product an image. Should the product look traditional, retro, or high-tech? At the same time, the consumer should understand how the product works, so that it is obvious how to operate it.

Form and function

The connection between form and function has been one of the most controversial issues in the

Figure 3.2.13 *Harley Davidson bikes can be customised*

history of design. When products were first mass-produced in Victorian times they were highly decorated to look like hand-made products, whether the decoration was appropriate or not. The development of 'reform' groups such as the Arts and Crafts Movement gradually brought about a change in the concept of design. The form of products was to be

Task

The Harley-Davidson has become a cult bike for many people (see Figure 3.2.13). The bike is sold in a standard form, with the opportunity of customising it by buying inter-changeable parts. Describe the image that such a product is perceived to bring to the user.

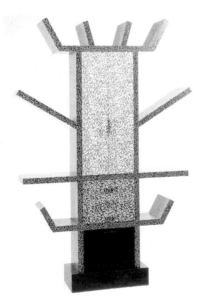

Figure 3.2.14 *The Casablanca sideboard designed in the Memphis style by Ettore Sottsass – balance involves the proportions of a product*

Figure 3.2.15 *Victorian decoration was often an afterthought or sometimes applied to disguise the function*

Figure 3.2.16 *A modern chair by Robert Venturi – from 1910, decoration became an integral part of the design*

simplified and the products well made from suitable materials. At the turn of the nineteenth century, developments in materials and technology enabled the production of innovative new products such as the telephone. Many of these products were so innovative that no one knew what they should look like! The development of mass production required that products be standardised, simple and easy to produce. The supporters of functionalism therefore suggested that the form of a product must suit its function and not include any excessive decoration.

From the functionalism of the early twentieth century and even up to the 1970s, the functional and technical requirements of mass production were used as a benchmark for the form of an industrial product. In other words, for a product to be mass produced at a profit, it needed to be simple and easy to produce.

The argument about form and function still continues. As a student designer, you are asked to simplify your designs for ease of manufacture. One of the many reasons for doing this is to give you time to manufacture your product. The concept of form and function is complex. Think about the design of modern products and of the development of styling to market them. Think about developments in the technical design of products. Think about marketing and image.

These days a product is thought to have three basic functions:

- its practical and technical function, if it works efficiently
- its aesthetic function, how it looks and feels, its styling
- its symbolic function, or the image it gives the user.

It is recognised today that design fulfils not only technical and aesthetic functions, but it is also a means of communication. Design tells the onlooker something about the user.

Aesthetics and function are clearly important to the success of products (see Figures 3.2.14 – 20). Making changes in a product's colour and styling is often a gradual process, because gradual changes make it easier to control development costs. It also means that the company can maintain its own 'personal' style. Where the changes are gradual, an existing product design specification is often used to develop a new one.

Task

Explain why image and style are increasingly important to the design of products. Choose two products that illustrate your answer. Explain why both are attractive to the consumer.

Figure 3.2.17 *The Drop chair, so called because of the shape of the seat, was designed by Arne Jacobsen*

Figure 3.2.19 *Shape and graphical images can be used to indicate control functions*

Figure 3.2.18 (top) *This functional light designed by Paul Henningsen in 1958 was inspired by the artichoke*

Figure 3.2.20 (bottom) *Texture used as decoration on a radio*

Culture

The design of a new 3D or graphic product does not emerge from a blank sheet of paper. Beyond the basic requirements of the brief are the available materials and production technologies of the time and the need to reflect the expectations of the consumer in terms of appearance, or 'style'. The styles of the day develop from the cultural ideas and concerns of the time, and often have their origins in the creative insights of an individual or small group of designers who challenge existing ideas and ways of doing things. The Dyson vacuum cleaner and the iMac computer are prime examples: familiar products appear in entirely new forms, gain public acceptance and are then applied to a much wider range of products. Indeed, most 3D or graphic products are derivative, in that they are a reworking, combination and further development of previous design ideas. Throughout history, designers have been influenced by what they see and the products they use.

The brief accounts of the major design movements and their leaders that follow provide only a brief overview of the way in which new products have emerged during the past century. As such, they provide a starting point for further investigation, and stimulation for the personal exploration and discovery of the fascinating way in which new ideas emerge and come to influence the design of the everyday objects that surround us.

Arts and Crafts movement

In 1890, the English designer William Morris attempted to revive interest in a return to the classic quality previously achieved using craft processes. His basic philosophy revolved around a love of simple things and a revolt against the over-decorated products of the Victorian age.

Morris was opposed to mass production and saw the effects of industrialisation as polluting the environment, giving poor working conditions and producing poor-quality, mass-produced products. He and the others in the Arts and Crafts movement wanted well-crafted, well-designed **consumer goods** to be produced. They promoted the idea of craft activity as being superior. They believed that design should be based on simple, organic forms from nature and stressed the importance of art in the practice of design.

Morris founded the Kelmscott Press, and several further private presses were established during the early twentieth century (see Figure 3.2.21). These private presses commissioned the design of the contemporary roman-style typefaces that are widely used today, as in this book, which uses a face called Latin725, designed around the 1870s.

Morris was also a talented designer of such products as furniture, wallpaper and carpets, as well as being a leading writer of prose and poetry.

Task

Find out more about the work of the following designers:

- William Morris
- C R Ashbee.

Art Nouveau

Art Nouveau was an important international movement that developed during the late nineteenth century. It is characterised by the use of organic, free-flowing lines and shapes.

Figure 3.2.21 *William Morris founded the Kelmscott Press*

Examples can be found in architecture, wrought ironwork, glass and furniture, jewellery, fabrics and wallpaper. The development of the advertising poster at the end of the nineteenth century was an ideal medium for graphic designers to apply the Art Nouveau style (see Figure 3.2.22). The design of the flowing, organic typefaces was interwoven into floral and decorative devices and swirling figures. The Art Nouveau designs of each European country and in the USA all have subtle differences.

Task

Find out more about the work of the following designers:

- Hector Guimard
- Victor Horta
- L C Tiffany
- Galle
- Toulouse Lautrec.

Figure 3.2.22 *An Art Nouveau advertising poster*

Modernism and the Bauhaus

The industrial designs that came to characterise the twentieth century originated towards the end of its first decade. The designs of the Scottish architect C R Mackintosh and the Viennese designers J M Olbrich and Otto Wagner began to use geometric shapes that had more potential for mass production. In 1907, the German Peter Behrens was appointed design coordinator to the electrical firm AEG, and was responsible for the design of the factory buildings, its products and publicity material. The logo he designed probably represents the first example of the implementation of a corporate identity scheme.

Dutch, Italian and German designers began to revolutionise typography and graphic layout. They sought to create images and forms that were appropriate to the machine age. Words became abstract shapes, and the symmetry and decoration of Victorian times became a thing of the past.

These new industrial styles were further developed and refined by the designers of the Bauhaus between 1919 and 1933. The Bauhaus was a school of art and design founded in Germany by Walter Gropius, who had earlier worked for Peter Behrens. It laid down many of the principles of design that are followed today and that are still taught in schools and colleges. One of the aims was to produce designs suitable for mass production, though it was not until the latter part of the century that products such as furniture and lighting were truly made cheaply in quantity. Another aim was to produce designs that were 'international', i.e. that would have appeal across all western countries.

Many of the Bauhaus's students went on to become leading designers in the fields of architecture, furniture design, ceramics, metalwork, textiles, stage design and photography. Laslo Maholoy-Nagy and Herbert Bayer were the most influential typographic designers, creating simple geometric typestyles such as bauhaus (which does not use any capital letters) and Universal that are in common use today. Many famous European artists also studied there.

Task

Find out more about the work of the following designers:

- C R Mackintosh
- Peter Behrens
- Walter Gropius
- Laslo Maholoy-Nagy
- Herbert Bayer.

Art Deco

From 1925 onwards, at the same time that the work of the Bauhaus was in full development, and not yet widely applied, the Art Deco style became extremely popular and influential. It was named after an international exhibition in Paris in 1925 in which all exhibits were required to be novel in their design. The style is characterised by many visual elements such as bright colours, images of the Sun, geometric shapes and zigzag patterns derived from Egypt. There was also an emphasis on the use of expensive materials.

During the inter-war years there was a considerable demand for eye-catching display typefaces for use on posters and packaging, and these often reflected the Art Deco style, and the early work of the Bauhaus.

The 1950s and 1960s

The first synthetic material was Bakelite, invented in 1902, and commercially used to make products such as radios and ashtrays in the 1920s and 1930s. It was during the 1950s, however, that plastics such as foam, nylon and polyester developed, arising out of war-time research. Plastic materials revolutionised industrial design and production as the moulding processes made it possible for the first time to create goods in considerable quantity at low cost. Images of new developments in science and technology, such as the splitting of the atom, caught the public's imagination.

During this time the clean, minimal and highly legible Helvetica typeface (designed in Switzerland in 1957) came to dominate typographic design, and became one of the most commercially successful fonts (see Figure 3.2.23).

Figure 3.2.24 *Psychedelic record cover from the late 1960s*

In contrast, during the late 1960s, 'psychedelic' posters and record covers led a revival in highly colourful and decorative typefaces and designs which often referred back to Art Nouveau styles (see Figure 3.2.24). This represents the start of what is sometimes called youth culture, and provided a distinctive new market, keen to establish its own identity and sense of individualism.

During the 1960s, the earlier interest in science gave way to images led by the Space Race and the first Moon landings, science fiction and fantasy, and 'futuristic' designs, applied to household products that were becoming increasingly automated. A rapid expansion of the early development of new forms of graphic products developed during this period, such as television graphics, information graphics and **corporate identity systems**.

The Helvetica typeface was designed in Switzerland in 1957

The Helvetica typeface was designed in Switzerland in 1957

The Helvetica typeface was designed in Switzerland in 1957

Figure 3.2.23 *Variations of the Helvetica typeface*

Task

Find out more about:

- The Festival of Britain
- Terence Conran
- Kenneth Grange.

Memphis

Memphis was the name of a group of designers who established themselves in Milan in Italy in 1981. The principal figure in Memphis was Ettore Sottsass, an architect who moved into product design and became a consulting designer for Olivetti, the typewriter manufacturer. The Memphis designers were interested in mass production, advertising and the practical objects of daily life (see Figure 3.2.25). They loved the fast changes brought about by fashion and their witty, stylistic design was influenced by comic strips, films and punk music. Memphis designers combined materials such as colourful plastic laminates (e.g. melamine and formica) as well as glass, steel, industrial sheet metal and aluminium. Many products looked like children's toys. The Memphis group introduced a new understanding of design and inspired many design developments in the 1980s. Much of the bright and interesting ideas since the 1980s have come about as a result of the influence of those involved in the Memphis group. The 1980s became a decade of design and the status of design itself grew. Design took over a key role in marketing, advertising and the development of individual life styles.

The design drawings of the Memphis group are of particular interest. They are often simplistic and childlike and, as such, contrast with the more formalised and realistic sketches of the industrial designers of the 1950s, 1960s and 1970s. The Memphis group did not in itself produce a great deal of graphic work, but graphic designers of the late 1980s applied the distinctive decorative colours and shapes to a wide variety of products.

Task

Find out more about:

- Ettore Sottsass
- George Sowden
- Michael Graves.

The New Design

The work of Memphis began a number of developments in Spain, Germany, France and the UK, which can be described as the New Design. The thinking of a number of designers in these countries was to move away from functionalism and to reflect the influences of daily life. Design in Germany and the UK developed along similar lines during this period, with designers working in materials such as concrete and steel and a simplification of form. In Spain, there was also a move away from functionalism, aided by the entry of Spain into the European Community.

In France during this period, the influence of Memphis was still felt, with unusual combinations of colour and materials being used in interior design. One of the most important French designers of this period was and still is Philippe Starck. He began a move away from the ideas of Memphis to designs which were marketable and relatively inexpensive. He has borrowed ideas from past styles and groups and favours unusual combinations of materials. Starck has been described as the 'super designer' of the late twentieth century. He has worked on interior design, furniture and lighting and is without doubt one of the most influential designers of this period.

Figure 3.2.25 *Kariba fruit bowl, circa 1982, Memphis group*

Task

Investigate one of the design groups described in this section. Produce illustrations and notes to show the type of products that were made by designers in that group. Describe the style and function of each product. Explain how the materials used in each product influenced its form.

The Marketing Age

The 1980s and 1990s represented a significant shift away from the focus led by the design of a new product towards the creation of the market for the product and the life-style brand image that would persuade people to buy it. The experience of shopping (sometimes known as retail therapy), the promotion and packaging of products were all important. Focus groups identified the needs of a target group of the population and the product, manufacturing and marketing strategy were developed in parallel in an attempt to get the right goods to the right place at the right time at a price the consumer was willing to pay.

At the same time, the domination of a particular 'style', unique to a particular time and culture, disappeared. Designers were free to draw on colours, shapes and forms from a rich variety of past styles and from countries across the world.

New materials, processes and technologies

The development of new materials, processes and technologies is a constant influence on the design of products. Advances in materials and manufacturing technologies tends to influence the style of products because they enable designers to experiment with the new materials and techniques.

New materials may be developed in response to legislation. In the USA, a recent directive banned the use of lead in mass-produced products. However, lead is an important constituent part of some mild steels since it gives a better finish – the steel is known as freecutting steel. University research has now developed a steel with a tin additive, rather than lead, which gives very similar properties, but reduces the lead hazard. An advantage of using this new process is that the new steel is cheaper to produce.

In recent years a range of so-called 'smart' materials have been developed. These materials have properties which alter in response to an input. Smart materials provide opportunities for the development of new types of sensors, actuators and structural components and can reduce the overall size and complexity of a device. Smart materials include the following:

- **Piezo-electric actuators**. These are used in greeting cards, which produce a sound from an electrical signal when the card is opened.
- Electrorheological and magnetorheological fluids, which change their viscosity within milliseconds when placed in an electric or magnetic field. In the future, they may be used in suspension systems and engine mounts.
- Shape-memory alloys return to their original shape after they have been plastically deformed by heat. They can be used to open greenhouse windows when it gets too hot or as heat-activated electrical cable fasteners.
- Optical fibres enable light to be transmitted over 200 km. They are extensively used in telecommunications systems, computer networks and in surgery.

The computer age

In many ways, printing methods had remained, in essence, the same for over 500 years. From the 1970s onwards, however, digital technology has revolutionised the design and printing of graphic products. Fonts are now created digitally, and their size and positioning defined to higher degrees of accuracy and flexibility than was ever possible using traditional typesetting processes. This has led to an explosion of new design approaches to typography and layout. The leading exponent during the 1980s was Neville Brody (see page 130), who freely mixed typestyles and sizes, unusual letter spacing and leading and shaped blocks of text.

During the 1990s, the rise of the personal computer and desktop publishing gave access to the design and production of printed products to people at home, and stimulated levels of public interest and awareness in the design of graphic products. Today, we are far more type-conscious and typographically literate than ever before.

On the web

Current developments in typography and layout are focused on the Internet. Designing a web page is very different from designing for the printed page or even the TV screen. At present, what can be delivered in terms of quality is highly restricted by image size and legibility, although many web designers have achieved remarkable results within these limitations. As the technology develops, however, the richness and interactivity of the web will increase dramatically, providing the next major step in the development of the communication of information using text, moving images and sound.

Design in the twenty-first century

It is becoming evident that the processes and disciplines of design that developed during the twentieth century are now undergoing a rapid transition. The old model of the role of the industrial designer specialising in 3D design, graphics, fashion, etc., as the singular creator of single objects, has been supplanted by a new approach in which the designer is becoming the orchestrator of complex systems in which information, materials, emotional sensations and technology interact together. Multi-disciplinary development teams comprise marketing experts, information specialists, computer programers, psychologists as well as production and materials technologists. Computer technologies and the possibilities of CAD/CAM systems are having an increasingly profound effect on the way in which 2D and 3D products are being designed and made.

2. Professional designers at work

The relationship between designers and clients, manufacturers, users and society

The role of the designer

Designers need to be aware of the many different considerations that must be taken into account when designing products. This means that designers have to fulfil many different roles which are usually interlinked. In large companies, designers usually work as part of a design and production team. A small company may employ only one designer, but very few designers work totally alone, since design and manufacturing is seen as an integrated activity. It is the responsibility of everyone in a company to get the product to market on time and to budget. Although the following roles of the designer are described separately, they form a seamless and integrated activity that is performed by the whole design and production team.

Artistic and aesthetic role

The first thing we notice about a product is how it looks – its aesthetic qualities. Many consumers are also concerned with the image the product portrays. With this in mind, the role of the designer is increasingly to give a product and the consumer an image. The artistic and aesthetic role is therefore one of creating and developing innovative, attractive products. This may be seen as the most creative aspect of the work of the designer, since influences such as shape, form, colour, pattern and style all need to be considered. Designers also need to be aware of current and future user and market needs, moral, cultural, social and environmental issues and the competition from other products. Successful products are market driven and market led.

Functional and technical role

All products must be designed with function and performance in mind. Some products, such as seat belts, are produced specifically for their functional performance rather than for aesthetic qualities. The functional role of the designer therefore is to keep up to date with technical information about materials and processes and to understand the competition from other functional products. In this role the designer is still being creative, but in a different way since the product's function, purpose, materials, systems, construction and finishing need to be taken into account.

Economic and marketing role

Consumers want to buy innovative, attractive products at a price they can afford. The economic and marketing role of a designer is therefore to design marketable products to a price point. In other words designers must work to target production costs, which are established at the start of a project. Target production costs are based on a study of the design, development and manufacturing costs of the product. They are also checked against the cost of existing similar products. In this role the designer needs to be aware of the market into which the product is to be sold and should have a clear understanding of production processes and costs.

Organisational and management role

In the past, the design development process was consecutive, with each department contributing to the overall process before handing over the product to the next department on its way to production. In this scenario, the organisational and management role of the designer is limited. Many companies these days use the concept of concurrent manufacturing. This brings together all the different departments (marketing, design, production, quality assurance, etc.) to work concurrently on product development. Designers are increasingly involved in the whole organisation and management of the product and generally work as part of a design and production team. This team shares all the information about the product, using software such as **Product Data Management (PDM)**. This enables fast and easy communication between design, production, suppliers and clients and results in a faster time to market of products that meet customer needs.

Professional practice relating to design management, technology, marketing, business and ICT

Design and marketing

Designing is done in response to a marketing need. Market research is often done by a marketing department, but all designers need to be aware of market trends. Sometimes individual designers establish a product need. This was the case with James Dyson, who in 1974 established the need for a new type of wheelbarrow – the Ballbarrow – and a new form of vacuum cleaner, first conceived in 1978 and finally marketed in the UK in 1993.

Marketing involves developing a product **marketing plan** aimed at the target market group. It involves developing a competitive edge through providing well-designed, reliable, high quality products at a price customers can afford, combined with the image they want the product to give them. This is sometimes called lifestyle marketing.

A marketing plan can involve the advertising and promotion of products and brands through retailers, newspapers and magazines, TV, radio, film and the Internet. A successful marketing plan uses market research to find out:

- consumer needs and **consumer demand**
- the age, income, size and location of the target market group
- the product type customers want and the price range they are prepared to pay
- trends affecting the market
- competitors' products and marketing style
- the estimated time required to develop and market the product.

Efficient manufacture and profit

Efficient manufacture is an essential part of design management. Unless operating in a non-profit situation all companies have to make a profit from the sale of their products after all manufacturing costs have been taken into account. The profit from manufacturing may be used for a variety of purposes including funding research and the development of new or improved products.

> ### Task
> Research examples of products that illustrate ways in which a designer has considered efficient manufacture and profit. Present your findings in the form of notes and sketches.

Aesthetics, quality and value for money

Aesthetics

All modern products are designed with aesthetics in mind, regardless of the use to which the product will be put or the environment in which it will be used. The food mixers shown in Figure 3.2.26 provide a clear example of the way in which aesthetics has influenced the design of a product.

Quality and value for money

Quality is clearly of great importance in the design of a product since no product will sell if it is of poor quality. Quality does, however, need to apply to the way in which the product functions as well as its appearance. Since all designing involves some compromise between function, appearance, materials and cost, quality has to be considered carefully, both by the designer and by those responsible for marketing the product. There will always be a need for products which vary in quality and hence in cost.

We expect a luxury car to be of better quality and to sell for a higher price than a standard mass-produced car. One of the difficult tasks facing the designer is to design in quality within the price range for the product. The customers for the product will want to feel that they are getting value for money but that they are also getting a product which is of good quality.

Value issues related to design

Value issues are inherent in designing and making and in many ways are a driving force behind design. One example is the increasing interest in the environment and the fact that one of the many roles of the designer is to design with recycling in mind.

As with other aspects of designing, there is often a conflict between values held by the designer, the client and the user.

- Cultural and social values are related to the way in which aspects such as fashion and

1918 model

1948 model

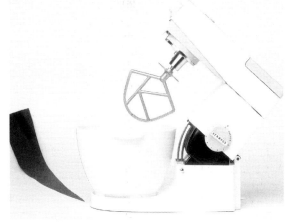

1960 model

Figure 3.2.26 *In the 1918 model of a food mixer, no attempt has been made to consider aesthetics; the machine is simply a smaller version of an industrial mixer. The Kenwood model of 1948 shows the use of aesthetics to improve the appearance of the product, although it still shows its industrial roots. Many machines designed at this time for production in different sorts of industries would have had similar design features. The 1960 model shows a radical rethink of the design with the use of modern materials and a much smoother look to the product as a whole*

lifestyle affect design. Trends in colours used for clothes, for example, influence colours in cars, furniture and interior design. Social influences from the media such as film, television and music also influence design. Exhibitions also stimulate a revival of interest in design influences such as a recent exhibition of Art Nouveau.

- Economic issues mean that there is a need to reduce costs and maximise profit if companies are to survive. This, together with developments in ICT have led to a shift in manufacturing away from traditional areas to LEDCs where labour costs are cheaper.
- Value issues are often very sensitive and companies need to approach these issues with great care. The clockwork radio, designed by Trevor Bayliss (see page 17), was designed to help those living in poorer countries of the world by providing a communications link at low cost. When production started, it was based in South Africa and was largely done by people with physical disabilities.
- When operating in the global market place companies have to be careful to use product names which do not cause offence to religious or cultural views. Consideration also needs to be given to the ethics of imposing values from the traditional industrialised nations on those countries which do not have the same tradition. For example, computers that are advertised on television world-wide are no use to people who do not have electricity, but who require the basic essentials of life. Is it moral to advertise such products and to create a demand for them?

The work of professional designers and professional bodies

There are many professional designers and companies that have made a significant contribution to the design of familiar 2D and 3D products. They have also influenced the work of other designers.

London Transport

The design of the graphics used for the London Underground developed in the 1930s and its subsequent updating in the 1990s provides an insight into the way in which successful design solutions develop as a partnership between a team of designers and the managers of a large commercial organisation, responding to the commercial needs and developments in production technologies of the time.

Frank Pick was the commercial manager of London Transport during the 1920s and one of

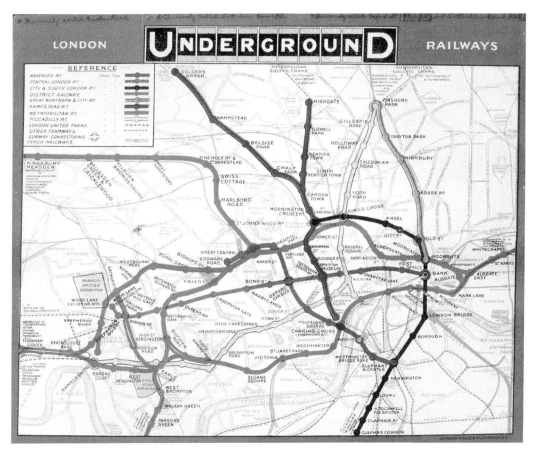

Figure 3.2.27 *An early London Underground map was a geographical representation of the network. As extensions were added in the 1920s, it became more difficult to fit the whole system on one map*

the founders of the Design and Industries Association. He initiated one of the earliest and most comprehensive corporate identity schemes in the UK. He employed the architect Charles Holden to oversee the designs of new underground stations and the typographer Edward Johnston to create a unified graphic appearance for the system. Under their direction, underground stations were transformed into open, bright, airy spaces. Escalators, lit by reflected light, replaced lifts and signs with clear lettering were positioned in a logical manner. Underground trains were fundamentally redesigned to be quieter and provide more passenger space. Similar changes were made to buses to make them sleeker. By 1939, every part of London Transport could be clearly identified as the property of a single organisation.

In terms of graphic products, a number of developments of the time are of particular interest and importance. The first was the changes made to the London Underground map. The 1924 map shows the stations in the geographically correct positions, following the general curves of the routes and the distances between the stops (see Figure 3.2.27). This created an organic layout, with graphic congestion in central London. The schematic map, introduced in 1931 and designed by Henry Beck, is significantly different, and has become one of the most widely accepted mental images of the city (see Figure 3.2.28). Inspired by electric circuit diagrams, the routes were organised along vertical, horizontal or 45-degree axes. This enlarged the distance between the stations in the central area, and reduced the distances in the outer area. Beyond the increased clarity, there is an element of deception in the map, as it suggests that long journeys are shorter than they actually are, and that the interchanges between lines at some stations are much closer than in reality. At the time, this led to growth in passenger usage.

The second development was the application of the typeface designed by Edward Johnston in 1916. Previous signs used a wide variety of unjustified type sizes; Johnston's sans serif type was clean, neat and efficient. It was not so much

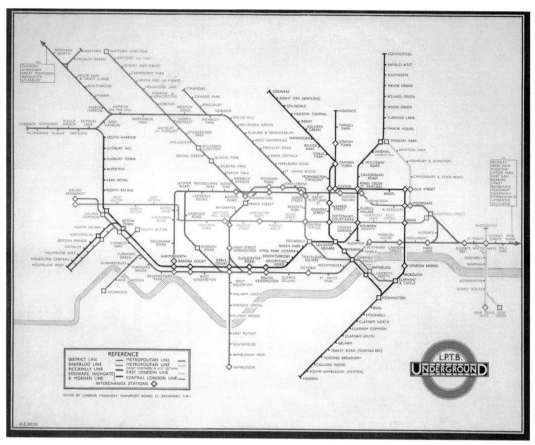

Figure 3.2.28 *Henry Beck's design solution was to abandon geographical accuracy in favour of clarity. Beck's idea was tested by an initially sceptical Underground in 1933, as a folding pocket map. The design is now recognised as a classic and is imitated all over the world*

the design of the type that was significant, but the fact that a series of standards for its use across a range of applications was established. The typestyle has since been adapted, but is still in use today. It also formed the basis of the widely used Gill Sans typeface designed in 1928 by Eric Gill.

Finally, mention must be made of the London Transport advertisements produced during this time, which formed a major element in Pick's drive to improve the images of London Transport. In common with many of the railway companies, leading artists of the day who were experimenting with new graphic styles and printing processes, such as E McKnight Kauffer, Rex Whistler, Edward Bawden, Paul Nash and Graham Sutherland, were commissioned to produce highly attractive lithographic posters that promoted idealised subjects such as the capital's attractions and their ease of access from the suburbs (see Figure 3.2.29).

From the 1950s to the 1970s, the London Transport corporate identity system remained largely unchanged, but began to lose the coherence and standards of quality of the original version. In 1984 the company of Henrion, Ludlow and Schmidt was commissioned to rationalise the signs, maps and other passenger information to present the London Underground as a more modern and responsible organisation. Further rationalisation was needed to help communicate the increased integration with other transport systems in the capital. The original distinctive Johnston type was retained, but further developed by the London consultancy Banks and Miles to improve its consistency and clarity in application. It is now known as New Johnston.

New signs are now produced using CAD (see Figure 3.2.30). The tradition of commissioning artists to produce London Transport posters remains, but these are now supplemented by the more frequent use of safety and information posters that use subtle verbal humour and simpler visual elements to attract attention and get the message across.

In recent years, developments in signage for transport systems has focused on the introduction of computer-controlled digital

Figure 3.2.29 *London Underground poster designed by Manner, 1929*

Figure 3.2.30 *The latest signage used by the Underground*

Figure 3.2.31 *From British Telecom to BT*

communication equipment in which constantly up-to-date information about running schedules and other information is presented to passengers waiting on the platforms or at bus-stops.

Wolff Olins

The company Wolff Olins was formed in 1965 by Michael Wolff (who left the company in 1983) and Wally Olins. Since then, it has pioneered the development of innovative corporate identity programmes for companies such as ICI, P&O and Prudential. One of its most famous projects was for BT in the late 1980s, following its privatisation. Market research at the time revealed that British Telecom was seen as being cold, bureaucratic and unhelpful. Its new

managers wanted it to be seen by domestic and international business customers as being an efficient global player and, at the same time, warmer and friendlier. It also wanted to get across the message that the company did not just provide telephones, but a wide range of telecommunication services.

Wolff Olins proposed a solution that completely replaced the old identity (see Figure 3.2.31). Although effective when launched in the early 1980s, the style of the old logo, with its use of dots and dashes and overall feel of being technology-led had been widely copied throughout Europe. Instead a fluid outline of a piper in red and blue was used, representing the act of communication and using the colours best associated with the UK across the world.

This was accompanied by the initial letters BT in a specially designed distinctive typeface.

Wolff Olins' work extended well beyond the basic logo, with designs being commissioned for stationery, bills, telephone books, vans, shops and offices and the companies sign systems, telephone boxes, staff uniforms and internal newsletters.

Aardman Animation

The Aardman studios are without doubt the biggest success story in the current wave of British animation. The initial success of the studios was with its 3D characters such as Morph produced for the BBC in the 1970s. Aardman animator Nick Park has enjoyed particular success with his *Wallace and Gromit* TV films, and more recently with the full-length *Chicken Run* made for the cinema. Previously he had worked on a successful series of animated TV commercials.

The early work of the company owed much to the traditions of the medium. There are essentially two types of animation – 2D, in which the characters' scenes are drawn and painted, and the less frequently used 3D, in which 3D scale models are constructed and moved. (In recent years digital computer technology has provided a third method of production.) The Aardman studios specialise in the traditional 3D method.

Their films start life as character sketches and storyboards. Even so, without good characterisation and an interesting storyline, an animated feature is unlikely to be successful, however technically accomplished it might be. The models created for Wallace and Gromit are surprisingly sophisticated, involving a jointed armature, a resin body and modelling clay (see Figure 3.2.32). A series of models of the same character are required for different actions, and at different scales. 3D sets need to be created, using ingenious methods of simulating objects at the appropriate scale. Lighting the characters and sets has to be just right too. Sound, including the complex process of synchronising the voices with the lip movements, also involves hours of detailed, methodical work. Finally, the films are edited in the same way as any other.

The particular strength of Nick Park's characterisation has resulted in a range of highly commercially successful 'spin-off' *Wallace and Gromit* and *Chicken Run* products.

Figure 3.2.32 *Nick Park works on the animation of the character Gromit*

Neville Brody

Neville Brody is currently one of the UK's best known graphic designers. He trained in fine art and graphic design at the London College of Printing, and designed record covers before becoming a magazine art editor.

Brody's innovative work in the early 1980s for *The Face* magazine was highly influential (see Figure 3.2.33). He began to develop and use typefaces as **graphic devices**, and experimented with unusual surface textures and decorative motifs which often ignored the conventions of layout grids and typesetting. His designs drew freely on the collages, photomontages and typographic experiments of the early movements of the twentieth century, such as Dada, Russian Constructivism, the Bauhaus and De Stijl.

Brody's work was often imitated during the late 1980s in packaging, magazines and record covers. In 1990 Brody set up FontWorks UK and continued to develop and promote the design of innovative typeface designs. He now enjoys an international reputation.

Figure 3.2.33 *Brody's influence on* The Face *magazine*

Michael Graves (1934–)

Graves is an architect who has also worked on the design of products – mainly furniture and ceramics. His early work was inspired by ideas from art history particularly classicism and cubism. Graves is one of the leading exponents of postmodernism and has designed in the Memphis style for Alessi (see Figure 3.2.34).

Philippe Starck (1949–)

The French designer Philippe Starck has been described as the 'super-designer' of the late twentieth century and he is one of the best-known designers in the world. In the early 1980s, he designed the interior of the Café Costes in Paris, which brought him to public attention. Starck designed all the fittings, including a three-legged chair, which has come to signify modern restaurant design. He designs saleable, relatively inexpensive products for mass production. These include many well-known products such as his Juicy Salif Lemon Press, designed for Alessi. Other work includes a range of consumer goods such as furniture, televisions, lights, water taps and toothbrushes (see Figure 3.2.35). Starck's work includes borrowed elements from past design styles such as

Figure 3.2.34 *The Graves kettle, produced in 1985 for Alessi is made from steel with a polyamide handle. The plastic bird mounted on the spout sings when the kettle boils*

Figure 3.2.35 *The M5 107 television and remote control designed by Philippe Starck*

including educational activities, a reference library, a register of designer/makers and a picture gallery.

The Crafts Council can help designer/makers by providing practical help and advice on business matters and as a source of information and reference materials. Potential clients can obtain information about crafts people and see illustrations of their work. Students can also use the same information to research contemporary design and craft work as a source of products to evaluate and develop ideas.

The Design Council

The Design Council was established in 1960 following earlier work through the Council for Industrial Design. The aim of the Design Council is 'to inspire the best use of design by the UK, in the world context, to improve prosperity and well being'. The Council does this through encouraging business, education and government to work together productively and communicate more effectively.

The Design Council supports design education in schools and colleges through publications about designing and making and collections of good product design.

It supports professional designers by advising businesses on the design and marketing of products. The Design Council can assist students of design by providing information and advice about its activities.

streamlined dynamic lines, organically formed handles and Art Nouveau-style chair legs. He uses unusual combinations of materials such as plastic with aluminium, plush fabrics with chrome or glass with stone.

The Crafts Council

The Crafts Council was founded in 1975 and grew out of the Crafts Advisory Committee. The main aim of the Crafts Council is to promote contemporary crafts in the UK and to provide a service to crafts people and the general public. The Crafts Council operates in a range of spheres

---

3. Anthropometrics and ergonomics

The basic principles and applications of anthropometrics and ergonomics

Anthropometrics

Anthropometrics is the name given to the study of human physical dimensions in relation to the objects which are used by people. The various dimensions of the human body – height, width, length of reach, etc. – are the most important aspects of anthropometric data which need to be considered when designing. Other factors, such as size of fingers and the force which fingers, hands or feet can exert, are also important.

When making use of anthropometric data, it is important to take into account the greatest range of sizes of users of a product. This applies particularly to the design of furniture and similar products such as seats in cars, trains and aircraft. It also applies to products with handles such as kettles, suitcases, tools, door handles, etc. Although anthropometric data cover the whole

of the population range, when designing, only part of this range of sizes are considered. If you are designing a chair which has to be comfortable for 100 people who have a range of heights, say from 1.5 metres to 1.9 metres, designing to fit the whole of the range would be difficult. Normally the smallest five people and the tallest five people are ignored, and the chair is designed to fit the remaining 90 per cent of the group. In anthropometrics, the range taken is on the same basis. Anthropomorphic data constitutes measurements taken from the 90 per cent of the population that is between the 5th and 95th percentile range.

Anthropometric data are available in chart form from a number of sources, including British Standards and text books for design and technology, and can be used in your designing and making. Sometimes, however, it may be necessary for you to collect your own data, where they do not already exist. This might be the case when you are designing for a particular individual or for a small group.

Ergonomics

Ergonomics is concerned with the relationship between people and the products which they use. Ergonomics makes use of anthropometric data to ensure that products can be used comfortably by the people for whom they are designed. Ergonomics is important in many areas of design, since products must be of a suitable size for the people who will use them.

Interacting with products

Almost all products need to be designed with ergonomics in mind. Jewellery needs to be designed to fit the people who wear it. Rings need to be adjusted in size to suit the finger of the wearer, and bracelets also need to fit the wearer. Other items of jewellery may not need such careful fitting.

Furniture is a very important area for ergonomic considerations, especially where tables and chairs are concerned. There are set ratios between the height of a table and a chair to be used with it, and there are optimum positions for certain tasks such as using a computer where the relationship between the user, the keyboard and the monitor are very important.

Interacting with users

The way in which a person uses a product is also an important ergonomic consideration. The size of a hand and the force of the grip are important in the design of handles for cutlery, tools and doors. A design may have to be modified to give increased leverage to allow sufficient force to be exerted. People who have difficulty holding some things, such as young children or those with some disability, may need to have adapted designs (see Figure 3.2.36).

Carrying handles such as those used on a toolbox or suitcase must be sufficiently wide to allow a large hand to hold them, but the diameter of the handle must not be too small or too large. Work benches need to be at the correct height to allow a range of tasks to be carried out.

Interacting with equipment

Control switches on equipment and machines have to be designed so that they can be operated easily and safely. Some switches may have to be positioned so that they cannot be operated by mistake and emergency stop switches positioned so that they can be operated by a foot or knee. Display readouts need to be clear, and glass or clear plastic covers must be positioned to avoid reflection of light.

Equipment and machines which have interchangeable parts, such as vacuum cleaners, are designed so that parts can be changed without the use of excessive force. Push buttons and catches needed to hold parts in place must hold with sufficient force, but also allow easy removal. Parts must be of a size and weight that can be easily held in the hand. For example, a Tour de France bicycle needs to be as light as possible (see Figure 3.2.37).

Interacting with environments

Environments where ergonomics is important include driving seats and controls in vehicles and aircraft, control spaces in machines such as cranes and excavators, checkouts in shops and control rooms in power stations and similar complex areas.

In vehicles, the seat and steering wheel must be adjustable to allow the driver to have the most comfortable position. This should allow the driver

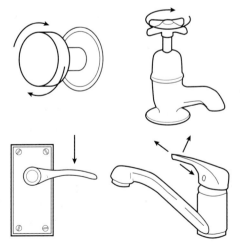

Figure 3.2.36 *The use of levers on door handles and taps makes them easier to operate, especially for small children and for other people who may have difficulty in turning the round knobs*

Figure 3.2.37 *Even a six-year-old should be able to hold a Tour de France bike over her head. Racers who use such bikes have them custom made to match their reach, width and length*

to see and operate all of the controls, display gauges and meters. Foot pedals must be arranged so that any driver can exert sufficient force to operate them over the full range of travel, or power assistance must be provided. Pedals need to be of an appropriate measurement to suit the size of foot.

In driving cabs of cranes and similar equipment, the operator needs to have a comfortable seat, since he or she may be operating the machine for long periods of time. The various controls must be within easy reach and the function of the controls needs to be clear to avoid the risk of mistakes. The load or working end of the equipment must be visible without the need to strain to see.

Checkouts in shops are arranged so that the operator does not need to lift heavy goods. Bar-code readers must be positioned so that they can scan the bar code on the goods without excessive lifting. The operator and the customer must be able to see the price display.

Control rooms in power stations and similar complex areas require operators to be able to see what is going on easily and to be able to quickly process complex information for gauges and displays. The layout and function of controls also needs to be clear so that they can be located and identified quickly when needed.

British and International Standards

The British Standards Institution (BSI) is now the world's leading standards and quality services organisation. BSI was formed in 1901 and is the oldest national standards-making body in the world. BSI is independent of government, industry and trade associations and is non-profit-making. It is recognised globally as an independent and impartial body that serves both the private and public sectors. BSI works with manufacturing and service industries to develop British, European and International standards. It operates in more than 90 countries including Europe, the USA, Canada, Mexico, South America, Russia, Singapore and China.

There are now many similar bodies that belong to the **International Organisation for Standardisation (ISO)** and the International Electrotechnical Commission. (IEC). The joint European Standards Organisation is called CEN/CENELEC/ETSI. **CEN** is the European Committee for Standardisation. It implements the voluntary technical harmonisation of standards in Europe, in conjunction with world-wide bodies and European partners. **CENELEC** is the European Committee for Electrotechnical Standardisation. **ETSI** is the European Telecommunications Standards Institute.

Setting standards

Most standards are set at the request of industry or to implement legislation. The setting of standards, testing procedures and quality assurance techniques enable companies to meet the needs of their customers. Tests are carried out against set standards for a wide range of products. These tests are applied to many products manufactured in the UK and overseas. Some British Standards are also agreed European and/or International Standards. Any product that meets a British Standard is awarded a Kite Mark, as long as the manufacturer has quality systems in place to ensure that every product is made to the same standard.

The relationship between standard measurements and the design of products

Whatever project you are working on, standard measurements probably exist in relation to the product you are making. Sometimes a single critical dimension is all that you need, whereas other designs may need the lengthy and complex calculation of data. When designing a one-off or custom made product, it may be necessary to use specific dimensions. For example, fitted furniture for an awkward alcove would need to be made to critical dimensions for it to fit in the space available. When designing mass produced products, however, it is usually necessary to match the product to a range of users who come in various shapes and sizes. Information about standard measurements may be found in the 'Compendium of Essential Design and Technology Standards for Schools and Colleges' and through the BSI website on www.bsi.org.uk/education.

Ergonomic considerations for designs and models

There are a number of considerations to be taken into account when designing.

- It is a fallacy to think that just because your product is the correct size for you, it will be right for everyone.
- Designing for the 'average' user doesn't mean that a product will be suitable for everyone. Since average dimensions only take account of 50 per cent of the population, the other half may find the product unsuitable or difficult to use.
- Although people are adaptable, it should not be used as an excuse for bad design. Problems like back pain, for example, are often the result of using furniture that requires unsatisfactory working positions.
- Many products are sold for their aesthetic properties and may be designed without reference to anthropometric data, because the data are expensive to buy.

Exam preparation

In order to revise successfully for this unit, you need to make sure that you have a good grounding in all of the topics included in the unit, so that you can apply your knowledge to the context of an exam question. Most questions will ask you to make specific reference to one or more products and give answers related to that product.

It is useful to cut out and keep items in newspapers and magazines that relate to design and technology, since these will keep you up-to-date with current product information. The Internet is also a good way to research products. You can gain more information if you get together with other students in your group and share information.

Make brief notes about each topic in the unit. Try to summarise key points on one or two sheets of paper and learn them.

Practice exam questions

1 All products are subject to aesthetic influences on their design.
  Identify two different products and compare their form and function.
  Give reasons for your views. (15)

2 a) There are a number of design movements that have influenced design since 1900.
   Two contrasting groups are the Bauhaus and Memphis. Compare the influence
   of these two groups on product design. (8)
  b) Describe the work of ONE designer from EITHER the Bauhaus or Memphis design
   movements. (7)

3 Value issues are inherent in product designing.
  a) Outline the importance of value issues in designing. (7)
  b) With reference to two contrasting products, describe how values have influenced
   their design and manufacture. (8)

4 Products can be manufactured as one-off, batch or mass production. Describe how
  each of these levels of production impact upon the design of products. (15)

CAD/CAM (G303)

Summary of expectations

1. What to expect

CAD/CAM is one of the two options offered in Section B of Unit 3, Examination Paper 3. You may have covered some of the option content in your GCSE course, depending on which focus area you studied. If, however, these materials are completely new to you, don't worry. This unit will take you through from first principles.

2. How will it be assessed?

The work that you do in this option will be externally assessed through Unit 3, Paper 3. There will be two compulsory, long-answer questions, each worth 15 marks. You are advised to spend 45 minutes on this paper.

3. What will be assessed?

The following list identifies what you will learn in this option and subsequently what will be examined.

- The impact of CAD/CAM on industry, including:
  - changes in production methods
  - global manufacturing
  - employment issues
  - trends in manufacturing using ICT.
- Computer-aided design, including:
  - CAD techniques
  - creating and modifying designs and layouts
  - 2D/3D modelling and prototyping
  - constructing accurate drawings
  - creating complex products

  - creating virtual products
  - creating total design concepts
  - common input devices
  - common output devices.
- Computer-aided manufacture, including:
  - CNC machines
  - using CAM for one-off, batch, high-volume/continuous production
  - advantages/disadvantages of CAM.

4. How to be successful in this unit

You are expected to demonstrate an understanding of what you have learned in the option. Examiners are looking for longer, more detailed answers that show a greater depth of knowledge and understanding at AS level.

There may be an opportunity to demonstrate your knowledge and understanding of the content of this option in your coursework. However, simply because you are studying this option, you do not have to integrate this type of technology into your coursework project.

5. How much is it worth?

This option, with Section A, is worth a total of 30 per cent of your AS qualification. If you go on to complete the whole course, then this unit accounts for 15 per cent of the full Advanced GCE.

Unit 3 + option	Weighting
AS level	30%
A2 level (full GCE)	15%

1. The impact of CAD/CAM on industry

The need for companies to develop competitive products or services is vital for their own economic survival, for the prosperity of their workforce and for the other businesses in their community or supply chain. Eventually, most products can be designed better or updated because of advances in technology or produced more economically by improvements in production methods. The purpose of any production system, including those in the graphics environment, is to ensure that the correct personnel, software, processes and systems are employed to ensure the best possible outcome for the client or customer and designers.

In all product sectors, designers are able to use the computer-based tools and features within a computer-aided design (CAD) system to create, develop, communicate and record product design information. Computer-aided manufacture (CAM) is a rapidly evolving set of technologies that translate design information into manufacturing information. **Computer integrated management (CIM)** systems are used increasingly to plan and control automated manufacturing processes at all levels from batch to mass or continuous production. The system of process planning combines a range of sub-systems including computerised sensing and control systems, robotics and computer-driven equipment; CIM is how the two systems CAD and CAM are integrated. The type and degree of integration will depend on the scale of production and other operational or commercial considerations.

Increasingly important in the printing industry is the use of computers and microprocessors to ensure that the printing stream works smoothly for the customer as well as the print room operator. Computer-based document management systems have been developed to allow automatic handling of the printing and finishing processes. It is possible to make easy and rapid interventions in order to prioritise printing jobs, choose printers and other output devices or take control of specialist printing jobs. Related developments include automated accounting functions that generate the data needed for tasks like cost allocation, reporting and resource planning. The easily available administrative and management data that are generated by these systems allows for the constant monitoring of manufacturing capacity and costs to generate management level reports and maximise efficiency.

Electronic document management

Once everything in a production process can be carried out electronically the strategy of electronic document management can be introduced. This is quite simply the scanning of physical documents and the importing of computer-generated documents so they can be searched and retrieved at a workstation anywhere, perhaps alongside an existing database system. This allows graphics and printing companies to offer their clients an immediate response to their queries or interest taking account of all existing correspondence. Search criteria can be met in a number of ways, either through document titles, built-in key words, or through **optical character recognition (OCR)** technology on words held within the documents. This procedure allows a company to research its documents and archives in many ways, allowing them quickly to draw together the information they require. In comparison with more labour intensive and traditional paper-based methods an electronic document management system is quick, flexible and cost efficient in time and resources, both physical and human.

The impact of CAD/CAM on design companies

In graphics related operations, whatever their scale, the introduction of CAD/CAM and related digital technologies means that they are able to offer a comprehensive service to their clients.

In large companies creative teams are increasingly multidisciplinary in order to reflect this convergence between the manufacturing technologies and the creative industries. Larger companies might need to have graphic and product designers, digital media specialists and other specialists such as interior designers and architects on their staff. Smaller sized companies are unable to employ such an extensive range of specialists so they tend to focus on particular market segments. It may be that a company employs both product and graphic designers on their staff or interior designers and architects could come together in a creative enterprise. Freelance graphic designers have to work closely with their clients and with their printing companies. If there is a print production problem, the designer can turn to these printers for advice based on their specialist experience. The key factor that unites these operations is the use of digital processes and technology to provide faster, more efficient customer service characterised by smooth workflow, short job turnaround times and the ability to meet special requests at any time.

Most business owners and marketing department heads do not have the time to stay on top of the technicalities of printing and website creation, let alone deal with coordinating the work of photographers, writers and illustrators. Whatever the organisational arrangement, these new computer-based operations all need the services of digital media specialists. These could be employed directly, but for many companies the most cost-effective way of enlisting specialist technical support is to make use of the skilled people employed in companies providing bureau services. These bureaux also have a vast range of specialist equipment that is too expensive for most companies who would not need it on a day-to-day basis. Typically, they can provide digital scanning or large format plotting and printing of electronic files that have been created from a variety of CAD applications. These files can usually be plotted or printed in monochrome or full colour on various media including paper, drafting film, vinyl for banners and signs, and textiles.

CAD/CAM – creating or responding to change?

CIM is an increasingly effective way of integrating CAD/CAM technologies into the production of a range of products including those that incorporate graphics such as product cartons or point-of-sale packaging. **Die cutting** is the manufacturing operation for the cutting and creasing of flat sheet materials so that that they can be formed into 3D cartons and packages. The **die-cutting tools** can be designed and tested on a CAD system. In high-volume manufacturing operations, computers and microprocessors provide the control systems that integrate the different production stages to ensure continuous production. Across all industrial sectors the development of computer-based design and production technologies is providing a driving force for change. The same technologies also provide an effective response mechanism by allowing companies to react quickly to changes in external factors such as a fluctuating market demand (see Figure 3.3.1).

The pressures for change

The twenty-first century shopping experience is fast, competitive and demanding. Producers and manufacturers of mass-produced items market and sell their products across the world. This creates pressure on the design processes, the

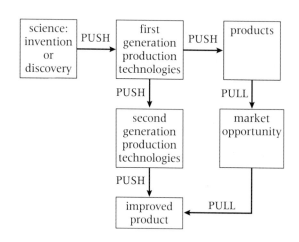

Figure 3.3.1 *Changes in production methods are either pushed by the development of new technologies or pulled by the demands of the market*

manufacturing processes and the organisation of the production and logistics facilities. A product may have to be remodelled to include different design features depending on where it is to be sold. For example, a point-of-purchase merchandising display for photographic film may retain the same shape and form but the language and styling of the text may have to vary from one international market to another. In some countries local legislation also imposes restrictions on manufacturers. Multinational manufacturers often face a challenge from local manufacturers of similar products. These product differences are classed as design variants. They range from the amount of recyclable materials used in the product, to whether a plug is fitted or not, to the fitting of different visual and graphic displays. Another pressure for change is that product lifespans are becoming shorter, especially in areas such as telecommunications and information systems.

In a fast moving sector like graphic design and its allied disciplines in manufacturing, the deadlines set by clients are often incredibly tight, creating further pressures on the design process. The application of computers has changed design practice from the traditional 'design for print' approach to interactive design. This new field of design encompasses many areas ranging from web design through to digital television. Web design involving designers in interactive communication design has emerged in a relatively short period of time. This means that designers are now faced with the prospect that they need to keep learning about new software and how to operate in new design arenas.

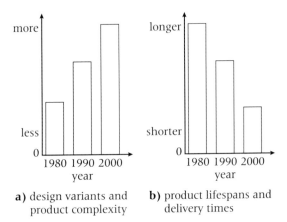

a) design variants and
product complexity

b) product lifespans and
delivery times

Figure 3.3.2 *Pressures to improve the design and
production process*

Keeping up with changes and upgrades to existing software also takes up an incredibly large amount of time and creates further pressures for change especially for those working in the freelance graphic design market. New technologies and the introduction of smart materials are also increasing the technical complexity of designs and how they can be manufactured. Figure 3.3.2 shows the effect of pressures to improve the design and production process.

Changes in design and production methods

A major design innovation has been the development of desktop publishing (DTP). This is a process whereby a designer sitting at a desk can create camera-ready copy and final film/artwork on a personal computer. Another significant change has been the continuing evolution of more powerful CAD systems for use on personal computers. For graphic designers in particular, a fully featured CAD system provides the means to develop design concepts and model or test ideas from the simplest to the most complicated product. DTP applications support presentation graphics and multimedia in the preparation of design for reproduction. These applications allow for the accurate positioning of pictures and blocks of text. Typography can be accurately controlled such as when editing and rotating text. Colour separation features allow the opportunity to specify percentage tints either as spot or process colours, which can then be output into separated film for printing processes.

Using software to design for manufacture
Design for manufacture software such as PowerSHAPE™ produced by Delcam allows users to take concept designs or designs prepared in other CAD systems and add specific and complex manufacturing features such as fillets or split and draft surfaces that are needed for the trouble-free manufacture of plastics-based products.

These fully featured CAD systems also allow the production of photorealistic images and 'virtual products' that can be used to 'sell' the products to potential clients or customers through different media such as magazines and the worldwide web.

Photorealistic images
The growth in the use of the Internet and the development of websites to advertise and sell products has increased the pressure on software designers to improve the 'real life' quality of product images. These digital images are described as photorealistic. Modern CAD systems allow greater control over the quality of these images, typically through the use of an object or image browser. One of the key features for graphic designers is the ability of desktop publishing software to manipulate images and represent different material qualities such as metallic, non-metallic and textured surfaces.

Page make-up and layout applications such as Adobe Photoshop™, Illustrator™, Aldus PageMaker™ and QuarkXpress provide interactive 'virtual studios' in which the designer can control the direction and intensity of light falling on a product image as well as the type of camera lens in order to create a photorealistic image. Designers can choose daylight conditions, spot or flood lighting. Camera lenses available include fisheye, wide-angle and telephoto. These determine how much of an image can be seen and from what distance. Images can be manipulated electronically; they can be retouched, masked, textured and resized to fit the space available.

Virtual products

The comparatively recent introduction of 3D CAD systems allows the creation of interactive virtual reality environments and products. A virtual product can be viewed from any angle at any distance. Products with moving parts can be seen in operation 'on screen' as animations. These animations are generally kinematic motions that do not take account of the mass of the object or the other physical forces acting on it. They do not represent a complete model of how the product will behave in reality.

We will return to the subject of photorealistic images and virtual products later in this unit.

Design management

In the more sophisticated CAD applications that require larger amounts of computer memory the software keeps track of what are called design dependencies. This means that when the one value or element on a drawing of a part or an assembly is changed, all the other values that depend on it are also automatically and accurately redrawn. This is known as **parametric designing** and is particularly useful in applications such as the packaging of consumables for special promotions such as '10% extra at no extra cost'. This means that although the relationships of the geometric shapes making up the package stay the same, all the dimensions have to be changed to allow the package to hold the increased volume of product. One simple change to a key measurement such as the height of the package causes all the other dimensions to be altered to the correct sizes. The graphics can then be added by cutting and pasting from the original drawing and resizing as required to allow room for the promotional text to be added.

The benefits of CAD/CAM product modelling

Combining 3D CAD systems with computer-aided modelling techniques such as **Rapid Prototyping (RPT)** allows the creation of physical models as soon as the 3D digital model is designed. This reduces potential communication problems in the product development team. A further benefit is that any potential errors or technical and tooling problems are found out more quickly. Changes to a product design in the later changes of its development are potentially very costly in terms of time and reworking costs.

Architectural model making is one area of product modelling to benefit from the application of CAD/CAM. Designs developed on screen can be quickly redeveloped to accommodate design changes arising from discussions with clients. Computers are now used to produce anything from a simple design study model costing relatively little to an illuminated, mechanised marketing model costing thousands of pounds. Model makers, such as Pipers of London, now use CAD/CAM in a variety of ways. The company is able to accept digital drawings direct from an architect and then use a range of computer-controlled cutting techniques to increase the accuracy, efficiency and quality of the final model.

Task

Construct a simple organisational chart indicating the main stages in the design process, as you understand it. Identify how you think computers could be used at each stage in the process indicating what features might make them useful or how they might have a limited application.

Computer numerically controlled (CNC) machines

CNC is in widespread use across industry and especially in automated production systems. Computer numerically controlled (CNC) machines are controlled using number values written into a computer program. Each number or code is assigned to a particular operation or process. In manufacturing equipment such as CNC milling and press machines, improvements in the machine-operator interface have made them easier and more intuitive to use. Computer programs, sometimes referred to as 'wizards', have largely eliminated the need to learn elaborate CNC machining or programing codes. Most CAD/CAM software is now capable of generating the required NC machining codes known as 'G' and 'M' codes from the digital data created from a drawing. Examples of 'G' and 'M' codes used by manufacturing equipment are shown in Table 3.3.1.

These codes can be shown on a screen as the product is either 'virtually manufactured' as a simulation or in reality on the CNC machine. The

Table 3.3.1 Examples of 'G' and 'M' codes

Code no.	Type	Description
G00	Rapid traverse	The tool moves from point 1 to point 2, along the shortest path available The feed rate (speed of movement of the tool) is usually set to run as fast as possible
G01	Point-to-point positioning	The tool moves from point 1 to point 2, in a straight line, with a controlled feed rate
M00	Stop program at this point	
M02	End of program	
M03	Spindle on clockwise	Code is followed by a number prefixed by the letter S, denoting the spindle speed
M05	Stop spindle	
M06	Change tool	Code is followed by a number prefixed by the letter T, denoting the number of the tool you wish to change to
M30	Program end and rewind to start of program	

numbers or codes are easily changed from within the CAD software when required. Most CNC machines also have the capability to be programed manually from an adjacent keypad.

The benefits of CNC machines are that:

- graphic images and products are produced accurately, quickly, with consistent quality
- they provide increased operational flexibility as they can be employed in batch and mass-production systems such as high- and low-volume print runs
- they can be used reliably in processes requiring continuous operation such as carton manufacturing
- they can used in conditions that are hazardous to human operators such as the guillotine cutting and trimming of large printed documents
- they are economic to operate over time even though they have a high initial cost.

Computer integrated manufacturing system (CIM)

In both small- and large-scale production processes, traditional approaches to designing and manufacturing involve a linear or sequential approach in which the image or product passes through a series of predefined stages.

In this system, often referred to as 'over the wall', process planning and the handling of production data are relatively straightforward activities, but design errors or manufacturing problems can occur at many points along the line depending on the number of people involved. This extends the time taken to design and manufacture a product and bring it to market.

As we have seen earlier, product life cycles are reducing significantly and with the globalisation of manufacturing there are severe competition pressures. These are forcing companies to look at more efficient ways of operating such as concurrent engineering.

In this manufacturing system, a product team is organised so that all the specialisms within a large company are represented. These multidisciplinary teams share their expertise right from the start of a product's life. They work together at the development stage to reduce errors and draw on the skills and expertise of others in the team. A print production engineer will immediately be able to tell a graphic designer whether what has been designed can be produced rather than waiting until the 'completed' drawing is received.

Desktop manufacturing systems
IThe role of computers in flexible manufacturing systems (FMS)

Software applications in a flexible manufacturing system (FMS) allow a central computer to process production data in order to sequence and control a network of machines and materials handling systems in order to meet the order book more efficiently. For instance, the printing industry is in flux; new market requirements are calling for new solutions to meet shorter deadlines, shorter production runs with reduced set-up times, last minute changes, improvements in quality and reductions in waste. At the most basic level the trend towards smaller, better-quality print runs means that it is very important to get the press inked up and achieve even inking conditions quickly. Several companies have changed to printing presses that use microprocessor-based systems to measure print quality, evaluate colours and use the data to automatically control the inking process.

The production data generated by computer-based production systems can also be used in many other ways. They can be used to judge progress against 'world-class manufacturing' quality criteria such as 'right first time every time'. The move to total quality management (TQM) means that product quality is no longer the responsibility of one department or one person. This new organisational culture of **continuous improvement (CI)** through concurrent engineering cannot exist without effective computer-based communications.

CAD and CAM systems generate vast amounts of digital data that can be used at different stages in an integrated manufacturing system. Relational databases and other data storage methods enable data to be tracked, stored and retrieved as required. Data can be used to plan CNC and other manufacturing operations – computer-aided process control (CAPP). These 'new' data combined with other data generated from the CAD model can be used to manage the production processes, a technique known as computer-aided production management (CAPM). Figure 3.3.3 shows a systems model of the use of software applications within CIM.

The role of computers in managing document archives

Speed, reduced cost of design creation and revision, accuracy, flexibility and the increased scope of drawing production have ensured that CAD has been rapidly integrated into design studios and drawing offices worldwide. Many graphic design companies, architects and manufacturers have large and difficult to store archives of past and sometimes present-day projects on media such as paper. Drawings were once produced on paper or other media and stored in plan chests and filing cabinets but they are now created on computers and are stored as electronic files. Many companies find it cost effective to use the technical facilities provided by

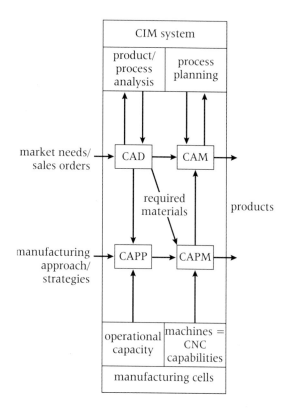

Figure 3.3.3 *A systems model of the use of software applications within CIM*

specialists firms called bureau services who can convert design information stored on paper copies into digital files that are easier and more convenient to store. These files can then be made available online in an information retrieval intranet or electronic data management (EDM) system.

Global manufacturing

We are currently living through a high-speed information revolution that is not only making the world a smaller place, but it is also increasing international competition and creating pressures on manufacturing companies to respond or die.

The trend towards cooperative and integrated working is increasingly more cost effective because of improved communication and data exchange systems that are not limited by geographical boundaries.

There are now many examples of products that are designed and engineered by teams working across the globe and assembled using component parts that, because of increased specialisation, are sourced from supply chains in different countries.

The London Eye is a feat of modern engineering and is an example of this method of

Figure 3.3.4 *The London Eye – an engineering project that drew on design disciplines from around the world*

design and manufacturing (see Figure 3.3.4). The project management team drew on specialist expertise and suppliers based in many different countries.

Employment issues

CAD/CAM technologies are directly affecting patterns of employment in all manufacturing sectors right across the globe. The numbers of people involved in manufacturing are declining as automated systems take over. For those that are left, their jobs have changed significantly (see Figure 3.3.5). For instance, as we have seen, CNC machines can be controlled directly from a central computer system. This means that some shopfloor jobs such as those in a large print room consist of nothing more than 'machine minding';

Figure 3.3.5 *A manufacturing cell at a Ford plant*

correcting faults that arise or replacing materials or resources that have run out. With developments in automation, robotics and artificial intelligence it is likely that there will be even less human involvement in the production processes of the future.

Employment trends on the design side

CAD systems reduce the need for large and labour-intensive drawing offices where historically drawings were hand-drawn by skilled technicians. However, the effective use of CAD requires workers who have high levels of computer and visual literacy alongside creativity and problem-solving skills. These desk-based workers do not necessarily need to be office based. With the growth in information communication technology they can work almost anywhere and if they do so they are classified as remote out-workers.

The increased use of CAD and CAM systems has also raised significant issues concerning initial and on-going training. For instance, the effective introduction of these new ways of computer-based working relies on the willingness of employees to adopt more flexible approaches to work and the recognition that they will need to retrain or update their skills as systems continue to develop. It also requires that people coming into this sector are more highly trained with much higher levels of basic skills than previously expected.

New opportunities for employment also arise with the adoption of CAD/CAM systems. For many small to medium-sized manufacturers, employing a specialist bureau service to provide professional CAD/CAM services to support a number of different design disciplines is an effective and flexible way to cater for the specific needs of their different customers.

Employment trends on the processing side

CAM operators now have to be trained to operate different machines and carry out many processing operations within their manufacturing 'cell' or work centre. They have to be good 'team players' offering help to others as required. Gone are the days of rigid demarcation and narrow specialisation when, for instance, lathe operatives were not allowed to set up and use another machine in the factory. This method of redeploying workers to the area of greatest need is a feature of modern production systems. These systems reduce queues for machines, remove potential bottlenecks on the production line and so reduce processing times. They are not viable without a skilled, trained and flexible workforce.

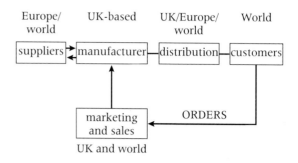

Figure 3.3.6 *The worldwide distribution of a UK manufacturing company*

In the UK there are severe shortages of these multi-skilled, multi-function operators in many areas of manufacturing. This, combined with the flexibility offered by computer-based systems, often means that companies can relocate their manufacturing operation to another part of the world. Design and development can remain in one country while manufacturing operations move abroad. This has a devastating impact on local and regional employment prospects. Until recently, large Japanese manufacturing companies offered 'jobs for life' but Japanese workers are now also feeling the pressure on their jobs from emerging manufacturing centres in Taiwan and other Pacific Rim countries. Figure 3.3.6 shows how the distribution of a UK manufacturing company has been spread worldwide.

The impact of CAD and CAM technologies on future industrial innovation

The globalisation of manufacturing industry means that considerable research and development time is being spent on developing innovative manufacturing systems that fully integrate the use of CAD/CAM with artificial intelligence or 'expert' systems, information systems and databases. Innovation is a process involving three types of technologies:

- **Critical technologies** are the 'building blocks' from which products develop. CAD/CAM approaches will be involved in the continued development of computerised sensing and control systems, materials handling, storage and retrieval systems and the development of industrial robots.
- **Enabling technologies** such as CNC machines are needed to make use of critical technologies.
- **Strategic technologies** are concerned with the decision-making process. In manufacturing this can range from decisions about capital

investment in new products, the most effective factory layouts and facilities, to the future use and potential benefits of systems based on artificial intelligence.

The drive for profitability

There is much research and development activity into data manipulation systems because a company's continued existence in both low- and high-volume manufacturing relies on the establishment of efficient, cost-effective and 'quick response' systems. Effective CIM systems that include CAD and CAM reduce the 'lead time' from product development to market; time to market is a key factor in profitability. The development of more sophisticated and powerful

DTP technology has improved the effectiveness and output of individuals, while developments in information and communication technologies now allow for the rapid and effective deployment of vast amounts of electronic data to virtually anywhere in the world. The development in mobile phone technologies is one area that will also contribute to the drive to profitability.

Task

Explain how DTP technologies contribute to increased profitability in both a small- and a large-scale graphics-based industry.

2. Computer-aided design

A computer-aided design system is a combination of hardware and software that enables designers working individually or in teams to design everything from promotional gifts such as pens or desk tidies to the International Space Station. CAD provides increased flexibility for designers and allows them more control over the quality of the finished product. In situations such as the production of a computer-animated feature film it allows digital image manipulation and the facility to combine computer-generated images with filmed images from real life.

In many manufacturing situations a CAD system will automatically cross-reference many drawings to one another to create a product inventory. This not only details the materials and components required to assemble the final product, it also informs the stock control systems in the manufacturing facility. The successful application of CAD is dependent on the amount of computing power that is available to the user.

In general terms, CAD requires a powerful central processing unit (CPU) and a large amount of memory. The images on a computer screen are made up of pixels. The term bit depth refers to the amount of data used to describe each pixel on the computer screen. Black and white line work is 1 bit deep. Greyscale is 8 bits deep. RGB is 24 bits deep. Images to be printed as CMYK separations should be 32 bits deep. In graphic terms, images can now be created and manipulated in 2D, 3D and video forms creating a need for greater processing speeds and increased amounts of computer memory.

Until the mid-1980s, all CAD systems operated on specially constructed 'dedicated' computers. The processing power and the speed of processors and graphics interfaces on personal computers has improved dramatically in recent years. 'Fully featured' professional quality CAD systems can now be operated effectively on **local area networks (LANs)**, wide area networks (WANs) and general-purpose office workstations and on personal desktop and laptop computers.

The components of a CAD system

The following are the main components of a CAD system; you will consider them in more detail later in this option.

Hardware

This term describes all the physical components of a CAD system including the input and output devices. There are three main categories of computer that can be used in a manufacturing situation:

- Mainframe computers have high processing speeds and the memory required for handling and storing the large amounts of data that are generated by high-volume manufacturing. A mainframe computer is accessed via a network of computer terminals connected to it. The networks can be local area networks, **intranets** or wide area networks.
- Minicomputers are smaller versions of the mainframe system used by large organisations mostly in various network configurations.
- Microcomputers, desktop or laptop personal computers (PCs) are used for individual computing needs or as machine control units (MCUs) for controlling a range of CNC machines.

Data storage devices include: hard and floppy disks, CD-ROM and DVD and other external storage devices such as Zip drives.

Input devices include: keyboard, mouse and tracker-ball, graphics tablet and stylus, digitiser, puck (cursor) and mouse, digital camera and video, 2D and 3D scanners.

Output devices include: monitor, printers of all types, plotters and cutters, CNC machines and increasingly Rapid Prototyping systems (RPT) for modelling 3D product ideas.

Software

The operating system (OS) provides the platform to run all the application programs. It is also used to manage the files in the computer. There are a number of operating systems available such as **Unix**, Windows, DOS and MAC-OS (Apple computers).

There are many CAD software applications based on these operating systems. All CAD software, whatever the operating system, provides the data for the graphic display and the other output devices such as printers and plotters.

A modern CAD program contains hundreds of functions and features that enable you to accomplish specific drawing tasks. A task may involve drawing an object, editing an existing drawing, displaying a specific view of the drawing, printing or saving it, or controlling any other operation of the computer. The software will contain a number of menus, commands, functions and features that enable you to specify exactly what you want to do and how you want to do it. For instance, the edit feature is a convenient electronic drawing aid that enables designs to be changed easily. The software may also have a number of specialised features such as providing animated 3D views or printing design information on different layers. This feature is useful in areas such as desktop publishing or the screen printing of a graphic design.

Drawings can also be plotted with specific colours, pen thickness and line types. In some CAM programs lines on a CAD design can be drawn in different colours to indicate the different depths of cut that a cutting tool has to make. This is a particularly useful feature when making folded card or 'creased' plastic products (such as folders and wallets to hold promotional leaflets) where one colour is used to indicate the folds or creases that are required to make the 3D product.

Graphical user interface (GUI)

This is a two-way link between the user and the computer. It provides an on-screen display giving

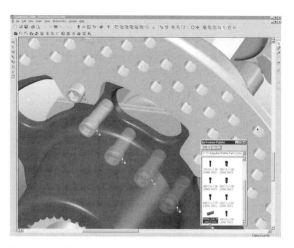

Figure 3.3.7 Example of a CAD window

visual clues to help the user communicate with the computer. The graphical user interface (GUI) allows the user to enter data through commands or functions that are selected from menus; by keyboard stokes (e.g. Ctrl+P); from toolbars, buttons and on-screen icons and from text or dialogue boxes. GUIs also allow a range of users with different operational requirements to set up their own preferred 'on-screen' working environment such as specifying dimensions and text that match industry standards such as **ANSI**, ISO and BSI. An example of a typical CAD window is shown in Figure 3.3.7.

Standards for CAD systems

A format, programing language, or operating **protocol** only becomes an officially recognised **standard** when it has been approved by one of the recognised standards organisations. However, you will find that many of the CAD/CAM standards are, in fact, accepted standards because they are widely used and recognised by the industry itself as being the standard. Some examples of such standards that you will come across include:

- Hayes command set for controlling modems
- Hewlett-Packard Printer Control Language (PCL) for laser printers
- PostScript (PS) page description language for laser printers
- Data Exchange File (DXF), a graphic file format created by AutoDesk and used on many CAD systems.

Managing CAD data

All CAD applications have a range of data management features and options. Drawings can be stored either on a hard disk, tape, CD-ROM or on a central network **server**. Drawing files are often large and take up a lot of memory. Some applications allow files to be compressed (zipped) to take up less of the system memory. Files are managed in directories and subdirectories. The software should also be able to translate drawings created by other CAD programs. Data Exchange Format (DXF) is one of the common data translation formats used by CAD programs. There are a number of other data formats available. Initial Graphics Exchange Specification (IGES) is an international standard, which defines a neutral file format for the representation of graphics data across different PC-based CAD systems.

The common formats for storing CAD data are:

- Windows Metafile Format (WMF)
- PICT – the standard format for storing and exchanging graphics on Apple computers
- Tagged Image File Format (TIFF)
- Hewlett Packard Graphics Language(HPGL)
- **Virtual Reality Modelling Language (VRML)**

Tasks

1 Find out and record what the initials ISO, ANSI and BSI stand for.
2 Give three examples of standards that apply to CAD/CAM resources, equipment or software.
3 Give three examples of standards that apply to graphic products.

The basic operating characteristics of CAD programs

- The user is able to interact with and manipulate images on screen.
- Displays can be divided into two or more windows, sometimes referred to as tiles or panes.
- Each window can contain a different view of the product or the different parts that make up the product.
- Models can be repositioned or edited independently in each window, the effects of one action affecting all relevant views.
- When working with multiple windows only one of the windows remains current or active, other windows are activated by clicking on them.
- The coordinates that make up the multiple views to create a 3D model or product are calculated automatically.
- Models or products can be displayed or viewed from any direction or viewpoint.

- Standard projections are supported including orthographic (1st and 3rd angle), oblique, isometric and perspective.
- Pull-down or pop-up menus contain the various commands that allow text, dimensions, surface finishes, textures and labels to be added. Features such as line styles and use of colour can be adjusted. Most CAD packages have keyboard shortcuts for drawing and other features such as Ctrl+G for grouping a collection of individual elements in addition to the standard word-processing shortcuts, for example Ctrl+P for printing.
- Primitives are predefined graphic elements. Most CAD packages have a library of 2D and 3D vectored objects that can be drawn, stretched and resized. Typical primitives are:
  - 2D: lines, polylines, arcs, polyarcs, circles, ellipses, splines, Bezier curves and polygons
  - 3D: cones, cylinders, prisms, pyramids and spheres.
- Predefined specialist libraries of graphic shapes can be provided to support specific design disciplines such as architecture, interior or textiles design.

Task

Logos often make use of primitives like the recently introduced British Farm Standard logo (see Figure 3.3.8). It is designed to help UK farmers compete with imported food supplies and it can be found on fresh product lines in supermarkets. Analyse the shapes in the British Farm Logo. Use a CAD program to create a copy of the logo showing clearly the primitives that can be used to construct it. Collect and create a display of examples of other logos that are drawn in this way.

Figure 3.3.8 *British Farm Standard logo*

Computer-aided design images and models

All 2D and 3D CAD systems are based on the fundamental need for a designer from any specialist discipline to work with a 'model' of a product design. These systems are classed as interactive graphics systems because the user has control of the image on screen; data can be added, edited, modified and deleted. The generation of digital models or virtual products removes the pressure to produce a physical model too early in the product development process. The virtual product becomes the designers' main means of communicating and talking about their ideas with others in their creative and manufacturing team. Creating a digital product model requires a system that is capable of processing mathematical functions and making complex calculations in order to produce models and manipulate graphic images in 2D and 3D. These images and models are either generated by vector or raster graphics.

Using vector and raster graphics in CAD systems

In systems using vector and object-oriented graphics, geometrical formulae are used to represent images as a series of lines. Raster graphics represent images as 'bitmaps' and the image is composed of a pattern of dots or picture elements (pixels). In design terms, the way a graphic image is generated or displayed affects the impact it will make or the message it conveys. Draw programs create and manipulate vector graphics. Programs that manipulate bit-mapped or raster images are called paint programs. Vectored images are more flexible because they can be resized and stretched. In addition, images stored as vectors look better on screen or paper, whereas bitmapped images always appear the same regardless of a printer or a monitor's resolution or picture quality. Another advantage of vector graphics is that representations of images often require less memory than bitmapped images do. CAD programs employ a combination of these two graphics. Vector graphics are used to draw lines and produce 3D shapes; raster graphics are used for the rendering of surfaces and textures.

The uses of CAD-generated images and models

Designers use CAD systems in a variety of ways depending on the properties of the image or the product to be modelled. They are concerned with how the image will sell the product, or in the case of a product, how it will function, how the parts go together (structure), its form (shape), as well as the materials, surface finishes, textures and functional dimensions. The design models that are generated on screen can be sent directly to a manufacturing centre. For instance, for folding box production CAD data can be used to produce the cutting dies for the carton and to prepare the labels in different colour variants and in different language versions if required.

Computer to plate

Traditionally, printing companies would receive the camera ready artwork from a graphic product designer and then use it to create the film and printing plates necessary for the printing process to proceed. Increasingly, a new technique called computer to plate is being introduced that removes the need to create films from the production workflow. Using computer-to-plate equipment designers can now use the digital artwork to output a set of printing plates directly from the desktop. There are no films, no chemical film processing and no operators tied up exposing plates so that make-ready times are significantly reduced. The quality of the final output is improved because a laser is used to cut the plates and the images produced have much greater printing definition.

CAD generated images on printed plastic are ideal for sales promotion and merchandising. Figure 3.3.9 shows plastic items that have been thermoformed, folded or die cut. The images are digitally printed directly on to the plastic, and this offers the opportunity to personalise marketing materials, such as mouse mats, by adding individual names. Each type of plastic has its own unique properties. In terms of visual design, plastics are often far more versatile and effective than either paper or card. It is possible to digitally print directly from a computer on to mirror effect plastic, plastics that are transparent or tinted, plastics with integral patterns and plastics with different textures.

Task

Polypropylene, PVC, polystyrene and polyester are all used in packaging and promotional materials that have images generated in a CAD program and that are digitally printed. Using the Internet and other sources, research the process of digital printing on plastic, illustrating your answer with examples of commercial products that use plastics.

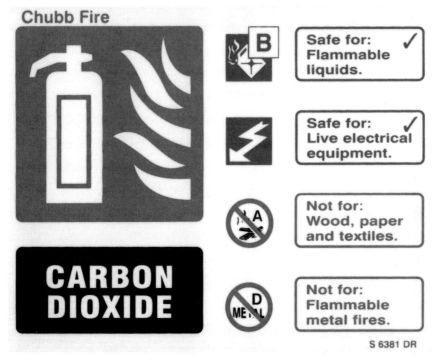

Figure 3.3.9 *This fire extinguisher sign has been digitally printed*

Comparing the benefits of 2D and 3D drawings

Historically, 2D CAD drawings have been used to develop, share and exchange product shapes and designs by electronic means such as email or local and global networks. The benefit of a 3D drawing is that it provides additional visual information about important things such as the form of the product. The development of affordable 3D CAD systems that do not rely on powerful mainframe computers is providing a means to communicate and exchange much more detailed design information. 3D facilities allow complex screen images or virtual products to be rotated, sectioned, measured and annotated to create a range of digital data. These data can be used via CAM to support an increasing range of CNC and automatic processing operations.

Using design data

In modern CIM systems design data can also be shared with the suppliers of materials, components and sub-assemblies, as well as the workers and machines on the production lines. Product data can be used directly or converted into other file formats for a variety of other uses, such as product and sales presentations, company reports, marketing materials and brochures that are made available at a company website on the Internet. The websites are designed to boost brand awareness and electronic retailing presence in home and international markets. Virtual Reality Modelling (VRM) and the use of 'knowledge-based' expert systems via the Internet are growing in importance for all the purposes described above. Multi-media approaches that create more flexible merchandising formats for retailers are covered in more detail later in this unit and in Unit 4.

The importance of CAD modelling

CAD modelling is now a key part of the industrial and creative design process for the following reasons:

- Designs can be developed and electronically shared with others which enables a fast turnaround of ideas. A team of specialists often design products such as point-of-sale displays. The first stage in the process of designing a 'shelf wobbler' or moving display (see Figure 3.3.10) might involve drawing the overall layout in 2D including all the crucial features of the brand or product being promoted. The

Figure 3.3.10 *Examples of 'shelf wobblers' and other kinetic displays*

design constraints can then be applied along with the way the different components that make up the display need to move (kinematics). The basic solid model properties such as colours and textures are added and the top-level layout is complete. The specialist teams such as those responsible for 'die cutting' the final shapes work on their particular components or features. If there are any design changes, these take place on the top-level layout as the reference point and all the design teams receive updated versions automatically.

- Ideas can be tested, evaluated and modified at all stages at any point in the process.
- The need to produce a range of costly prototypes or samples is reduced. Photorealistic images can be created and modified without the need for physical models and expensive photography.
- Products and processes can be simulated or animated and then evaluated on screen. This means that development time, design costs and the use of resources is significantly reduced.

2D and 3D geometric models

Geometric modelling is concerned with describing an object mathematically (algorithm) in a form that a CAD or graphics program can display visually. Geometric models are subdivided based on the amount and kind of information they store. The three divisions are wireframe, surface and solid models (see Figure 3.3.11).

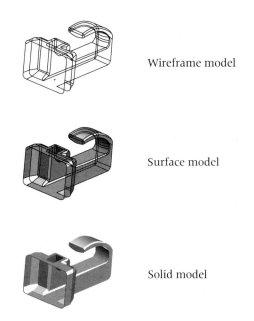

Wireframe model

Surface model

Solid model

Figure 3.3.11 *Applications of 3D modelling techniques*

Wireframe models

Wireframe models in 2D or 3D are most effective for sheet metal products and simple frame constructions without a great thickness of material. In a wireframe model an object is represented as a collection of points, lines and arcs. Wireframe models can be ambiguous and difficult to 'read' or interpret. A realistic form is only achieved by the generation of a lot of data, which increases image-processing times. Dimensions, annotations and other 'attributes' of the object may be stored but there are no visible surfaces. This means that surface or solid properties cannot be computed and rendered images cannot be generated. Additionally, 3D-wireframe objects lack information about points inside the object and the geometric data that are produced are incompatible with the requirements of CNC programs.

Surface models

Surface models in 3D can provide more machining data than wireframes and generate a more realistic 'picture' of the model. This technique is an alternative to solid modelling but provides fewer data. Surface models, either flat or curved, are created by 'patches'. Polygons define contours and surfaces. As with the wireframe, the surface model contains no data about the interior of the part.

Solid models

Solid models in 3D can produce full digital mock-ups and a comprehensive dataset including product assemblies. Solid models are clear; there are no visual confusions as with the other models described. They provide complete representations of the properties of the solid.

Rendering

Rendering is the process of adding realism to a computer model by adding visual qualities (see Figure 3.3.12a). These include colour; patterns and textures; surface shading with or without light sources; hidden line removal and hidden surface removal. Hidden line removal is an important drawing function as it removes lines from the drawing that would normally not be visible from the chosen viewpoint making the model less ambiguous. The semi-hidden function displays 'hidden' lines in the 'dashed' line style (hidden detail), which you will be familiar with in engineering drawings. Hidden surface removal is a technique for filling shapes on the model with colour to improve visual understanding.

Rendered images are also used in advertising, sales literature, assembly illustrations, operational

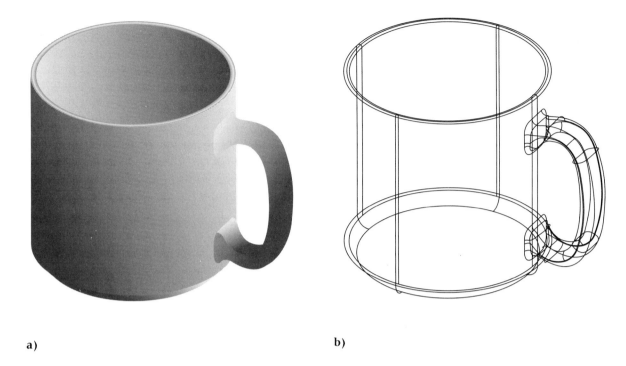

a)

b)

Figure 3.3.12 *Rendering and sweep techniques*

instructions and other information sources. The degree of realism that can be achieved is dependent on the quality of the available software and hardware, the designer's creative and visual abilities and the time available.

Shading
CAD software has a range of available shading options. The three most common are as follows:

- *Flat shading* is quick and simple. The surface of the object is divided into small polygons that are all shaded uniformly. This type of shading gives the object a faceted appearance. The curved surfaces are represented as a series of flat surfaces rather than a smooth curve.
- More realistic effects are achieved by *graduated shading*, which removes the sharp edges created by flat shading and replaces them with a gradually changing shading pattern.
- *Phong shading* is the most accurate as it incorporates 'highlights'. Each pixel on the shaded portion can be assigned a brightness value. As a result, the rendering quality and visual realism are very good but Phong shading is time consuming and slows down the processing of images.

Sweep techniques
Sweeping refers to a class of techniques used for creating curved or twisted solids (see Figure 3.3.12b). Sweep techniques involve drawing a profile along a path. The profile is usually a closed geometric form such as a 'D' shape. The path indicates how or to where the profile will be 'swept'. Moving the profile along the path that can be linear, circular, radial, spiral or some other configuration then creates the solid. Handles on a cup or a threaded part are examples of profiles that can be generated by sweeping.

Textures
Textures can have the 2D qualities of colour and brightness and they can have the 3D properties of transparency or reflectivity. Textures can be mapped electronically around any 3D model, a technique known as 'texture mapping'. Textures are an important part of creating ever-more realistic images but they use lots of memory and image processing can be slowed down.

Task
Produce a series of drawings for your project folder to demonstrate your capability in using the techniques described above using a CAD program that you have access to.

Constructing accurate drawings within a CAD system

To describe an object for manufacture accurately all the appropriate 2D orthographic views and 3D visualisations must be drawn. Production drawings communicate all the information necessary for the production of products and assemblies. All production drawings, whether CAD or manually generated, can be classified into two major categories:

- *Detail drawings* are drawings of single parts and include the additional information such as dimensions and notes relating to materials, finish, weight or calliper of paper, or standard colour options such as Pantone numbers that are required to produce the parts.
- *Assembly drawings* document all the necessary parts needed to assemble a product and how they fit together. The dimensions in an assembly drawing, such as a point-of-sale display, usually refer to the spatial relationships of different parts to each other rather than the size of the individual parts. An assembly drawing may be a multiview drawing or a single profile view. Ballooned letters or numbers are attached to leader lines to 'reference' or identify the parts in the assembly. The letters or numbers also identify the part in a list which is usually placed to the bottom right of the drawing. The 'parts list' provides information regarding the name of the part, what material it is made from and the minimum number of each part that is required. Standard 'off-the-shelf' parts and components like fixings and fastenings are also included in the parts list.

Dimensioning and annotating a drawing

Dimensioning and annotating a drawing provides accurate information about the size of the product and its component parts. Different types of features require the use of different dimension formats. CAD systems can provide linear, angular, cylindrical and radial dimensioning. Notes are added to drawings to provide additional information about the project. They are used to indicate specific surface finishes or materials; or other special manufacturing requirements such as the size or depth of holes.

The importance of dimensioning standards

In any manufacturing system, it is important that all the people reading a 'drawing' interpret it in exactly the same way. However, dimensioning practices may vary from company to company or from country to country. A set of international standards (ISO) has been developed to specify acceptable dimensioning practices.

CAD systems offer standard dimensioning formats allowing the fundamental principles of dimensioning to be followed. Some are listed below:

- Each feature of an object is only dimensioned once in the view in which it is most clearly seen.
- Each dimension should include an appropriate tolerance.
- Dimensions should be located outside the boundaries of the object wherever possible and there should be a visible gap between the object and the start of a dimension line.
- Crossing of dimension lines should be avoided wherever possible.
- Dimensions should refer to solid rather than hidden lines.
- Dimensions should be placed as close as possible to the feature they are describing.
- When dimensions are 'nested', the smaller dimension should be placed closer to the object.

Figure 3.3.13 shows some standard dimensioning formats.

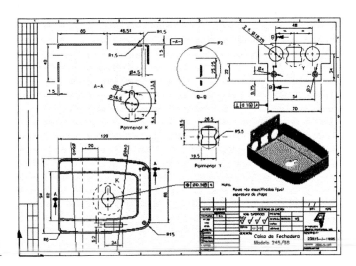

Figure 3.3.13 Standard dimensioning formats

Sections

Sections are an essential aid to understanding the complexity of a product. They should make a drawing easier to understand. Standard views show all the exterior features of objects, but if the interior features are shown just as series of dashed lines (hidden lines) it can cause confusion to less expert readers of the drawing. In a CAD drawing, as in manually produced drawings, sectioning cuts the object with an imaginary plane (cutting plane), making interior features, which were hidden, visible. The generation of sectioned views is quick and relatively easy when using a CAD program. The solid parts of the object in contact with the cutting plane are cross-hatched. CAD systems will allow many types of section view to be drawn. The choice of method depends on the internal complexity of the object.

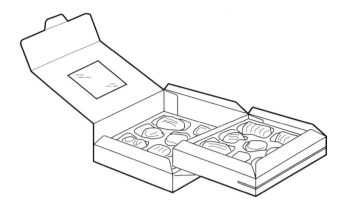

Figure 3.3.14 *An exploded view of a box of chocolates*

Tasks

1 Using a CAD program, generate some simple block shapes with holes, recesses and cavities.
2 Using either the views you generate above, or views supplied by your teacher, investigate how to produce:

   a) a full-section view, i.e. show an entire orthographic view as a section view, with half the object removed
   b) a half-section view, i.e. show one-half of the orthographic view as a section view
   c) an offset-section view, i.e. a type of full section using two or more cutting planes that meet at 90-degree angles
   d) a removed section view – this is similar to a revolved section but it is not drawn within the view containing the cutting plane line but is shown displaced from its normal projection position.

Exploded views

Manufacturers in all design disciplines use exploded views to provide a visual explanation of how a product is assembled. A typical example is shown in Figure 3.3.14.

Virtual reality (VR) techniques

Virtual reality is an emerging technology that combines computer modelling with simulations in order to enable a person to interact with or be immersed in an artificial 3D visual or other sensory environment. Three-dimensional 'virtual products' can be created and viewed from different angles and perspectives.

Virtual Reality Modelling Language (VRML) is a specification for displaying and interacting with 3D objects on the **world wide web** using a **web browser** with a VRML plug-in or one that supports it. The development will have a significant impact on all industrial sectors but especially manufacturing. For instance, potential customers will be able to download virtual products and examine all aspects of them offline anywhere in the world. There are many other exciting developments that are beyond the scope of this book, but here are two examples for you to find out more about:

• Virtual manufacturing is a rapidly developing technology being pioneered by research teams all over the world. One of these teams, based at the University of Bath, produced this definition in 1995: 'Virtual manufacturing is the use of a desk-top virtual reality system for the computer-aided design of components and processes for manufacture'. In their system, a user wearing a helmet with a stereoscopic screen for each eye views animated images of a simulated manufacturing environment. The illusion of being there (telepresence) is caused by motion sensors that pick up your head movements and adjust the view on the screens accordingly, usually in real time. Real time is the actual time during which something takes place. Simulations of manufacturing processes are usually accelerated to save time.

• Denford Ltd has created an innovative 3D website that uses virtual reality worlds to create a tour of its Professional Training and Development Centre where visitors can control a robot and try out trial versions of the company's CAD/CAM products (see Figure 3.3.15).

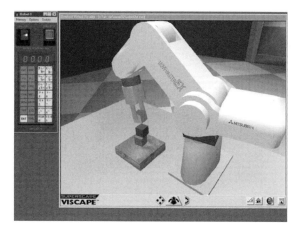

Figure 3.3.15 *3D website*

Input devices used in CAD systems

As we have seen, CAD systems have to be both interactive and graphical in use. In addition to the keyboard such systems need input devices and a user **interface**, operating as an input/output device, that allows interaction with the computer-generated model. There are two types of interface in common use:

- *Command-driven interfaces* require specific commands or codes to be used in order to make something happen. These interfaces process data quickly and are flexible in use but they rely on operators trained in a particular code or command set. Incorrectly entered codes make it difficult to edit or revise the drawing. These types of interface are being gradually replaced by graphic user interfaces.
- *Graphic user interfaces (GUIs)*, first developed in the 1970s, take advantage of the computer's graphics capabilities making the CAD program easier to use. Most GUIs use WIMP format, (Windows, Icons, Mice and pull-down/pop-up menus operating environment). The user does not need to learn complex command codes. GUIs are effective in allowing the user to:

  - control the system by using set commands; by selecting functions via a series of windows; by a menu system; by screen icons or by direct actions such as 'clicking and dragging'
  - receive information and feedback relating to what the system is doing, for example displaying an hourglass icon or progress meter
  - enter data that will be used by the system in constructing the model
  - select relevant data or parts of the model for the system to manipulate.

Input devices position or locate, point or pick or combine these functions. To signify that an action has to take place the user presses a button or switch that is provided on the input device. Many input devices produce an **analogue signal** (A) that needs to be converted into a **digital signal** (D) in order for it to be processed by computers. This (A to D) process is completed using a device called a digital signal processor (DSP) which can also be used to produce an analogue signal for use by an output device (D to A).

The mouse

Invented in 1963, at the Stanford Research Centre, the mouse is the most common input device. It operates either mechanically or optically to control the movement of a cursor or a pointer on the graphic display. It allows the user to 'point' to a function and 'click' to 'execute' it. All mice contain at least one button and sometimes as many as three, which have different functions depending on what software is used. Some also include a scroll wheel for 'scrolling' through large documents. The big disadvantage of using a mouse is that a positional error will occur if the mouse is lifted from the surface that it is running on. It is also virtually impossible to trace a drawing from a paper sketch or drawing (see digitiser).

The trackball or tracker-ball

The trackball or tracker-ball is basically a mouse lying on its back. The cursor on the screen moves as a thumb or fingers, or even the palm of the hand rotates the ball. As with the mouse there are one to three buttons which are used like mouse buttons. Unlike a mouse the trackball remains stationary so has the advantage of not requiring much space and it will operate on any type of surface. For both these reasons, trackballs are popular pointing devices for personal and laptop computers.

Table 3.3.2 *Types of mice and computer connections*

	There are three basic types of mice:	Mice and tracker-balls connect to computers in different ways:
	Mechanical – has a rubber/metal ball on its underside that can roll in all directions. Mechanical sensors within the mouse detect the direction the ball is rolling and move the screen pointer accordingly	Serial mice connect directly to the RS-232C serial port or a PS/2 communication port. This is the simplest type of connection
	Optomechanical – similar to a mechanical mouse, but uses optical sensors to detect motion of the ball	PS/2 mice only connect to a PS/2 communication port
	Optical – uses a laser to detect the mouse's movement along a special mat with a grid that provides a frame of reference. Optical mice have no mechanical moving parts. They respond more quickly and precisely than mechanical and optomechanical mice, but they are also more expensive	Cordless mice are more expensive and do not physically connect to the computer. They rely on infra-red or radio waves to communicate with the computer On Apple Macintosh computers the mouse connects through the ADB (Apple Desktop Bus) port

Table 3.3.2 shows the three basic types of mice and identifies the different ways in which mice and tracker-balls are connected to computers.

Digitiser

Digitising tables have a large working area, typically over A0 paper size, enabling users to enter large-scale drawings and sketches into a computer. They operate to a great degree of accuracy and avoid the positional problems described previously when using a mouse. A digitising table consists of a reactive electronic surface and a cursor (also called a puck) that has a window with cross hairs for pinpoint accurate placement and it can have up to 16 buttons to execute various functions. Each point on the table represents a fixed point on the computer screen. To determine the exact position of the puck the surface may have a grid of embedded wires, each carrying a coded signal. The puck has an electronic device that can pick up these signals that are then digitally translated to give an exact position on the screen.

Graphics or digitising tablet

A graphics tablet is a tabletop digitiser consisting of a rectangular board (tablet) and pen (stylus) electronically connected to a computer (see Figure 3.3.16). The board contains electronics that detect movement of the pen and in the more sophisticated tablets a pressure sensitive stylus enables the user to simulate the marks that a pen or brush would make when it is pressed harder on the paper. The advantage of this device is that the designer can use hand-drawing techniques in order to 'draw' ideas electronically. The drawing can then be manipulated in the normal way by the CAD software.

Figure 3.3.16 *A graphics tablet*

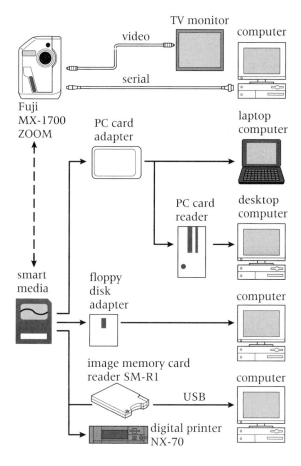

Figure 3.3.17 *Connecting a typical digital camera to a computer system*

system for processing and image manipulation within minutes of shooting. Digitally enhanced images now appear in a whole range of media ranging from product presentations such as photo montages on product labels to the Internet and print-based product catalogues. If the requirement is to get images in electronic form in the fastest possible time, then a digital camera is the best choice.

PhotoCD™

PhotoCD™ is a useful way of getting images from an ordinary camera into a computer fitted with a CD-ROM drive. The film to be developed and converted is sent to an authorised Kodak PhotoCD™ developer who provides a CD-ROM containing the compressed digitised images. The quality of the images is relatively good and the CD often contains several sizes of each individual image for various uses, the largest size being 2048 x 3072 samples per inch (spi). One of the advantages of this method is that CD-ROMs are an effective way of keeping image files organised without occupying expensive disk storage space on the computer system. However, in some cases, a dedicated slide scanner, although a more expensive and less convenient option, provides a higher quality image that has not been compressed.

Digital camera

Digital cameras store images digitally rather than recording them on film. Once a picture or image has been 'captured', it can be downloaded as data into a computer system for manipulation within a CAD program, stored on a photo CD or printed on a dedicated digital printer (see Figure 3.3.17). The quality of a **digital image** is limited by the amount of memory in the camera, the optical resolution of the digitising mechanism, and by the resolution of the final output device. Established printer technologies have limitations but there are three printer technologies, 'thermo autochrome', 'dye sublimation' and digital printers that produce better images. We shall look briefly at all three in the section on output devices.

The main advantages of digital cameras for designers are the operating costs; the speed of data conversion (there is no film processing at the start of the process); and the image manipulation and editing that is possible. A digital camera allows an image to be put in a CAD

Figure 3.3.18 *A **sleeve** for a food product that combines text, graphics and photographic images*

Digital video (DV)

A new generation of video cameras that are entirely digital are now on the market. In a digital video (DV) camera the output from the camera is already in a compressed digital format, so no analogue to digital conversion is required as with analogue video cameras. Images can be taken straight to a PC-based video editing system. Digital video systems allow the communication, control and interchange of digital, audio and video data. The benefits of applying these DV systems within CAD other than for video conferencing are an area of considerable research.

Scanner

A scanner is a device that converts analogue data, into digital data that can be read by computer software for desktop publishing (DTP), CAD and other programs that combine text and graphics. The data can be edited, stored and used to develop design ideas and specifications.

The quality of a scanned image depends on the resolution of the scanner. The more dots per inch (dpi), the sharper and more detailed the image on screen. As is the case with most computers generated text and graphics, the final appearance of the image will depend on the quality of the output device – a laser printer produces better quality images compared to a low-quality inkjet printer. High resolution scanning is important for line artwork to ensure that the lines retain their smoothness when printed on a commercial high-resolution output

device such as a 2500 dpi imagesetter.

Storing scanned images requires large amounts of random access memory (RAM) on the hard disk or other storage space. A colour image measuring 200 mm x 250 mm at a low resolution of 300 dpi would require in excess of 7 Mb (megabytes) of RAM. The higher the resolution required, the more memory intensive is the image. Graphic designers use scanners mainly for:

- scanning in line artwork for reproduction or positioning on a graphic image
- scanning images to be used as positional guides on a page or other graphic layout
- inputting text copy by means of optical character recognition (OCR) software
- scanning images that will be digitally manipulated to achieve a particular visual effect.

Scanners are able to scan both 2D surfaces and 3D objects (see Figure 3.3.19). Developments in the light sensitive scanning head called the charge coupled device (CCD), the control technologies and the processing software mean that scanners are now much easier to operate. The images produced range from simple black or white (line images) to greyscale and full-colour images with in excess of 256 colours.

Types of scanner in common use in CAD

A scanner may be monochrome, greyscale or colour and it may be one of the following:

- *Flatbed scanner* – the most versatile and popular desktop device for scanning reflective, line and greyscale artwork. It looks and works like a photocopier as the image is placed face down on a flat glass screen and a scanning head passes across the image. There are three types of flatbed scanner:
  - entry-level flatbed scanners generally share the following specifications:
    216 mm x 280 mm scanning area, low to medium resolution and are low cost. They often come with entry-level image manipulation software. These machines frequently offer excellent price/performance ratio
  - mid-level flatbed scanners differ in three important ways: they cost much more; they are targeted at a professional market and often come with more sophisticated image

manipulation software; they also have significantly better technical specifications resulting in scans of correspondingly higher quality. Some mid-level scanners may also offer a larger scanning area compared to flatbeds

- high-end flatbed scanners are often seen as a practical alternative to drum scanners offering sophisticated design features that graphic design requires. They are expensive but offer a noise-free design, large scanning areas and very high image resolution.

- *Sheet-fed or edge-feed scanners* – similar to flatbed scanners, they operate differently as the original is moved under a fixed scanning head. They are used for high-volume work.

- *Handheld scanners* – portable, low cost, low resolution and small capacity devices with restricted scanning widths.

- *Large drum scanners* – capable of scanning both opaque documents and transparencies at high resolutions of over 400 dpi; expensive.

- *Transparency/slide scanners* use a scanning camera to produce high-quality colour images when the transparency or slide is placed between the camera and a light source. Slide scanners cost a lot more than the relatively inexpensive flatbed transparency option. For those who may need only an occasional transparency scanned, a flatbed with transparency adapter is the way to go. But if you scan a lot of transparencies, then the only equipment that offers the best quality scans are dedicated transparency scanners. Expensive, multi-format transparency scanners aimed at the professional market can scan everything from 35 mm slides to 100 x 130 mm high-quality transparencies. As scanning technologies improve these high-end transparency scanners are presenting an alternative to drum scanners by offering more features, better software, and faster scanning time.

- *Overhead scanners* – scanning cameras mounted above a copy table on which the original such as a photographic image is placed. They can also be used for scanning transparencies that have been mounted on a lightbox on top of the copy table.

- *Drum scanners* – desktop scanners in which a laser beam acts as the light source to capture the image from the original attached to a spinning glass drum.

- *Video Digitisers* are mostly used for multimedia purposes, especially in the creation of *QuickTime*™ movies, but they can be used to capture still images for print. Video cameras utilise the same digital CCD arrays found in flatbed scanners. These CCD arrays produce an analogue signal (50 or 60 MHz) that either drives other analogue devices such as VCRs and television sets, or is captured on to videotape. Video cameras are not technically scanners in the truly digital sense of the word, but the analogue video signal can be digitised using specialised hardware and software. Video image capture software is very similar to traditional scanning software, while the hardware is usually a board that fits inside a computer. Video cameras provide an inexpensive way to get images into a computer, but the image resolution will be low (only 640 x 480 pixels) and the colour accuracy is often poor.

3D scanners

3D scanning or digitising creates a series of profile curves that define the surface characteristics and the physical geometry of a three-dimensional object. The digital data collected and recorded allow a 3D representation or model to be created within a computer. CAD and graphics software can then be used to blend and render surfaces to add colour, reflections, texture and other visual techniques to add visual realism. There are two different methods of scanning a 3D object in order to produce a digital file, contact or non-contact.

Contact scanning systems make physical contact with the object with a probe that can be passed around and over the surface of the object. The tracking is done by a probe that is machine driven or manually operated. A typical manual system uses a mechanical arm that has digital sensors in each joint (see Figure 3.3.19). A typical use of these devices is in converting a concept model made in any suitable medium such as clay into a set of electronic data that can

Figure 3.3.19 *An arm scanner*

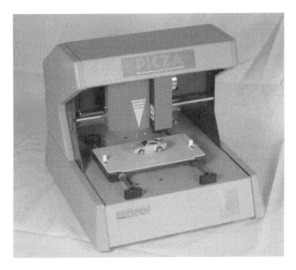

Figure 3.3.20 *A Roland scanning machine*

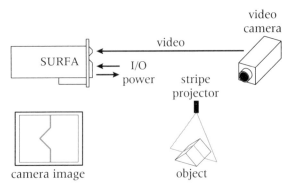

The SURFA board uses Digital Signal Processing of video data to capture surface shapes in real time at over 14,000 points per second.

Figure 3.3.22 *Laser stripe triangulation*

be used by a CAD program or a CNC machine. An operator moves a probe over the surface of the object clicking and recording the positional data. These points generate the required electronic profile curves. A machine-driven device common in UK schools and colleges is the Roland model, which can operate as a low-cost 3D scanner or as a small capacity CNC milling machine (see Figure 3.3.20). The scanned data can be exported in 3D DXF or VRML format for use in other software.

Non-contact scanning systems mostly use a geometric technique known as triangulation to create the 3D shapes. Laser scanners are non-contact and high-speed devices but they are expensive. These devices scan by either directing beams at various points to create a profile curve or by generating a 'laser stripe profile' of the object (see Figures 3.3.21 and 3.3.22). The beams or stripes are reflected back to a series of sensors or into a video camera. In both cases an accurate surface representation is built up by scanning

from different planes and angles and mathematically processing the digital data.

Ultrasonic scanners are also non-contact devices that 'bounce' sound waves off the surface of the object and triangulate the points in 3D. The devices are not portable or easy to set up. They are noisy in operation and do not provide a high level of accuracy.

Magnetic tracking scanners work on a similar principle to ultrasonic devices using magnetic fields to create the triangulated points. The major drawback of such systems is that they cannot scan any object with a metal part.

Common output devices

An output device is what makes a computer capable of displaying and manipulating pictures (graphics) or machining a variety of materials. Laser printers and plotters are 'graphic' output devices because they permit the computer to output pictures. They produce 'hard copy' on paper or film. A monitor is an output device that can display pictures. A CNC milling machine is an output device that can produce physical objects in a range shapes and materials.

Linking CAD and CAM

CNC machines provide an interface between CAD and CAM. Numerically controlled (NC) machining processes were first developed in the USA in the 1940s. The motors and machining tables on the NC machines were originally driven from instructions provided on a punched card system and then from magnetic tapes. Programs generated and transferred from computers now control the operation of many NC machines, hence computer numerically controlled or CNC

Figure 3.3.21 *Laser scanners in operation*

machines. There are a host of CNC machines including milling machines, lathes, drilling machines, press engravers and die cutters used for producing cartons and packages. New classes of 'tool-less' CNC machines are in use in industry, for example CNC laser devices are increasingly being used in cutting and machining applications.

Transferring data to output devices

Both CNC machines and graphic output devices are generally connected to the computer by either 'series' or 'parallel' cables, but some devices such as printers can be operated by infra-red sensors and are therefore 'wireless'. When cables are used they are connected into the computer via communication ports, the size, shape and number of connecting pins varying according to the specific make of computer. Almost all graphics systems, including CAD systems and animation software, use a combination of vector and raster graphics. Most output devices are raster devices.

Displaying graphics

The graphic capabilities of a monitor or display screen make it an important output device in any CAD system. There are many ways to classify monitors. The most basic is in terms of colour capabilities, which separates monitors into three classes:

- Monochrome monitors actually display two colours, one for the background and one for the foreground. The colours can be black and white, green and black, or amber and black. They are often used on the displays of a CNC machine where a colour image is not so important.
- A greyscale monitor is a special type of monochrome monitor capable of displaying different shades of grey. They are also often used on the displays of CNC machines.
- Colour monitors can display any number from 16 to over 1 million different colours.

The importance of screen size for CAD work

The choice of screen size, or viewable area, is particularly important in a CAD system. Monitors that measure 400 or more millimetres from corner to corner are often called full-page monitors and are considered more suitable for graphics work. Large monitors can display two full pages, side by side.

The resolution of a monitor indicates how many pixels are on the screen. In general, the more pixels (often expressed in dots per inch – dpi), the sharper the image and the better the resolution.

Printers

A printer is an output device that prints text or graphics on to paper, card and film. Not all printers can produce high-quality images. Printers are classified by the following:

- The quality of type they produce – either letter quality (as good as a typewriter), near letter quality, or draft quality. Inkjet and laser printers produce letter-quality type.
- The speed at which they print. Printing speeds are measured in characters per second (cps) or pages per minute (ppm). The speed of printers within a particular class can also vary widely. Laser printers print at speeds ranging from four to 20 text pages per minute.
- The effectiveness of the methods used to create colour and more realistic images by employing techniques such as dithering, colour matching between the screen and paper, half toning and continuous tones.

The following printers are considered below:
- inkjet
- laser
- dye sublimation
- thermo autochrome
- digital
- offset.

Inkjet printers

These non-contact printers produce low-cost, high-quality text and graphics in colour by combining cyan, magenta, yellow and black (CMYK). A typical commercial use of a larger inkjet printer would be in the production of exhibition panels or short-run or one-off posters and photographic enlargements that require clear and colourful images at a comparatively low cost. The images are drawn on-screen to produce digital artwork for printing on to A2, A1, A0 and custom paper sizes paper before being encapsulated in clear plastic, for added durability, and mounted on foamboard or some other suitable substrate.

Inkjet printers all use some sort of thermal technology to produce heated bubbles of ink that burst, spraying ink at a sheet of paper to form an image. As the ink nozzle cools a vacuum is created and this draws in a fresh supply of ink for heating and spraying. The print head prints in strips across the page, moving down the page to build up the complete image with a resolution that can range from 300 to 1200 dpi. They are sometimes referred to as 'bubble jet' printers, which is actually the trade name of Canon's own inkjet technology.

Disadvantages of the inkjet printer are that the ink cartridges need to be changed more frequently and expensive specially coated paper is necessary to produce really high-quality images, which significantly raise the cost per page. Choosing the right paper for inkjet printing is important. Some images, especially those with large areas of colour, can 'bleed' if the paper is too absorbent or too much ink has been applied. Bleeding causes images to blur as the colours merge and run together.

Another disadvantage of this system of printing is that images are easily smudged as some types of ink take a little time to dry. A further drawback can be that the diameters of the nozzles used in inkjet printers are very small and can easily become clogged.

Laser printers

These printers use the same technology as photocopier machines. The printer receives data from the computer, which is processed and used to control the operation of a laser beam directing light at a large roller or drum. Altering the electrical charge wherever laser light hits the drum creates the required image. The drum then rotates through a powder called toner. The electrically charged areas attract the powder and the print is made when it is transferred on to the paper by a combination of heat and pressure. Laser printers produce very high-quality text and graphics.

Dye sublimation printers

These low-speed devices produce relatively expensive but high-quality graphic and photographic images. The four coloured inks or dyes (CMYK) are stored on rolls of film. A heating element turns the ink on the film into a gas. The amount of ink that is put on paper correlates to the temperature of the heating element. The temperature varies in relation to the image density of the original drawing or artwork. The reason that it produces images of such high quality is that the ink is applied as a continuous tone rather than as a series of dots and special paper is used that allows the dyes to diffuse into the paper to mix and create precise colour shades.

Thermo autochrome (TA) printing

This method is used to print high-quality images generated by digital **photography**. The process is more complex than either inkjet or laser printing. TA paper contains three layers of coloured pigment, cyan, magenta and yellow, each of which is sensitive to a particular temperature. Three passes are needed to get the three colours to show. The printer is equipped with both thermal and ultraviolet heads; the heat from the thermal head 'activates' the colour in the paper, which is then 'fixed' by the ultraviolet light.

Digital printing

Digital printing is a system of printing based on rapidly evolving digital technology, which involves linking state-of-the-art printing presses and computers, bypassing the traditional route of making printing plates. High-speed digital printers use laser technology to provide 600 dpi print quality for text, photographs and graphics on a wide range of media. These printers combine ultra high resolution with over 4000 adjustable levels of shading and can print at a rate of 250 copies per minute. For printing from hard copy the original is scanned in once reducing wear and tear on the master copy. If a digital file is used every print is a 600 dpi original. There is no deterioration in print quality. The printers come with a full range of image handling capabilities including image clean-up, reduction and enlargement, photo-screening, cropping, masking, rotating and the moving of text and graphics. The advantages of digital printing include:

- digital job preparation and processing is controlled from the end user's desk without having to save data on disks, CDs or other traditional media for later delivery to the printer
- reliability as end users can download their printing jobs via a modem, **ISDN** or network without worrying that they will output incorrectly; the software allows users to specify finishing and media types and sizes from the desktop
- what you see is what you get, as documents can be seen on-screen as they will be printed
- print proofs can be made at the end user's terminal for checking
- the ability to select media finishing such as stapling and gluing
- less operator stress as different printing jobs can be sorted so that those requiring the same media can be printed together
- more productive print operators thanks to streamlined job flow and lower document production costs
- printing on demand as digital technology allows images to be scanned or transmitted electronically into memory to be stored for future use
- the smoothing out of production runs which is important when market demands are fluctuating; for example, you can print 700 jobs one week and then increase production to 1000 copies the next with no additional set-up

charges as all the files and processing settings are stored digitally.

In your school or college you may come across specialist types of digital printers that are used to produce photographic quality images from a digital camera without having to transfer data to a computer. The printers can be connected to a monitor for viewing and editing images and layouts. The widespread introduction of digital printers is limited at the time of writing because there are no agreed standards for them. Manufacturers of digital cameras will supply digital printers that match their range of cameras.

Offset or offset lithography (litho) printing

Offset printing is currently the most common commercial printing method, in which ink is offset from the printing plate to a rubber roller, then to paper. This type of printing technique can be divided into two parts: duplicating or standard printing; and commercial quality or special printing.

Standard printing is a cheap alternative to photocopying. Any reasonably typed, hand-drawn or computer-generated original can be used to produce long copy runs. It is often offered as a 'same day' or an 'over the counter' service for all small- to medium-run jobs, i.e. 1–20 originals with 100 copies of each.

Special printing describes high-quality and multicolour commercial type printing. In addition to paper and card, a wide selection of materials can be printed such as plastic, foils and other laminated paper finishes. Increasingly, a graphic artist provides the designs, producing digital data that can be sent directly for computerised phototypesetting. These computer-based production methods provide considerable savings and can be made with no loss of quality when compared with the traditional commercial printing methods.

The use of computers in specialist printing situations

Screen process printing, sometimes referred to as silk-screen printing or serigraphy, is very common. The introduction of computer technology in the 1980s, in the pre-press side of screen printing, particularly in image capture and manipulation, was a major step forward in the scope and range of images that could be applied to objects. With this process it is possible to print on to a great variety of surfaces such as plastics, wood, metal, glass, textiles, paper and board.

The process is typically used to place images, many of which are digitally produced, on to products such as binders, document wallets, mouse mats, window and car stickers as well as posters and all types of point-of-sale and promotional materials. Other uses of screen printing include displays on computers and the badges and control panels on electrical equipment. In the home you will find that many textiles and items of clothing, sports bags and T-shirts have all been screen printed.

Artists have also used silk-screen printing, especially since the days of 'Pop Art' in the sixties; Andy Warhol, Rauschenberg and Hamilton are a few examples. Some artists and print makers who want to produce multiples of their work, limited editions and other projects often like to provide their own colour separations prepared on film or as camera-ready artwork, while others prefer to supply working drawings, sketches, together with their instructions and colour specifications. There are no set rules to producing a print and every artist works differently so printing houses have to be able to operate flexibly. Typically, they might offer a photo stencil-making service to printmakers wishing to screen print their own work. For those that are now using digital imaging techniques the company must be able to offer the production of computer-cut stencils.

Continuous stationery covers products such as invoices and statements and involves the original being created in a word processing or desktop publishing application before direct or indirect transfer into the printing process.

Flexographic printing (flexography) provides economical printing on unfinished surfaces, wrapping paper, cardboard, plastic film, and it is also used for printing self-adhesive and fabric labels. Digital images can be drawn from a range of sources and then transferred to an image setter or to a plate setter for production of the printing plate.

Other uses for computers in the commercial print process

Computers are playing an increasing role in each of the three stages of the commercial printing process pre-press, press and post-press:

Pre-press includes all the various printing related services performed before ink is actually put on the printing press and includes image capture and manipulation, scanning and rasterising, colour separating and proofing. A typical pre-press department concerns itself with pasting-up camera ready artwork to produce final films of a specified area, usually a page, which contains all the images, text and tints in position, necessary for the production of a printed plate. Images and text produced on a computer are sent directly to a Raster Image Processor (RIP) which converts the electronic information into a format that an image setter can understand and process. An image setter is a high-resolution device that prints directly to film or bromide significantly improving print quality and reducing preparation time. Increasingly, pre-press also involves electronic publication direct to devices such as a plate setter which is a device to create printing plates directly from a digital page, missing out the need for a film stage.

Press is the stage of printing the image on to the chosen media in which automated, CNC and microprocessor-controlled presses are being increasingly used. The market for short four-colour runs is one of the fastest growing and dynamic of all. It is characterised by short runs with extremely fast delivery and the need to be able to receive and act on last-minute changes. With changing market requirements, new approaches have to be taken and digital technology is increasingly at the core of workflow. The technology enables companies to become complete service providers from digital data to the final product. In the new digital printing presses the printing plates are imaged using data direct from a computer which reduces the number of production steps, increases speed and improves reliability. Computers provide profitability with a high degree of automation, user-friendly operation and enormous flexibility for job changes.

Post-press/finishing is the stage at which the printed item is cut to size, folded as necessary and finished as appropriate by binding and stitching.

Plotters

A plotter is a high-quality impact-printing device that draws images on paper or any other suitable medium directed by commands from a computer. There are two classes of plotter, vector and raster. Vector plotters produce an image as a set of straight lines, fill patterns are clearly visible and they operate comparatively slowly. All raster plotters generate an image as a series of points. The way that the points are printed varies and the methods are beyond the scope of this book. These plotters produce very large, full-colour drawings with a high degree of quality.

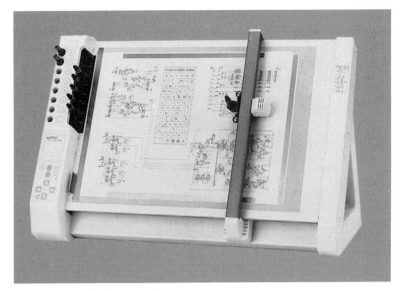

Figure 3.3.23 *An XY plotter*

XY plotter

The term XY refers to the axes along which the plotting pen can travel. Plotters differ from printers in that they draw lines using a pen. As a result, they can produce continuous lines, whereas printers can only simulate lines by printing a closely spaced series of dots. Multicolour plotters use different-coloured pens to draw different colours. Pens can be picked from a bank of penholders or changed individually as the different colours are called for. In general, plotters are considerably more expensive than printers. They are used in engineering applications where precision is essential. An XY plotter is shown in Figure 3.3.23.

Plotter-cutters

Plotter-cutters can plot drawings in the same way as the XY plotter described above but they can also produce cut shapes in card, vinyl and other sheet materials using thin blades that can be adjusted for depth and pressure of cut. This allows the plotter-cutter to undertake finely controlled cutting techniques such as 'scoring' in which a card that is to be folded is only cut to a certain depth to make the subsequent folding easier. Computer-controlled vinyl cutters are used in the production of advertising and promotional products as well as in exhibition screens, sign and banner making and many other similar processes. Figure 3.3.24 shows a typical plotter-cutter.

Engraving machines

Engravers also have a variety of uses ranging from sign and print plate making to the

Figure 3.3.24 *A plotter-cutter*

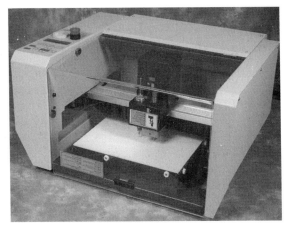

Figure 3.3.25 *A CNC engraver*

production of jewellery, medals and 3D reliefs (see Figure 3.3.25). They can be relatively small devices that sit on a desktop or larger floor-standing machines. The machines can operate in the x, y and z axes. This allows the engraving of 3D surfaces and curves as well as lettering. Cheaper versions of CNC engravers for schools and colleges have less control over the z axis of the cutting tool. This reduces the overall flexibility of the machine because it has to engrave at a fixed depth. One of the characteristics of an engraver is the high spindle speed that is needed because of the very small diameter of the V-point engraving tools. We shall return to CNC cutting speeds and feeds later in this unit.

3. Computer-aided manufacture

The production benefits of using CNC machines

CNC machines are widely used in a range of industries. Because they can operate in more than one axis they can generate shapes ranging from straight lines to very complex curves. The technology and functionality of CNC machines has continued to evolve since they were first used in the 1970s. CNC machining centres allow a single machine to carry out more than one manufacturing operation, for example a plotter-cutter can perform two main manufacturing functions. It can be used for high-quality, large-scale colour plotting on a range of media. If the cutting knife is fitted it can be used to cut and perforate card and other media such as vinyl to produce 2D shapes, such as packaging nets, screen printing templates, exhibition screens and advertising banners.

Processing flexibility and improved output quality is the key to the future development and profitability of computer-based manufacturing systems. Where they can be used, computer controlled machines offer significant reductions in processing times. In volume production, such as commercial printing, set-up or make-ready times are significantly reduced, allowing greater flexibility supporting 'quick response' production capable of handling production runs ranging from under 100 to print runs in the thousands.

CNC prototyping

The development of new 'tool-less' cutting technologies such as the use of computer-controlled lasers is further extending the range and complexity of products that can be prototyped or manufactured such as new designs for shampoo or perfume bottles. Rapid Prototyping (RPT) is an emerging CNC application that creates 3D objects using laser technology to solidify liquid polymers in a process called stereolithography.

Boxford Ltd in the UK has produced a relatively low-cost RPT system for schools and colleges based on a prototyping system called layered object modelling (LOM) (see Figure 3.3.26). Models are imported from a CAD program as in the industrial systems but the

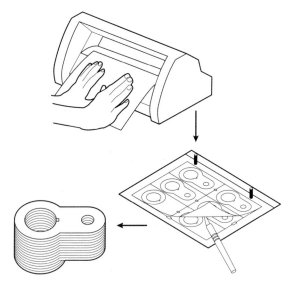

Figure 3.3.26 *Rapid Prototyping using CNC machines*

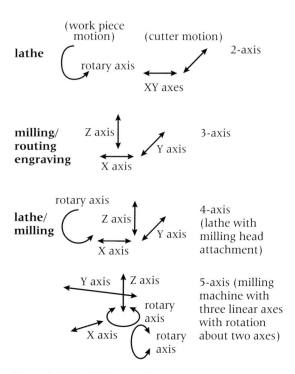

Figure 3.3.27 *CNC machines can move and operate in up to five axes*

models are assembled from thin sticky backed paper. A CNC vinyl cutter cuts the slices (in the industrial process a laser is used). The layers of sliced paper are built up on a pegged jig to make the 3D prototype.

> ## Task
> Use the Internet and other information sources to find out more about RPT. Produce an information sheet to record your findings. How could you use Boxford's RPT system in your design work?

Operating characteristics of CNC machines

On CNC machines such as plotter-cutters and embroidery machines either the tool or the work piece is able to move in up to five axes to generate the required point-to-point, straight or contoured tool paths, as shown in Figure 3.3.27.

On larger scale manufacturing machines such as plotting and cutting devices a computer-generated cutting or plotting program is fed into a machine control unit (MCU) which includes a manual control keypad and in some cases a display screen. The MCU reads and converts the digital data it receives in order to control the analogue machining movements that the machine has to make. Movement is controlled by a series of servo systems or, in low-cost devices, stepper motors, called actuators. As the machine is operating the MCU is constantly receiving feedback information from sensing devices

(**transducers**) or encoders on the machine. This information is used to correct any errors of spindle speed, feed rate or cutter position. This sensing and response mechanism is called a **closed-loop control** (see Unit 4, Section B, Option 2). On smaller desktop CNC devices such as engravers, plotters and vinyl cutters the cutting program is transferred from the computer by cable or infra-red to the CNC machine. The machine has the circuitry and computing capacity to generate the codes needed to move the machine in each of its three axes. Figure 3.3.28 shows a block diagram of a typical CNC machine.

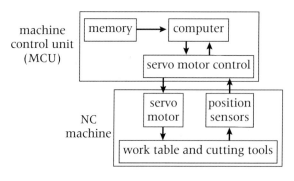

Figure 3.3.28 *A block diagram of a typical CNC system*

Toolpaths, cutting and plotting motions on CNC machines

There are three type of motions employed on CNC machines:

- The cutter or plotter moves from 'point to point' as it moves between two specified positions. However, the path between the points does not have to be a straight line; it can be an arc or a series of curves.
- Moving a cutting tool or plotting pen parallel to one machine axis is known as 'straight cutting'.
- Contouring allows both point-to-point and straight movement in more than one axis, allowing complex curves and shapes such as spirals to be generated (see Figure 3.3.29).

On an engraver or another cutting machine the depth that the tool cuts in any one pass depends on the amount of material to be removed and the surface finish required on the finished article. The inside of a mould for a precision plastic product to be made by injection moulding has to have a finer surface finish than is needed on the outside. Roughing cuts remove large amounts of material, set out the basic profile and cut close to the different depths that are required. Finishing cuts provide the required final shape and surface finish. Finishing cuts are made in a number of ways. Depending on the functions available the cutter can:

- move in one direction, constantly lifting back from the end of the cut to the start
- constantly cut as it moves backwards and forwards
- follow the contours on the surface
- follow parallel paths over the surface of the object (see Figure 3.3.30).

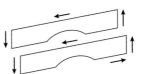

a) The cutter moves in one direction constantly lifting back from the end of the cut to the start

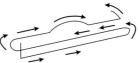

b) The cutter cuts constantly as it moves backwards and forwards

c) The cutter follows the contours on the surface

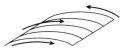

d) The cutter follows parallel paths over the surface of the object

Figure 3.3.30 *Tool paths for machining surfaces*

Task
Determine what types of movement that the CNC machines which you have access to in your graphics and practical work are able to make. Compile a table showing the results of your investigations using one of the four toolpath models to describe the movements you observe.

Types of CNC machine
- Lathes for operations in metal and plastics used to produce 3D product shapes and moulds for plastic products.
- Milling machines for operations such as mould making, profiling, pocketing and surface milling used to produce the cutting dies used in the print industry.
- Routers and engravers for producing printing plates and 2D and 3D graphic designs.
- Flatbed cutting and perforating machines used in the post-press stage of commercial printing.
- Pressing, punching, bending and die cutting machines for processing sheet materials such as card and plastic sheet to produce products such as flat carton packs and document wallets.

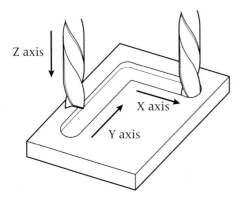

Figure 3.3.29 *Point-to-point cutting of a printing plate in three axes*

Speeds, feeds and rapid traversing

The speed at which the cutting tool or device such as a plotting pen or the work piece moves in relation to one another on a CNC machine is critical to its efficient operation. The plotting pen has to have time to make its mark on the paper. On most XY plotters it is usually automatically programmed. The feed or cutter moving rate is also programmed and is measured in millimetres per second (mm/s). It determines the rate of progress along a cutting or a plotting path. The feed rate has to be precisely controlled on CNC devices such as embroidery and knitting machines to avoid machining problems such as snagging and jamming. On engraving machines metals are machined with slow feed rates and small depths of cut in comparison to engraving plastics. Rapid traversing describes the way the cutter, plotting pen or printing head moves when it is not engaging with the material being worked. On a plotter, for instance, different lines are drawn apparently at random but, in fact, the plotting software has calculated the most efficient way of drawing and building up the image and when different pen changes need to occur.

On a CNC engraver or other cutting device the cutter revolution or spindle speed is measured in revolutions per minute (rpm) and will vary according to the material being machined and the diameter of the cutter. As a general rule, the bigger the cutter diameter, the slower the spindle speed and the softer the material, the faster the spindle speed. On a typical engraving machine spindle speeds usually range between 5000 and 15 000 rpm. To give an example, acrylic is engraved at a spindle speed of 10 000 rpm with a depth of 0.2 mm at an XY feed of 15 mm/sec and a Z feed of 5 mm/sec.

The types of cutting tools on CNC machines

CNC machines such as engravers and millers use cutting tools that are similar to those used in their manual equivalents. The tools are held in collets. The choice of cutter depends on the geometry or shape required and the cutting loads it has to withstand. Cutters are typically made from high-speed steel (HSS), tungsten carbide or ceramic materials. These materials can operate at high cutting speeds without losing their hardness. They can resist the heat generated as a result of friction during the cutting process. Tungsten carbide and ceramic materials are relatively expensive so these materials are only used on the cutting edges of a 'tipped' tool. Some cutters have removable blades for ease of maintenance and for sharpening purposes.

The geometry of the cutting tool determines what shape the tool generates and how the waste materials are moved away from the cutting edges. On a twist drill or a milling cutter 'flutes' force the materials up and away from the cutting operation. On a lathe tool the angled surfaces combine to move the cut material away. Routing cutters are small in diameter and can be half round or three-quarter round in shape. Engraving cutters are V-shaped.

Task

Slot cutters, end mills, bull nose cutters and face mills are some of the cutters you might find on a CNC milling or routing machine.
a) What features do they have in common?
b) What are their respective advantages and disadvantages?
c) What are the benefits of a collet rather than a chuck for holding the cutting tools?

Scale of production and the use of CAM

The choice to use CAM will depend on the intended scale of production, the number of identical components to be produced and the cost benefits that it can bring. There are essentially three categories of production:

- one-off production (sometimes called 'jobbing')
- batch production
- mass-production systems that include high-volume 'runs' or continuous operation.

One-off production

This means that just one product is 'made to order', such as an architectural model or a prototype. Whatever the product, the company making it has to have a highly flexible production facility and in most cases a highly skilled team of workers. Products made by this process are invariably more expensive to buy because they are individually designed and made.

Batch production

This is a flexible production system that is used to make a relatively small number or batch of components or products. It is found in areas such as commercial printing where print runs are of different lengths and degrees of complexity. A key feature in batch production is the ability to respond quickly to demand. In modern commercial printing houses once a batch run has been completed the computer-controlled machines can be easily reprogrammed and quickly made ready for another printing job.

Mass production

Set-up and break-even costs can be higher than in the other two systems but the high volumes of product generated in continuous operations such as those in the printing industry significantly reduce manufacturing costs. In an industry such as carton making the products are mass produced and though they vary in their complexity they all are highly standardised, which means that processes can be highly automated, are less labour intensive and more cost efficient.

Task

Use the information you have been given about different scales of production to describe how they might be applied in different types of printing operation. Give examples of products that incorporate graphic elements to support your answer.

Recent developments in production systems

Mass-production companies, such as carton manufacturers, are looking to develop more flexible systems based on the batch production principle. Earlier in this unit, we looked at the pressures for change including the demand for a greater variety of products and the move towards products that are 'customised' to meet the demands of a particular market sector. In the advertising and manufacturing sectors both print and carton manufacturers are under pressure to reduce the unit costs of their products but at the same time they are expected to offer their customers a greater range of options as 'standard' rather than as 'optional extras'. To meet these demands all manufacturers, whatever their sector, need flexible production or manufacturing systems to allow products that can be 'made to order'. Flexible manufacturing systems (FMS) are particularly important for companies like those involved in merchandising, promotional or point-of-sales display when the scale of production cannot justify a fully automated production line but where there is a need to have a 'quick response' system. Outsourcing becomes an option in an FMS as it allows a company to take advantage of the IT revolution, reduce and control costs by freeing up capital that would otherwise be invested in costly specialist equipment.

An example of outsourcing is in the growth of CAD bureau services. These are specialist companies that have a vast range of specialist equipment aimed at the design and reprographic needs of the graphic market. Typically, these bureaux provide large format plotting and printing of electronic files created from a variety of CAD applications. These files can usually be plotted or printed in monochrome and full colour on various media including bond paper, tracing paper, drafting film, vinyl for banners and signs and textiles in much greater widths than is possible with most common output devices. For smaller companies, a bureau service offers the most cost-effective way to use expensive specialist equipment for jobbing or small batch production such as high-resolution full-colour printing up to 1440 dpi or digital scanning to copy, enlarge or reduce hard copy originals into an electronic file format. Additionally, bureau services might offer a range of computer design services such as slide and multimedia presentations, Internet websites, logo design, business stationery including business cards, letterhead and envelope design, brochures, flyers and posters and advertising design.

Advantages of CAM

Control of the production process to remove risks

CAM allows production processes to be automated so that manufacture can be more precisely monitored and controlled. Raw materials such as corrugated packaging materials and additional components such as card dividers in composite packages can be moved safely so that they are at the right stage of production at the right time. Automated lifting, handling and carrying systems allow people to be taken off tasks that are potentially hazardous or dangerous to health.

Flexible scale of production

CAM systems allow the scale of manufacturing to range from batch and mass to continuous production. For example, in the carton manufacturing industry the advantage of CAD/CAM is a reduction in make-ready times. The CAD programs not only generate the carton graphics and labelling, they also generate cutting programs that are transferred to the die cutting machines online leading to extremely fast and flexible order processing. The introduction of computer-controlled systems of manufacture such as 'just in time' (JIT) means that:

- CNC machines can be operated more flexibly allowing manufacturers to develop quick

167

response systems to meet changes in design or marketing strategy such as the addition of information on an existing package design as part of a sales campaign for the product

- production levels can be linked directly to the size of the customer order book because manufacturing can be speeded up by a reduction in make-ready or set-up times
- the company is less likely to be damaged by sudden changes of demand in its particular market
- orders for raw materials can be generated automatically as and when required, just in time, so avoiding large and costly stockpiles of materials and components that are not adding value to the process
- the inventory or levels of finished products ready for distribution to other manufacturers or retailers can be kept to a minimum which keeps costs to the minimum
- production increases and quality is improved.

Reduced manufacturing times

CAM can reduce the manufacturing time from days to just a few hours. Complex 2D and 3D shapes and forms are easily developed, edited and tested for ease of manufacture through on-screen simulation. As an example, computer-based design teams in a promotional merchandising company can now provide advice on all areas including structural design, they can provide rough visuals and graphics including coloured samples, complete camera ready artwork and cutter drawings. The technology now available also allows the production of computer-cut samples and direct access to suppliers and customers via modem and ISDN links.

Operational reliability

CAM improves levels of operational reliability and the finished quality of products. Modern CAM software can generate and simulate cutter and plotting paths that highlight potential production problems. In specialist areas like carton packaging, the bending allowances for producing 3D forms from 2D sheet materials using creasing and scoring rules can be calculated by the CAD software and applied directly to the cutting or creasing presses.

Consistency in repetitive situations

CNC machines only need to be 'trained' once and they can do repetitive tasks rapidly with fewer errors than human operators. Machines can operate 24 hours a day, do not need rest breaks or holidays and are not affected by other human constraints.

Improved productivity

Production 'throughput' is improved with lower processing costs and less waste of materials and resources. Manufacturing costs can be estimated with a greater degree of certainty, as production rates are more consistent than is possible with human-operated machines.

Disadvantages of CAM

Costs

The cost of buying and installing computer-controlled machines is high when compared to manually operated machines. For really high-volume mass production such as carton making a purpose-built automatic machine may be a more cost-effective solution to the use of CNC machines.

Employment

With developments in automation, robotics and artificial intelligence it is likely that there will be even less human involvement in the design and production processes of the future.

Worker involvement

For some people CAM can create emotional and other psychological problems at work. When machines and processes such as those found in the printing industry are controlled directly from a central management system, some jobs consist of nothing more than 'machine minding' leading to poor job satisfaction and reduced productivity. Many companies are having to work hard to devise systems and develop employee schemes to maintain the interest, enthusiasm and cooperation of their workers to maintain high levels of productivity.

Task

You might want to refer back to the first section of this unit to consider the employment issues surrounding the introduction of CAM in more detail. Explain what you consider the key issues to be.

Exam preparation

You should now have completed the work and covered all the sections in this unit option. Much of the work covered will possibly have been done in a theoretical context, with some practical or experimental work using the CAD/CAM equipment and software that are available to you. You will probably have answered many of the questions and problems set in the text throughout the unit. These will have enabled you to apply what you have learned in that particular section.

There are some additional questions for you to do below. They are similar in format and style to those that you will find in the exam paper.

Here are some tips to help you prepare for your exam:

- plan your programme of revision
- do not leave it until the night before
- little and often is better than a lot all at once
- try condensing all you know about a specific topic on to one side of A4. Then make notes under key headings or by using bullet points.

Practice exam questions

1 a) The use of computers as drawing aids has changed the way designers create their ideas. Explain the following terms related to the use of computers within the design process:
   i) digitiser (2)
   ii) 3D scanner (2)
  b) Explain how computers can be used at each of the main stages in the design of a point-of-sale display for a sweet or confectionery product. (5)
  c) Giving three product examples from your chosen area, explain the advantages of using CNC machines in manufacturing the products you have chosen. (6)

2 a) Explain what the following terms mean:
   i) virtual product (2)
   ii) photorealistic image (2)
  b) Explain, using two examples of products from your chosen area, why CAD modelling is a key part of the industrial design process. (5)
  c) Explain how different scales of production might affect the choice of CAM as an appropriate manufacturing process for two graphic products of your choice. (6)

3 Explain the impact of information and communications technology on the design and manufacture of five products in your chosen materials area. (15)

4 Use examples of products from your chosen area to explain the computer-based methods which manufacturers use to respond quickly to the needs of a rapidly changing global or local market for their products. (15)

Part 3
Advanced GCE (A2)

Materials, components and systems (G401)

Summary of expectations

1. What to expect

Unit 4 is divided into two sections:

- Section A Materials, components and systems
- Section B consists of two options, of which you will have to study and be examined in the same option that you studied at AS level.

Section A is compulsory and builds on the knowledge and understanding that you learned during Unit 3.

2. How will it be assessed?

The work that you do in this unit will be externally assessed through Unit 4, Paper 1.

Section A consists of general knowledge questions from within the overall unit content. There are six compulsory questions, each worth 5 marks. You are advised to spend 45 minutes on Paper 1.

3. What will be assessed?

Although your answers should be short and concise, the examiner is looking for a greater depth of knowledge, understanding and application to that demonstrated in Unit 3.

You must be able to clearly demonstrate that you have a thorough knowledge and understanding of the topics covered in each question.

4. Content

The following list summarises the topics covered in Section A. You are expected to be familiar with the concepts and work covered under these headings:

- Selection of materials, relating to:
  - quality
  - manufacturing processes
  - material limitations
  - wear and deterioration
  - maintenance
  - life costs.

- New technologies and the creation of new materials, including:
  - the creation and use by industry of modern and smart materials
  - the impact of modern technology and biotechnology on the development of new materials and processes.

- Value issues, including:
  - the impact of value issues on product design, development and manufacture.

You should apply your knowledge and understanding of materials, components and systems to your project work in Unit 5.

5. How much is it worth?

This unit, with the option, is worth 15 per cent of the full Advanced GCE.

Unit 4 + option	Weighting
A2 level (full GCE)	15%

1. Selection of materials

Everyone who is involved with either designing new products or improving existing products must have an understanding and working knowledge of materials in order to be able to select, process and finish the materials.

- A car designer strives to produce aerodynamic, efficient machines that are aesthetically pleasing with improved fuel economy and reduced emissions.
- An aircraft designer must try to balance the weight of the plane with the requirements that are set out with respect to strength, temperature fluctuations and the repeated stresses and strains imposed on the structure when taking off, landing, in turbulence or adverse weather conditions.

Materials choice

When choosing any material for a specific application, full consideration must be given to the material, its working properties, the relevant manufacturing processes that can be used, appropriate finishes and dimensional accuracy. Cost is also important in the selection process since it has a significant bearing on the purchase price of the raw material, subsequent processing methods in manufacture and the retail price of the product. If any one of these variables were to be changed it would affect many of the others and may ultimately end up making the material unsuitable for its task. It is therefore essential that designers or engineers fully understand the working constraints and limitations of the material that they are considering using.

Think about this!

As a designer, you need to consider the choice of material very carefully in your coursework project and the earlier that you can do this the better. You also need to consider the manufacturing processes that are going to be used since they will play a significant part in the project's cost-effectiveness.

When trying to make associations between materials, processes and available **components** a designer should be aware of the differences in terms of availability and different manufacturing processes. For example, consideration could be given to the production of sections in aluminium and plastic by extrusion compared to rolling and drawing in the production of steel sections. This would enable a designer to have some understanding of how those materials may behave or in what forms they may be available.

In the early stages of designing, a general overview is taken about what can be achieved with a material, whereas in the later stages a much more detailed analysis is needed when specifying manufacture. The selection of appropriate materials and manufacturing processes is quite a complex issue, with one influencing the other. Materials choice has become more difficult in recent years due to the rapid growth in the availability of new materials. For example, in addition to the more familiar woods and rubbers available, there is an ever-increasing range of new materials in the form of alloys, ceramics and composites. Production engineers have also continued to develop and refine manufacturing processes and procedures. With the widespread use and availability of computers, the control of machines and processes has meant that new demands are now able to be met with increased precision and lower **overhead costs** in terms of a human workforce.

The prime concern remains that of the relationship between materials and processes and the final form of the product. In order to examine this relationship, it is useful to compare several products that fulfil the same **function** but are made from a range of different materials.

Manufacturing inevitably involves a production process which is in direct response to materials, people and money. Before we go on to consider the relationship between properties, characteristics and materials choice in relation to a series of products, it is worth reviewing the various levels of production, whether it be a one-off diamond ring or the continuous flow production of electrical wire. The level of production is one of the most important aspects in determining why a product is the way that it is.

Task

What are the limits of the materials that you are considering using for your coursework project for Unit 5?

Levels of production

The **level of production** can be categorised in the following three areas.

One-off production

This is where a single item is required, often in response to an individual customer's need or requirements. It may be a stadium for the World Cup finals or a wedding dress or a very specialised piece of equipment for a disabled person. As a result of this low level of production the associated cost tends to be much higher because of the materials used, labour required, skills, design and production costs. This presents in some cases, such as the piece of equipment needed for a disabled person, a conflict of interests as to whether the project is financially viable. The project may be rejected on the grounds of cost and investment of time. The social issues in terms of providing a better quality of life should also be considered.

Batch production

This level of production makes products in specified quantities. They can be made in one production run of up to 1000 items, or just ten items depending upon the scale of the project. This level of production has the advantage of freeing up machines once a batch has been completed so that with new tooling it can be used to make other products. If a batch of the original product is required by the customer, the original tools can be re-set once again. This level of production is generally capable of responding very quickly to customer demands and once in production the batch size can easily be increased or decreased. Workers who operate this type of machinery are likely to be more skilled because of the versatility of the machines and tools. One disadvantage of batch production is that serious problems may arise if appropriate planning is not carried out.

High-volume production

Large quantities of products are produced, sometimes on a 24-hour basis of continuous production. The machinery is highly specialised, and therefore expensive, but the variety of products manufactured is kept low. This results in minimal changes to tooling and setting costs which take time and money. Most of the operations that take place are repetitive and components are assembled on assembly lines. In some cases robots are used to assemble products and this eliminates any risk of incorrect assembly or complacency by human labour where it can become very tedious working on repetitive tasks.

> ### Task
> Compare and contrast the labour implications and the tooling costs involved in the production of plastic disposable pens and a wheelchair.

A comparison of how costs change as the level of production is increased is shown in Figure 4.1.1.

Case study 1: chair design

The chair is an example of how designers have made statements concerning their personal design philosophy through the look, construction and materials used in the chair. In the examples selected overleaf it is worth considering the date each chair was designed and constructed, the materials used and the manufacturing technologies and processes available at that time. The material influences, to some extent, the form of the product, but obviously manufacturing processes have been exploited, in some cases in order to achieve the desired outcome.

One-off		Batch		High volume
high		labour costs		low
low		efficiency		high
general		tooling		specialised
high		unit costs		low
high		labour intensity		low
low		capital investment		high

Figure 4.1.1 *A comparison of how costs change as the level of production increases*

Figure 4.1.2 *Mass-produced wood chair designed by Michael Thonet, 1902–3*

Figure 4.1.3 *The Wassily chair, made of tubular steel, designed by Marcel Breuer, 1925*

Michael Thonet

Michael Thonet was very successful at mass producing chairs. At the time, 1902–3, the process of steam bending solid wood was revolutionary. He mass produced the individual components and stored them ready for assembly at a later stage. The limitations of the material have been fully exploited both by being able to bend and also as a structural member since the chair has been reduced to its simplest form (see Figure 4.1.2). There is a fine balance in this example between the development of mass-produced standard components and the **aesthetic** qualities of the timber even though it has been largely manufactured by machines.

Marcel Breuer

The B3 chair from 1925 became more widely known as the Wassily chair (see Figure 4.1.3). The designer Marcel Breuer was an architect who had graduated from the Bauhaus, a German art school. The basic philosophy of the Bauhaus was to create products that avoided historical styling, and that used modern industrial materials. The products were to be designed using just the basic elements and were not to be over decorated.

The Wassily chair has many of the characteristics of the Bauhaus principles. It has been stripped down to its basic structural geometric form and it has been widely recognised as one of the first and finest examples of modern tubular steel furniture.

Tubular steel is a material that is widely available today at relatively low costs. It is manufactured by either rolling it and drawing it through a series of ever-decreasing sized dies, in which case it is seamless, or it can be rolled from a sheet and electronically resistance welded (ERW). Once degreased, it is capable of taking a variety of surface treatments such as plating, dip coating and spray painting. Structurally, tubular steel can be regarded as being very stiff with respect to torsional forces and reasonably strong when subjected to bending forces.

The combination of resistance to bending and torsional forces make steel an ideal choice of material for this application. The simple geometric construction has meant that the manufacturing processes required to make it have been kept to a minimum. Simple angle bends are easy to achieve on a pipe bending vice in the workshop, however in industry the frames would be produced on a larger scale, and different techniques used. The pipes are internally filled with oil under pressure to stop them from caving in when bent, and an example like this would be fixed into a jig and then subjected to movement provided by hydraulic rams.

Task

Discuss the properties of the materials used in the manufacture of both Thonet's chair and the Wassily chair.

Gerald Summer

Gerald Summer's lounge chair of 1933–4 was specifically designed for use in the tropics (see Figure 4.1.4). Maintenance and deterioration were uppermost in his mind when he designed it. The humidity would affect the dimensional stability of any natural timbers used. Any joints used could open and subsequently result in failure of the product.

Plywood, which is a manufactured board, has excellent properties with respect to dimensional stability. It also has great strength due to the nature in which the alternate layers have been bonded together at right angles to each other. Various grades are also available, with some types being more resistant to moisture and therefore even more suited to this type of environment. What is quite extraordinary about the chair is that it has been manufactured from a single sheet of plywood. With cuts in the appropriate places, the rear legs and arm rests have all been formed without the addition of extra components. Therefore, all possibility of joints having to be cut have been eliminated.

Today, this sort of structure would be manufactured by sticking much thinner sheets together with decorative veneers on the outside surfaces over a former. This process is known as laminating and it is used as an alternative to steam bending to shape timbers. The life expectancy of Summer's chair has been greatly extended as a result of him using these materials in this way. It is an excellent example of where the designer has fully considered the environment in which the product is to be used.

Verner Panton

When Verner Panton designed his stacking chair he was given the credit of having created the first, single-piece plastic chair (see Figure 4.1.5).

Figure 4.1.5 *Stacking chair made of plastic, designed by Verner Panton, 1960*

The plastic is formed in such a way that a cantilever-type structure is created, which makes it structurally sound in terms of taking the weight of the person sitting on it.

Although the original versions were produced in 1960, it was not until seven years later that the technical difficulties had been resolved and the chair was put into production. The availability of bright colours in a glossy finish are trademark signs of a plastic product. Plastics generally require little surface finishing and because of the nature of the material once plasticised, it can be injected or blown into the most complex shapes. Panton fully exploited the materials to create an icon of design in the 1960s.

Jasper Morrison

The Thinking Man's Chair is by Jasper Morrison (see Figure 4.1.6). He is a new designer who has emerged in the UK since the 1980s. He very much believes in and considers the importance of design in relation to the look, manufacturing processes and costs of the product. His chair is constructed from spray-painted tubular steel and flat steel for the backrest and seat. The flowing gentle shape and curve of the seat are in contrast to what appear to be very cold and heavy engineering materials. The materials used possess great properties as far as strength is concerned and they can be easily joined by brazing or welding. Jigs would have been used to ensure that all pieces

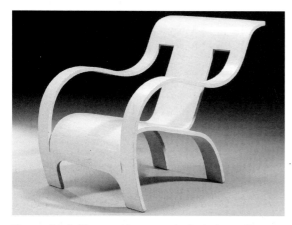

Figure 4.1.4 *Plywood lounge chair designed by Gerald Summer, 1933–4*

Figure 4.1.6 *Thinking Man's Chair made of tubular steel and flat steel, designed in 1987 by Jasper Morrison*

Task

Items of furniture can be manufactured as one-offs or as high-volume components from either single materials – wood, metal or plastics – or in any combination of these. The manufacturing processes, however, govern the limits the material can be put through.

Choose two contrasting pieces of furniture and analyse the materials used and the manufacturing processes in relation to the shape and form of the piece.

Case study 2: product packaging

The design and manufacture of packaging for products has become almost as important as the product itself. The presentation of a product reflected in its packaging allows the manufacturer to target a specific market for its goods. Therefore, the packaging of a product must fulfil two major functions:

- The aesthetic presentation of a product for **marketing** purposes:
  - to influence the consumer to notice and buy the product
  - to identify the product and **brand**.

- Meet the functional requirements of its contents:
  - to contain and protect the product during distribution and storage
  - to maintain the hygiene and safety of the contents
  - to communicate information and usage.

These requirements will be important factors in the design of the product packaging. The

were held in the correct place before any joining took place. The chosen finish would be easily and quickly applied to leave a hard-wearing surface requiring little or no maintenance.

The general overview that needs to be taken by the designer of any product is represented in Figure 4.1.7 which looks at the interaction needed between the material choice, design implications and manufacturing processes.

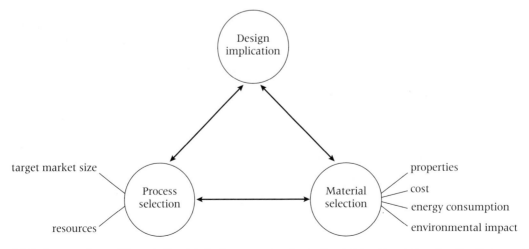

Figure 4.1.7 *An overview of the relationship between manufacturing processes, design implications and the choice of material*

designer will have to select a suitable material for the product and its target market. The cost of the material and the production process are also determining factors in the selection of materials as they will have to be incorporated in the retail price (see Figure 4.1.7).

As packaging is designed to be disposable, it is a major environmental concern. It contributes around a third of household waste in Europe and North America. This has led to European Union legislation on waste management, targeting packaging as a specific area for improvement.

The use of **non-renewable energy resources** in the production process and the raw material itself, along with excessive packaging used for some products, have been targeted by environmental groups campaigning for reform. Environmental groups have applied pressure to manufacturers to change how their products are packaged and the type of materials used. This awareness of environmental responsibilities has changed the way product packaging is designed and manufactured. Manufacturers have placed a greater emphasis in selecting materials and production processes to improve their environmental record. The following examples show a few of the ways in which organisations have responded.

Re-use
- Refilling plastic bottles with the product rather than purchasing another when only the contents are needed.
- Supermarkets may reward customers with small discounts when plastic carrier bags are re-used or may offer strong reusable bags for customers to purchase.
- Secondary use of packaging such as soap packaging being used for a durable soap dish.

These examples reduce the volume of waste and reduce the consumption of plastic.

Recycle
- The use of recycled paper/card, metal and plastic in the production of packaging (see Figure 4.1.8).
- Replacing original materials with ones that can be recycled after use.
- The use of shredded paper and biodegradable material that might otherwise be disposed of for packing and protecting products in transit.

Reduce
- Reducing the amount of packaging to the minimum requirement.
- Developing and producing CFC-free packaging material.
- Making the product smaller, for example condensing washing powder into small tablets

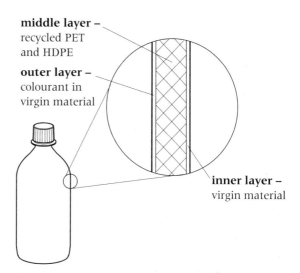

Figure 4.1.8 *New and recycled plastic material used in the production of a detergent container*

which require less packaging.
- Using biodegradable and indestructible plastic such as ICI's Biopol (see page 185).

As well as environmental concerns, the function of the packaging in relation to the product inside is a major consideration. The overall cost of producing the final item along with the materials innate properties are the two determining factors in material selection. The following, along with glass, are the three main materials used for packaging.

Card and paper
These materials are commonly used for packaging as they are inexpensive, easily formed and offer a wide range of properties. When card is shaped in a 2D form, as a net, it has benefits such as higher volume storage and transportation. It also enables common commercial printing methods to be used such as **offset lithography** without the need for specialist machinery.

To reduce costs card used for packaging is often produced from low-grade unbleached pulp, one side is then coated or a layer of higher grade paper is applied for better results during printing.

The weight of card used ranges from 120 gsm to 220 gsm. Card that exceeds this weight would not be used because of the high cost. Corrugated board would provide a cheaper alternative – this is used mainly for products that are not purchased from off the shelf such as white and brown goods that do not require such a high level of printed presentation. Corrugated board also offers the product greater protection because it can be produced in a greater thickness and, due to its construction, offers a slight padding.

As card and paper are naturally absorbent they are susceptible to water and grease. A polyethylene coating can be applied to the surface of the card to give it a resistant finish. In doing this, the card cannot be recycled.

Card packaging can also be produced in a 3D shape. Egg cartons and electrical product packaging are examples of pulp formed in a one- or two-part mould. The initial tooling cost for this production method is high and provides a crude surface finish. However, it does allow for intricate shapes to be produced and delicate products can be effectively supported. This type of packaging is an environmentally friendly alternative to polystyrene and thermoplastic.

Plastic

Thermoplastics are widely used in packaging, particularly for food. This is because they provide a non-absorbent, hygienic surface and an air-tight seal can be created to prolong the shelf life of the food inside. These materials also allow the customer to view the product before purchasing it without having to open the packaging. Two common forms of thermoplastic used are PET (polyethylene terephthalate) and HDPE (high density polythylene), which are both recyclable materials (see Figure 4.1.9).

The two main production methods used are as follows:

- **Vacuum forming/blow moulding**. This is an inexpensive method that can be used for both mass and batch production. Production costs of the moulds are low in comparison to injection moulding. This is especially true when large, intricate, departmentalised forms are required for products such as toys and food. The resulting form may lack rigidity and durability so card may be used for support.
- **Injection moulding**. The tooling costs for injection moulding can be very expensive, especially for packaging inexpensive products like food. Items such as bottles and tubs have to be produced in high volumes to warrant outlay costs. Companies will often use generic products produced by packaging manufacturers and apply their own labels. The advantages with injection moulding methods are the creation of more rigid and accurate shapes.

Figure 4.1.9 *The Body Shop uses bottles made from HDPE and also offers a refill service*

Metal

Metal is used predominately in packaging food and drink. Aluminium and tin-coated steel are the two main materials used because of their non-corrosive properties when vacuum sealed. This permanent seal maintains freshness, longevity and protection for the product.

In the manufacture of cans a high proportion of reclaimed and recycled material is used. Mechanical magnetic sorting at waste sites can detect and withdraw the steel to be melted down. Reclaiming aluminium is more difficult so manufacturers have offered economic incentives for people to recycle. This has resulted in the use of aluminium in drinks cans becoming more common. Extracting aluminium is an expensive process with high energy costs so recycling is an economically viable alternative. The properties of aluminium are ideal for packaging as it is a lightweight, malleable material which can be easily formed and shaped.

Tasks

1 Identify a range of products that are over-packaged.
2 Brainstorm ideas of how one of these products could be packaged in a more environmentally friendly way.
3 Sketch ideas of how the product packaging could be reduced.

Case study 3: kettles

Kettles are good examples of how materials have changed and brought about new developments in manufacturing technologies (see Figure 4.1.10). The Victorian kettle made from cast iron pushed the material to its limit as a hollow product at the time. As a metal, it would naturally have the required thermal conductivity properties where it could be heated over a fire or on a range. The kettle would require careful handling with gloves or some insulating material because the handle would also get very hot.

As materials became more readily available, copper replaced cast iron kettles. Copper kettles could be fabricated by soldering or formed from a single piece or spun. The spouts would be made from a single piece and soldered into place. Copper has a much higher coefficient of thermal **conductivity** than cast iron and so makes it much more efficient in terms of the amount of energy required to heat the water inside. Great care still needs to be taken though, since the handle gets even hotter due to the thermal conductivity of the copper.

With the advent of plastics and injection moulding, it became possible to mass produce components quickly and cheaply. Plastics are an ideal material for use in kettles since they are excellent insulators of both heat and electricity. Due to new manufacturing processes, new and improved shapes were designed with enhanced ergonomic performance. It also became possible to build in new features such as water-level indicators.

Designers such as Michael Graves and Philippe Starck challenged the concepts of shape and form in the 1980s and 1990s with the 'Graves kettle' and the 'Hot Bertaa' aluminium kettle respectively. These have become modern-day icons of design and yet they still retain the basic function of the kettle – that of heating water.

Victorian cast iron kettle

Copper kettle

Plastic injection-moulded kettle, 1980s

'Graves kettle' designed by Michael Graves and produced for Alessi, 1985

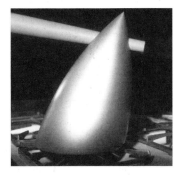

'Hot Bertaa' aluminium kettle, designed by Philippe Starck, 1991

Figure 4.1.10 *Kettles through history depicting how materials have allowed designers to challenge both shape and form*

Task

Working in small groups, consider the following aspects of kettle design:

- How will kettles develop in the future?
- What will customers want?
- Are there any new technologies that can be incorporated?
- Produce some concept sketches for a new kettle.

Present your thoughts to the others in your class in a formal presentation.

Life costs

The cost of a product can be expressed in many ways and it has to take account of many factors. Even before a product takes its place on the high street shelves, the cost of research and development has to be met and inevitably recouped in the final price. Advertising costs, production, packaging and distribution also have to be considered and added into the price paid by the consumer.

The cost implication, however, far outreaches the ticket price that is displayed on the product. Certain products and materials have hidden costs attached with regard to life costs and the various implications associated with them.

Environmental costs

The production of almost everything requires materials that affect the environment in one way or another. For example:

- the depletion of our forests for timbers that take hundreds of years to replace
- the extraction of metals and ores from the Earth's surface which requires mining and open-cast mining
- the exploration for and extraction of oil for the production of polymers.

These natural resources are all being used at ever-increasing rates and we are not able to replace them at anywhere near the same rate. In fact, as far as oil is concerned we are not able to replace it at all. Forests can be managed and to a certain extent we are able to replenish stocks. It does, however, take considerable time for a tree to reach maturity before it can be felled and converted into useful timber.

The amount of energy needed to produce products through their manufacture also needs to be considered. The production of steel and any subsequent processing requires an enormous amount of energy to heat it up to temperatures in excess of 1200°C. The casting of aluminium and alloys requires high temperatures, as do forging and any related heat treatment processing needed to change or enhance the mechanical properties. The environmental cost also must be considered since emissions from factories have to be monitored and treated as appropriate. Emissions can take the form of gases and vapours as well as miscible oils and solvents which need to be treated as industrial effluent.

The deforestation of areas has led to soil erosion in certain parts of the world and the land has now become infertile as the rains and storms have washed away the topsoil. This has led not only to environmental damage but also to an economic downturn for those timber-producing regions which can no longer work the land to generate income.

Finally, the disposal of the product must be considered when it reaches the end of its effective working use. The designer should have regard for how it can be recycled or how parts can be disposed of. In 1992 an environmental scheme was established to monitor, assess and encourage manufacturers to make products that cause less damage to the environment. It provides consumers with information which allows them to make better-informed decisions about the products they buy.

Task

Consider items of furniture such as a school desk and chair and a car battery and analyse each of them in terms of the following:

- use of natural resources and energy
- disposal of waste
- impact on air, water and soil
- how it can be recycled or disposed of
- whether the materials are renewable or non-renewable
- how much energy was used in the manufacture.

2. New technologies and the creation of new materials

The creation and use by industry of modern and smart materials

Liquid crystal displays

Liquid crystal displays (LCDs) are used as numerical and alpha-numerical indicators and displays. As they require much smaller currents they have replaced LED (Light Emitting Diode) displays because they prolong the life of batteries by using microamperes rather than milliamperes.

Liquid crystals are organic, carbon-based compounds, which exhibit both liquid and solid characteristics. When a cell, containing a liquid crystal, has a **voltage** applied across its terminals, and on which light falls, it appears to go 'dark'. This is caused by the molecular rearrangement within the liquid crystal. A liquid crystal display has a pattern of conducting electrodes which is capable of displaying the numbers 0 to 9 via a seven-segment display. The numbers are made to appear on the LCD by applying a voltage to certain segments which go dark in relation to the silvered background.

Smart composite materials

'**Smart materials**' is a term that has been applied to a broad range of materials whose physical properties can be varied by an input. These materials are now being used to replace devices and components that once had separate sensing and actuation components. A single smart material can reduce the overall size and complexity of a device. Smart materials have opened up enormous developments for new sensors, actuators and structural components.

Smart materials can be subdivided into various groups and probably the best known of these are the piezo-electrics. These devices produce a small shape change when a low voltage is applied and because of their response times they are being utilised in high frequency devices such as ultrasound generators and audio speakers.

Piezo-electric actuators work in two ways; they either produce movement in response to an applied voltage or they produce a voltage in response to an applied pressure. The voltage generated as a result of the material being deformed is sufficient enough to light an LED as shown in Figure 4.1.11. Piezo-electric actuators are used in greetings cards that play tunes when opened. They produce a sound from an electrical signal as a result of the card being opened.

When piezo-electric **transducers** are used as sensors they are capable of picking up small signals which can then be amplified and processed. They are used in a wide variety of applications including burglar alarms. The transducer produces a voltage in response to a loud sound, such as a breaking window, or to a movement such as treading on a mat or stair tread.

The transducers are made from minerals, ceramics or polymers. The piezo-electric film is bonded to a base material once coated with a metallic film and contacts are then attached. In some cases amplifier circuits are built in to the whole transducer but in most examples they are left like the examples in Figure 4.1.11.

Shape memory alloys constitute another group of smart materials. During the late 1970s, metallurgists discovered shape memory alloys. When made into components, these alloys can be plastically deformed at a certain temperature. They will retain their shape while held at this temperature. When the temperature is removed, the component reverts to its original shape. It was later discovered that by processing alloys in certain ways some alloys could be 'trained' to show memory where the components demonstrated total reversibility on heating and cooling.

Early applications saw shape memory alloys used in greenhouse window openers. They consist of a coil spring of copper-zinc-aluminium alloy which opens and closes the window on a hinge in response to the changes in ambient temperature. A more recent development has seen shape memory alloys used as thermally activated fasteners for use as electrical cable fasteners and hydraulic pipe connectors. The connectors are prepared by making them expand, normally by chilling them, and then placing them over the area to be joined and heating them making the shape memory alloy return to its original shape.

Shape memory properties occur in several alloys including nickel-titanium, gold-cadmium and iron-nickel-cobalt-titanium with the most common of these being nickel-titanium, called nitinol.

The cycle of being able to straighten and bend when heated can be repeated millions of times. The heat needed to make the shape memory alloy move is generally provided by passing an electrical current through it. Because shape memory alloys have a relatively high electrical resistance heating it in this way is an appropriate method.

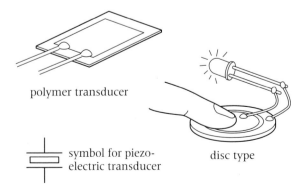

polymer transducer

symbol for piezo-electric transducer

disc type

Figure 4.1.11 *Piezo-electric actuators and transducers*

Task

Many piezo-electric transducers are used in greetings cards. Analyse such a card by considering the following:

- how the transducer is activated
- the voltage that is generated across it
- the commercial availability of such transducers.

Common shape memory alloy wires are available in a range of different diameters from 5 mm to 5/1000ths of a millimetre. Nitinol has been developed to 'remember' that it has a shorter length when it is heated above its transition temperature which is between 70 and 80°C.

Task

It has been suggested that shape memory alloys are to be fitted into bras and trouser creases. Discuss what the designers are trying to achieve here and how would the temperatures be controlled.

Some examples of other smart materials include the following:

- *Piezo-electric ceramics.* These materials expand and contract in response to an applied voltage. They are being considered as replacements for the large heavy iron wound speaker drivers which would reduce weight and the overall size and bulk.
- *Electrorheological and magnetorheological fluids.* These special fluids can change their viscosity within milliseconds when placed in an electric or magnetic field. The automotive industries are carrying out research on these fluids to ascertain their suitability and application for use in suspension systems and engine mounts.
- *Optical fibres.* Fibre optics allow light to flow along them since the glass used has a high refractive index. This technology makes it possible for light to be transmitted over 200 km. The advantage of glass fibres over copper conductors in telecommunications is the small diameter of the glass fibres and their greater load carrying capacity. They are now extensively used in telecommunications systems, computer networking and in surgery where they can be inserted into the body for exploration purposes. Currently, fibre optics are being used by the US space agency NASA to measure the strain in composite materials. When a fibre optic breaks due to failure of the material around it, the light signal fed through it becomes interrupted indicating failure.

Task

Chris Boardman rode a bike with a single-piece frame made from carbon fibre to win an Olympic Gold medal. Detail the benefits that he would have gained from this type of bike frame compared to the more conventional and traditional metal-framed bikes.

New materials used in the computer and electronics industry

Semiconductors are a special type of material that have an electrical resistance which changes with temperature, i.e. they conduct electricity better as the temperature rises. The resistivity falls as the temperature rises. Semiconductors include both compounds and elements. Perhaps the most common type of element is silicon.

Silicon is the second most abundant and widely distributed of all the elements after oxygen. It is estimated that about 28 per cent of the Earth's crust is made up from silicon. It does not occur in a free elemental state but it is found in the form of silicon oxides, such as quartz, sand, rock or in complex silicates such as feldspar. Silicon is used in the steel industry as an alloying element but its main use is in the computer technology industry. In particular, it is used in the manufacture of silicon chips and other electronic components such as **transistors** and **diodes**. It is also used in the manufacture of glass, enamels, cement and porcelain.

Like other semiconductors, silicon behaves as an electrical insulator or conductor depending on the temperature. At absolute zero, the material is an electrical conductor. When heat is applied to a piece of silicon, some of the electrons gain enough energy to jump from a valence band to a conduction band. The movement of these electrons can carry charge through the material when a potential difference is applied across it. The higher the temperature becomes, the more electrons jump across the bands, and the resistance falls making it an electrical conductor.

The production of integrated circuits (ICs) involves cutting single crystals of silicon into thin wafers. The surface of each wafer is then coated with a photosensitive polymer so that it can be etched by exposing light to predetermined areas. Several masks of different layouts are needed in order to construct the integrated circuit. As the accuracy of photolithography, (the process used to transfer the patterns) has improved, accuracy has improved to within less than one micrometre. Due to the individual devices being so small, the integrated circuits are produced on a larger wafer up to 20 cm in diameter that can contain up to one thousand million circuit elements (see Figure 4.1.12).

Computers have their entire central processing unit (CPU) made from a single integrated circuit containing more than one million transistors. The same computers also require memory chips and their capacity has increased and continues to increase at an ever-increasing rate.

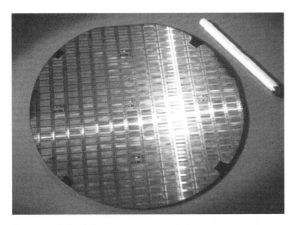

Figure 4.1.12 *Silicon wafer containing many smaller ICs*

Task

Silicon has led to rapid developments in the world of technology. Produce a time-line over the past ten years looking at some of the developing technologies and products that you have been exposed to such as CD players, increasing computer power and laptops, and more recent developments such as DVDs and MP3 players.

Impact of modern technology and biotechnology on the development of new materials and processes

Genetic engineering in relation to woods

Recent developments in science and technology have given rise to a whole new culture of genetically modified foods, plants, materials and animals.

Biotechnology is now at the forefront in the production of new materials and research, and experimentation into genetically engineered and modified timber is growing rapidly. Genes are currently being investigated with a view to providing faster growing trees. This would enable forests to be managed more efficiently in terms of replacing trees at the same rate at which they are cut down. The new trees planted would therefore reach maturity much quicker in terms of the useful bulk of timber that can be obtained from it.

Investigations are also taking place as to how wood can be engineered to be more resistant to wear, rot and animal infestation. Obviously, there are benefits to be had in relation to these improvements. What though are the impli-cations beyond this as a result of timber not rotting? If it was not for decaying and decomposing timber millions of years ago, we would have no coal or oil today. If timber does not rot, then maybe it will have to be burnt to be disposed of. The consequences of burning the timber will result in emissions causing damage to the ozone layer and to acid rain.

One major area of development as far as genetically engineered timber is concerned is in the production of paper. The process of making paper is an environmentally damaging one which involves the use of some very toxic chemicals. The chemicals have to be used to remove lignin from the wood pulp. Lignin is a natural tough polymer-like material that gives the tree its strength and rigidity. In the USA, scientists have discovered a way of reducing the natural lignin content and also producing trees that grow faster. It is thought that this will lead to the reduction of toxic chemicals used in the paper-making industry.

The same scientists when working on aspen trees, (the trees traditionally used for pulp production), made some remarkable discoveries. As they peeled back the bark from the saplings, they found that the timber was not its usual whitish colour. The timber was in fact a salmon reddish colour. Each sapling exposed was slightly different in colour and some in fact were spotted in appearance. Its use is now being considered beyond that of pulp and paper production. It is not inconceivable to think that we may end up with coloured timbers being used in furniture, panelling or even external cladding that would require no maintenance.

The scientists involved in this work are now considering the study of these aspens in a natural environment. As with any work of this nature, however, they are having to apply to the United States Department of Agriculture for a licence to do so.

Task

What benefits might the impact of genetically engineered timber bring to the following?

* pests and timber infestations
* use in the construction of outside furniture
* the long-term future of a managed timber supply.

Environmentally friendly plastics

A plastic is basically any material that can be heated and moulded so that it retains its shape once it cools. Today, a wide variety of plastics exist as polymers and resins. Synthetic plastics are derived from oil and hydrocarbons form the basic building blocks. The carbon-based molecules bond together through a chemical reaction and the bonds made determine the physical properties and characteristics.

One disadvantage of plastics generally has been the issue of how they are disposed of. Recent advances in biotechnology and bioengineering have seen the development of **environmentally friendly plastics**. 60 years ago the world's first biodegradable polymer, Biopol was created (see Figure 4.1.13). Much has been made of the need to create plastics by other means than from petro-chemicals. This led to the development of fibres such as nylon from petrochemicals and PVC.

Biopol has the advantages of both synthetic polymers and those of natural polymers. It is bio-degradable, can be produced in bulk and the raw materials are readily available. Scientists discovered that a natural polymer produced from food stuffs could be mass produced by fermentation. Once the polymer had been extracted it was substituted into the production process to produce plastic products. The major advantage of the 'plastic' once it had reached the end of its life was that it could be disposed of by biodegradation.

'Green' credit cards made from Biopol have now started to find their way into the market place as a genuine replacement for some of the 20 million credit cards in circulation today.

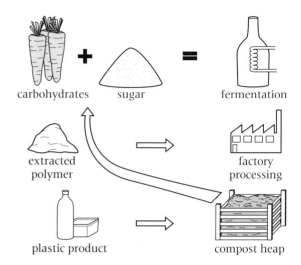

Figure 4.1.13 *The production of Biopol*

Task

Discuss the environmental benefits in terms of recycling plastics that materials such as Biopol have brought.

Special effects on television

Television production companies are beginning to incorporate the computer technology that has been developed by the film industry to create sophisticated special effects. Two main types of visual effects used are:

• blue screen
• computer-generated image.

Blue screen

This is a popular and widely used effect as it enables actors or scale models to be imposed on a separate filmed or computer-generated background. This technique can be used to create scenes that would be impossible (or too dangerous!) to film or to impose actors in a futuristic computer-generated scene.

The background and foreground of a scene are shot separately as two pieces of film and then combined in a process known as compositing. This can be achieved by projecting the two films simultaneously, frame by frame on to a third film. However, with the advancement of digital technology it is more likely that the two films will be digitised and the **composite image** made up on a computer before being processed as film.

First, the background is filmed or created on computer – this is called a background plate. Then the actor or model is filmed against a blue background or screen, which can be passed through a red filter to make it appear black. Silhouettes are created of the model/actor from the blue screen footage; one is black on a white background and the other is white on a black background. There are now four pieces of film – two originals and two mattes. They are:

• the background
• the actor/model
• the actor/model in black silhouette on a white background
• the actor/model in white silhouette on a black background.

These pieces of film are layered over one another and combine to make a composite image. The black silhouette is placed on the background plate and this creates a 'hole' into which the footage of the actor/model can be accurately placed. For each movement of the actor/model a separate frame must be produced.

Computer-generated image

Computer-generated image (CGI) is a development from blue screen technology in which computer-generated images are combined with film sequences in layers. The techniques used to manipulate existing images are part of a stage called **post-production**.

The scenes are filmed and scanned into a machine at a resolution of 12 750 000 dots per frame. Once this has been done, the scene can be manipulated in a number of ways.

- Rotoscoping – elements within the scene can be outlined and lifted out so that they can be replaced by other images.
- 2D painting – this technique can be used to add separately produced computer-generated images or elements to the scene. In addition to enhancing elements, this technique can also be used to remove unwanted elements from the scene, i.e. wires, safety equipment.
- Compositing – the process of layering the separate images together to produce a final sequence.
- 3D tracking – a 3D model of the scene is created in computerised form that incorporates the position of the camera and where it will move. This computer-generated replica of the scene allows the 3D elements that are added to be positioned accurately and realistically in relation to the rest of the scene and the camera.
- 3D **modelling** – figures and objects are computer generated in a 3D model form. The movement of any figure or object will have to be incorporated to make it appear realistic in the context of the film. For example, human movements can be replicated from motion capture data. An actor is fitted with a suit with light reflective markers positioned on every joint. A number of 3D cameras capture the light from the actor's movements from a variety of angles and the information is transferred into digital data, processed by the computer and used to develop the model.

Digital photography

Digital cameras employ a 4.4 mm x 1.6 mm sensor that converts light into electrical charges. This sensor is either a charge coupled device (CCD) for high-quality cameras or a complementary metal oxide semi-conductor (CMOS) for more basic models.

The CCD (see Figure 4.1.14) is a group of miniature light sensitive diodes called photosites. These diodes convert light (photons) into an electrical charge (electrons). The brighter the light that hits the photosite, the greater the electrical charge will be. The charge is converted from analogue to digital as each pixel value is recorded as a digital value.

The photosites do not register colour so to obtain a full multicolour image the sensor filters the light into its three basic primary colours. This process can be achieved in a variety of ways depending on the model of the camera. The most practical method to capture a full-colour image is the installation of a permanent filter positioned over each individual photosite. Most cameras use a Bayer filter pattern, which alternates a row of red and green filters with a row of blue and green filters. There is twice as much information from the green filters as the red and blue combined. This pattern is used because the human eye is not equally sensitive to the colours blue, red and green, more green pixels are required to create an image that will be perceived as true colour by the human eye. In effect, four separate pixels determine the colour of a single pixel by forming a mosaic.

Digital cameras use 'demosaicising algorithms' to convert the mosaic of separate colours into true colours. A true colour is formed for a pixel by averaging the colour value of the pixels that are closest to it.

To enable the camera to store the vast amount of information needed for these images, files have to be compressed. This can be achieved in two ways; repetition or irrelevancy.

Repetition

This process relies on the fact that certain colour patterns develop on a digital photograph and some shades of a colour will be repeated. The basic information needed to reconstruct the image is stored. This technique may only reduce the file by 50 per cent or less.

Irrelevancy

A digital camera records information in such detail that not all of it is detectable by the human eye. Therefore, any unnecessary information is discarded; the camera will offer different levels of compression by allowing the user to opt for

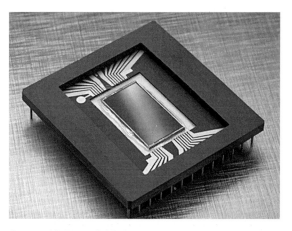

Figure 4.1.14 A CCD from a digital camera

varying grades of resolution. Higher resolution equals less compression and vice versa.

The information for photographic images is stored internally to be **downloaded** on to a computer via a serial or parallel port. Removable flash memory devices such as memory sticks are used to store files but some models store the file on a standard floppy disk.

As with standard film cameras, digital cameras use aperture and shutter speed to control the amount of light reaching the sensor. Most digital cameras are set for optimal exposure and focus automatically. However, some models do offer adjustment options allowing the user more creative control over the final image.

Internet website design

All websites are constructed using **HTML (hyper text mark-up language)**. This language or code forms the common structure of the **world wide web**; therefore, to place a site on-line all the information will have to be coded as such. For large commercial sites HTML code will be written for quicker navigation and downloading. Software applications for creating web pages will automatically apply HTML code, allowing web pages to be easily formed.

When creating web pages a combination of text and image is applied. The layout can be formed in a similar way to **desktop publishing (DTP)**, but with the added advantage of animation, sound and scrolling pages. A number of pages create a site so the design should allow for easy navigation between them, as well as consistency in appearance. Connections from one page to another can be achieved using a:

- hyperlink – text
- hotspot – picture, graphic, designated area.

When the mouse pointer passes over a hyperlink or hotspot it will change into a different icon such as a pointing hand. This indicates a link to the user. Hyperlinks and hotspots can be created in the website design software with the appropriate tool. Once a link has been created, it must be assigned. This can be to:

- another page on the site
- another section on the same page
- another website on the Internet
- an email address.

When creating a website consideration must be given to file sizes. The inclusion of large files such as scanned images will extend downloading time. This may prove tedious for users, so compression formats such as JPEG can be used to minimise size.

Colour printing in newspapers

Colour printing in newspapers is achieved by an offset lithographic printing method (see page 100). The **layout** of images and text is prepared on a desktop publishing (DTP) application; and a film negative is created from these digital files in order to produce the lithographic plates.

Colour images are separated into tonal black, cyan (blue), magenta (red) and yellow (see Figure 4.1.15) for individual plates to be produced. This separation of colours takes place at the film output stage. Registration marks are placed on the image to denote each colour area. These marks enable the image to be aligned accurately and prevent it from becoming blurred. A blue line print is made from the negatives to check the position before printing begins.

During printing, a continuous roll of paper sheet is passed through a bank of rollers that add the colours separately. The process begins with black, then cyan, magenta and yellow are added in sequence to build up the full-colour image. Registration alignment can be adjusted during this operation. The register control system (RCS) works in conjunction with a strobe light and camera to read the alignment of the registration marks. Adjustments are made to the colour rollers throughout the process to ensure the colour registration is correct. The amount of ink released into the units is also controlled, blending the colours to achieve the desired look. Prior to being placed on the press, the plates are scanned and the data are transferred to the main control unit. These data provide information regarding the levels of ink to be released to maintain the integrity of the colours. The registration marks are removed when the product is cut, folded and bound.

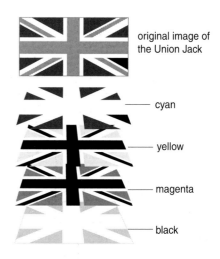

Figure 4.1.15 *The four colour layers*

original image of the Union Jack

— cyan

— yellow

— magenta

— black

Holographic images

Holography is the science of producing 3D pictures. Photographic images and television pictures are viewed on a planar medium and only capture the irradiance of the subject. Holographic images in contrast include the irradiance and phase of the object.

Holographic science is constantly evolving and inventing new methods of application. It can be utilised for a variety of purposes such as embossed surface images on credit cards or cockpit instruments for planes.

To produce a basic holographic image a photosensitive material needs to be used to record the image. This can be specific holograph film or other materials such as photochromatic thermoplastic. The holographic image is not recorded like a camera image. In photography differing intensities of light are reflected by the object and imaged by a lens. In holography the phase difference of the light waves are captured after bouncing off the object to give them depth.

A reference beam or laser light emits a 'plane' wave. By using a beam splitter two beams are formed. The one (reference) beam is spread with a lens and aimed at the film. The other beam (object) is spread with a lens and aimed at the object. When the beam hits the object it is changed from a plane wave. The wave is modulated according to the physical dimensions and characteristics of the object. This light which deviates with intensity and phase hits the film along with the reference beam (see Figure 4.1.16).

The two beams interfere with one another as they pass through each other. The crest of the one plane wave meets either the crest of another (constructive interference) or a trough (destructive interference). Both types of interference are needed to visualise the image.

The film records the wavefronts. It is not a point-to-point recording like standard photography but a recording of the interference between light that hits the object and light that does not. It acts like a complex lens, reconstructing the image so it is perceived as if the object were really there.

Use of high-wattage lighting for projecting images on to buildings

The large-scale projection of images on to buildings has been used by the advertising industry and by visual artists such as Krysztof Wodiczko. There are a variety of systems produced for this purpose (Pani, Xenon, PIGI, etc.), but the equipment used to project large images is based on the same principle as slide projectors.

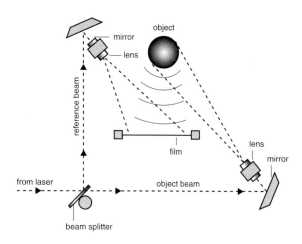

Figure 4.1.16 *How holographic images are produced*

The size of the slide used is usually 18 cm x 18 cm or 24 cm x 24 cm and is individually mounted or placed in a carousel. The slide film can easily be damaged by environmental factors such as warmth, light and pollutants and can result in the image losing its intensity. High-performance slide films are used and hardened glass can protect it from pollutants, but a steady temperature must be maintained.

Filters prevent bleaching of the slide which is caused mainly by exposure to UV radiation. The high wattage light output, often halogen, passes through the slide and the lens. The lens and light output required are dependent upon the size of the image to be projected and the distance between the projector and the building.

Mail-merge software

Mail-merge software is available with most word processing packages. It is used to create personalised documents, letters and electronic mail from a standard template. There are two key files needed to perform a mail merge:

- Data list – name, address or any specific information to be included in the template.
- Document or letter template – the standard document to be sent to the individuals on the data list.

Special place holders are inserted into the template to mark where the information is to be placed, for example 'Dear Sir/Madam' will be replaced by 'Dear [name]'. This is then assigned to the correct field in the data list. Mail merge will replace this generic entry with the actual names of the recipients on the list. This process can be repeated throughout the document for various fields of information.

Non-carbon reproduction

Non-carbon reproduction (NCR) or carbonless copy paper permits multiple copies to be made without intervening layers of carbon paper. The paper translates pressure, such as that made by a pen, into a dye reaction which transfers the image to the copy. Carbonless copy papers are mainly used for continuous form sets, covered pay slips, delivery invoices and payment receipts.

The recycling of materials

Once the function of a product has been performed or it reaches the end of its useful working life, it has to be disposed of. This is becoming a prominent issue that society is having to face. In Germany, for instance, the supplier and manufacturer are responsible for the return and disposal or recycling of packaging such as expanded polystyrene and cardboard. The consumers have to return the packaging to the store where it is then collected and managed by external waste handling companies.

It is difficult to recover materials from certain products for a number of reasons, the most significant being how the materials are used within a product. Sometimes base metals have been plated, sprayed or even galvanised using large quantities of zinc, tin, silver or gold. It is very difficult to recover these materials commercially with the exception of the more precious types.

One of the most complicated pieces of engineering and manufacturing which combines a multitude of components and materials is the motor car. Recently, there has been pressure on the automotive industries to consider future manufacturing implications with regard to recycling and the reuse of materials and components. Companies such as Rover and BMW have pledged to make this a key consideration for all future designs. Rover has set up an auditing system that documents what the car and its various components are made up from in order to make the recycling easier. It is also considering how to reduce the complexity and variety of materials used so that it becomes commercially possible to recover a greater amount of the car's material.

The UK government has set some very tough targets for the automotive and recycling industries to achieve. The average proportion of car parts by weight that has been declared as waste for disposal by shredding has to be decreased to 15 per cent by 2002 and to 5 per cent by 2015. Any materials that are disposed of rather than recycled also have to be included within this figure. Companies involved in the dismantling of cars should reuse components and fluids such as brake fluids as far as is technically and economically viable. Non-recyclable parts will need to be disposed of properly or they will need to be passed on to a certified disposal company.

The basic materials used in the manufacture of cars (see Figure 4.1.17) are:

- steel
- aluminium
- glass
- rubber
- plastics.

Other materials are used in solders and plating but in quantities which are too small to justify their recovery economically. Laminated glass is used in windscreens as well as for the headlights. Laminated glass is difficult to recycle due to its laminar structure as is the glass used for headlights since they tend to be made as sealed units and therefore they are combined with many other materials. In each case, the glass is ground up into fine particles where foreign bodies can be removed by various sorting methods and the remaining glass cullet is sent off for recycling where it is simply introduced into the furnace and remelted.

It is currently estimated that two billion tyres are scrapped every year and although some are used as planters in the garden, most are burnt or recycled. Some, however, are re-treaded, i.e. the worn tread is stripped off and fresh rubber is moulded on to form a new tread. Re-treaded tyres have to undergo some tough tests to ensure that they are safe to use before they can be sold.

Plastics are widely used and they present enormous problems in terms of recycling. They are used in textiles coverings for seats, foam in the seats, switches, dashboards and most of the interior trims. It is estimated that about one tenth of the weight of a car is plastic and it is very difficult to sort, separate and recycle any of it because of the wide variety of plastics used. If any of the plastics are recycled they have to be separated into their various forms. They are then shredded into tiny pieces, melted and fed into an extrusion machine where the output is chopped into small pellets to be used again in the production of new products, such as garden hose pipes. It goes without saying that the only type of plastics which can undergo this recycling process are thermoplastics.

Aluminium is used for engine blocks or in the form of alloys for wheels. These large items can be easily removed for recycling but smaller parts can be separated out once the remaining parts

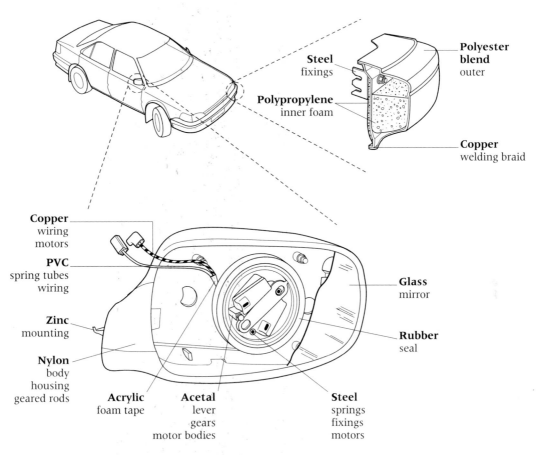

Figure 4.1.17 *Materials used in the manufacture of a car*

have been shredded. The reclaimed aluminium is taken back to the foundry where it is introduced into the furnace and melted down to make 'new' aluminium.

Steel is the world's most recycled material and that used in the production of cars is no exception. The majority of recycled steel is used in the construction industry and large engineering structures. Car bodies and chassis are cut into shreds, chopped up and passed over a magnetic separator. The ferrous metals are held in the magnetic field and the non-ferrous metals are deflected into a separate container. The shredded steel is then introduced back into the furnace in the production of new steel. This recycling process of metals, be it aluminium or steel, makes enormous savings in terms of energy consumption compared to the making of steel from scratch.

The overall impact of recycling at this level has enormous benefits:

- by reducing the demand on valuable resources
- by reducing the amount of energy consumed in the production of the car
- by reducing the overall environmental impact.

Tasks

1 Choose a household product such as a washing machine or a fridge. Describe how parts of it could be recycled.
2 Discuss some ways in which the general public can be encouraged to undertake more recycling.

Modification of properties of materials

Metals

An alloy is a combination of one metal with one or more other metals and in some cases, non-metals. Steel is an alloy of iron with carbon; brass is an alloy of copper and zinc. There are thousands of alloys in existence but they are strictly controlled by the **International Standards Organisation (ISO)** and similar bodies. There are very stringent guidelines and specifications laid down that dictate the maximum and minimum limits of composition and the mechanical and physical properties required by each.

An average car may contain as many as ten different alloys, including:

- mild steel for the body panels
- heat-treated steel alloys for gears
- cast iron or aluminium alloys for the engine block
- lead alloys used in the battery, etc.

Alloys are prepared by mixing two or more metals in the molten state where they dissolve in each other before they are paired into ingots. Generally, the major metal is melted first and the minor is then dissolved into it.

Steel has been covered in some detail in Unit 3 but it is worth considering a few of the details once again. Iron as a pure metal is very soft and ductile, and carbon is very brittle. Yet when the two are combined they form a metal that exhibits none of the properties that the two original ingredients possessed.

As would be expected, the ratio in which the two materials are combined also affects the mechanical and physical properties of the new material. Figure 4.1.18 gives an indication of how the increase in the percentage of carbon affects the hardness of steel. It should be noted, however, that as a direct result of increase in hardness, the ductility (the ability to be drawn out) is decreased.

Engineers have to make compromises at times by having to balance how the increase in one property decreases another. Other properties can be further enhanced by alloying. Resistance to corrosion can be increased by the introduction of chromium into the alloy mix.

A material's properties can also be modified by **heat treatment**. The basic processes are hardening, **tempering**, **annealing** and normalising. A summary is given in Table 4.1.1.

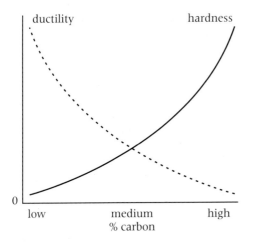

Figure 4.1.18 The effect of carbon content on hardness and ductility

Table 4.1.1 Heat treatment processes

Process	Description
Hardening	The hardness, resistance to indentation and abrasion can be increased in steels with a carbon content of 0.4 per cent or greater. The process involves heating the steel to just above its upper critical temperature and holding it there while it soaks before it is quenched in water. (Other types of hardening are discussed on p192.)
Tempering	After the material has been hardened it is very brittle, and to remove this property it must be tempered. This involves heating the material to a certain temperature depending upon its eventual use before it is quenched in water.
Annealing	As the structure of the metal is deformed by cold working, it becomes harder. To restore its original grain structure, heat is applied to the piece which allows the strain to be dissipated throughout. Annealing takes place at relatively low temperatures.
Normalising	This process is confined to steel and it is used to refine the grain structure after a component has been forged or work hardened.

Work hardening

If you were to take a thin piece of metal, such as a paper clip, and bend it backwards and forwards, it would eventually break. However, there are two noticeable changes that occur before fracture; the metal becomes harder to bend and it gets hot: we say that this metal has been 'work hardened'.

Many manufacturing processes rely on this change. Prior to processing, mild steel is ductile allowing it to be pressed easily in large dies, resulting in components whose strength has been increased slightly by work hardening.

Age hardening

Age hardening of materials was discovered by accident. In an attempt to improve the strength of cartridge cases made from aluminium, a German research metallurgist created an aluminium alloy containing 3.5 per cent copper and 0.5 per cent magnesium. The initial result of heating and quenching the metal was not successful. He returned several days later to discover that the hardness had increased due to its age. Special alloys have now been created specifically for their age hardening properties.

The hardening of metals can be achieved in many different ways. It is not possible to harden carbon steels with less than 0.4 per cent carbon content by conventional means. This does not mean, however, that they cannot be hardened in any other way.

Case hardening

Case hardening is a process that is used to form a hard skin around the outside of a component. The metal has to be heated to a cherry red colour before it is dipped into a carbon powder and allowed to cool. The process needs to be repeated two or three times before it is finally heated and quenched in water. This type of hardening is used on axles where only the surface is subjected to wear, but a soft core needs to be retained that is capable of withstanding shock and sudden impacts.

Induction hardening

Induction hardening also creates a similar pattern with a hard outside and a soft core (see Figure 4.1.19). Again, axles and shafts are hardened in this way.

An induction coil is used to heat a localised area of the work piece and water jets cool the work rapidly. This results in just the outer surface of the axle being hardened.

Flame hardening

A similar process, flame hardening, is used on flat surfaces and complex profiles (see Figure 4.1.20). The surface is again heated to a tempera-

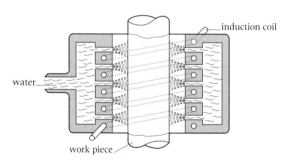

Figure 4.1.19 Induction hardening

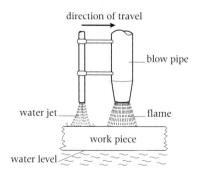

Figure 4.1.20 Flame hardening

ture above its upper critical temperature, normally by an oxy-acetylene torch, and it is followed immediately by a water cooling jet. Both flame and induction hardening can only be used on steels that have more than 0.4 per cent carbon content.

Sintering

Materials that have very high melting points are often processed by sintering. Sintering is a process associated with powder metallurgy and it is an economic way of shaping materials with minimal waste. Metal powders are mixed in the proportions required to form alloys, for example copper and tin to make bronze. Graphite and lubricating oils can also be added at this stage before they are compacted and sintered to form bronze bushes. The graphite and oils make them ideal as oil retaining bushes for use in washing machines and vacuum cleaners where they can be expected to last for the duration of the product's lifespan.

The process of sintering gives the powdered mixture its strength by heating and fusing the particles together.

Plastics

It seems somewhat strange to be thinking how plastics can be changed and modified but it is possible. In the same way that separate metals can be alloyed, separate **monomers** can be altered. Two or more monomers can be combined to form a new material and this process is known as **co-polymerisation** (see Figure 4.1.21), and the new material is known as a co-polymer. An example of this would be a mixture of vinyl chloride and vinyl acetate. The combined co-polymer is known as polyvinyl chloride acetate and is shown in Figure 4.1.21.

Cross-linking is another way of increasing strength. In the vinyl chloride monomer, one atom of hydrogen has been removed and replaced by an atom of chlorine. The new links have formed to create a new polymer, polyvinyl chloride (PVC), also shown in Figure 4.1.21.

Additives can also be used to increase the mechanical properties of plastics. Plasticisers are added as liquids to improve the flow of plastics when being used in moulding processes. They also lower the softening temperature and generally make them less brittle.

Fillers and foamants are added to plastics in an attempt to reduce bulk and overall costs because they are less expensive than polymers. They improve strength by reducing brittleness which also makes them more resistant to impact from shock loading.

One of the early problems with plastics was

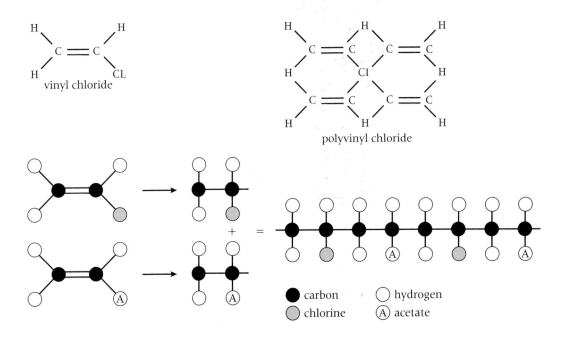

Figure 4.1.21 *Co-polymerisation*

their inability to resist deterioration and exposure to ultra-violet light. The addition of stabilisers has made them more resistant to ultra-violet light and they no longer yellow or become transparent when exposed for long periods.

Tasks

1 Produce an information sheet on one side of A4 paper only for a GCSE student who needs to revise the basics of heat treatment.
2 Why is it sometimes necessary for certain components to have a hard surface on the outside but to retain a softer central core?

Woods
Seasoning
Natural air **seasoning** is the traditional method of removing the excess moisture from newly felled trees. Although it is a very cheap system to operate it is dependent upon the weather conditions. The timber slabs are stacked with sticks between them allowing free air to circulate. This results in the evaporation of moisture, but this method has two major disadvantages:

• It is slow and inaccurate, taking on average one year to season 25 mm of slab thickness.

• The moisture content can only be reduced to that of the surrounding atmosphere (15–18 per cent).

Kiln seasoning, however, results in a much quicker, more reliable method. This time, the timber is stacked with sticks between, and is placed into a sealed chamber. Steam is pumped into the chamber and is absorbed into the timber. The humidity is then drawn out by an extractor fan, the temperature is raised and hot air is circulated. Very careful monitoring and recording takes place in order to attain the precise moisture content level. Moisture content can be calculated using the following formula:

Percentage of moisture content =

$$\frac{\text{Initial weight} - \text{Dry weight}}{\text{Dry weight}} \times 100$$

Overheating can result in a form of a case-hardened timber which is brittle on the outside. Kiln seasoning has a number of advantages:

• It only takes between one and two weeks per 25 mm of slab thickness.
• Less space is required since the process is quicker.
• There is improved turnover of stock.
• It kills insects and bugs in the process.
• Accurate moisture content can be achieved.

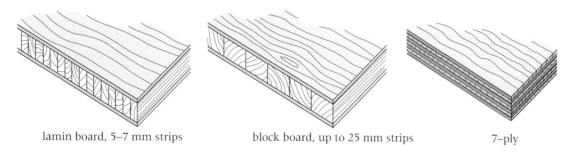

| lamin board, 5–7 mm strips | block board, up to 25 mm strips | 7–ply |

Figure 4.1.22 *Laminated forms of manufactured sheet timber*

Laminating

The development of plywood was an important process that improved the physical properties of timber. Plywood is made from thin layers of wood, and veneers, about 1.5 mm thick, called laminates. They are stuck together with an odd number of layers, but with the grain of each layer running at right angles to the last. This means that the two outside layers have their grain running in the same direction.

The interlocking structure gives plywood its high uniform strength, good dimensional stability and resistance to splitting. It is also available with a variety of external facing veneers and it can be made with waterproof adhesives which means it can be used externally.

These veneers or laminates can also be stuck together over formers to produce curved shaped forms. This process is known as **laminating** and since different shapes can be formed, the strength of the material can be further enhanced by the shapes into which the material is formed.

Other types of manufactured boards are also available that possess qualities of strength: block board, lamin board and batten board all use the laminating, joining together, approach to increase the overall strength (see Figure 4.1.22). These three boards are also clad on their external surfaces with thin laminates for decoration and strength reasons.

Particle boards such as hard board, chip board and medium density fibreboard (MDF) also exhibit properties of strength although they are made in a different way. Again, they can all be prepared so that they are waterproof and can be used externally. In particular, due to its uniform structure, MDF is particularly strong as well as having an excellent surface which is capable of taking a variety of finishes.

3. Value issues

Value judgements are made each and every day of our lives and, as designers, we make evaluative judgements when making and taking decisions about what we should pursue and develop or cast aside.

'Is it worth doing?' Most companies will invest enormous amounts of money into research and development to gauge consumer interest and demand. They will proceed with a project purely on its commercial viability. Design projects also come to fruition because of people's needs. They may well be genuine needs such as a stair lift for someone who has arthritis in his or her legs or they may be **market-led** needs, for example mobile telephones.

Impact of value issues

Every need is a value judgement, if only for that person. Not all needs though will be met because of the economic and commercial viability for the required product.

An example of the decisions that have to be made and the implications considered is the situation faced by a company that designs and manufactures motorised wheelchairs. The views of physiotherapists and engineers who had been working closely on developing **prototypes** at hospital level were taken into consideration. They had produced a series of one-offs where chairs had been individually moulded around the patient. The chair is a means of providing physical support that can eliminate the chance of the medical condition becoming worse.

At this point, a company specialising in the design and manufacture of motorised wheelchairs was contacted. From its point of view the problem was to develop a solution that could be mass produced rather than made as one-offs.

Although the company was a multimillion-pound concern, it needed to ensure that the product was commercially viable.

The base, controls and drive mechanism already existed but the company had to redesign the seat so that it could be mass produced. The marketing manager was concerned with 'Will it make a **profit**?'. The marketing manager worked hard trying to assess and generate sales by canvassing opinions from patients and specialists alike. 'What are you prepared to pay?' was the question asked of the potential customers.

The product was developed to a final prototype stage and fully tested. It was regarded as a success by the potential customers but after significant investment of time and resources, the company decided to shelve the product because it could not guarantee commercial viability. The sales that it had predicted fell well below what the company considered to be acceptable levels. Although the company felt a sense of priority and value about the product in that it improved people's quality of life, it was ultimately concerned about its profit levels.

If private money is not invested, should government provide funds to allow this sort of product to be manufactured? Sometimes company profits are put before the genuine needs of some people.

As a result, the SAM (seating and mobility system) wheelchair has been redesigned and relaunched by the company. The design maintains the same principles as the first using the same seat. This supports the body in a forward leaning position to align body posture for users with spinal deformities (Figure 4.1.23).

The modification in the design comes from adding a motor to allow the user mobility and directional control through a joystick. The base material has been changed to plastic to ensure that any moving parts are covered. The cost of the wheelchair is, therefore, higher but in doing so it has become a more commercially viable product by offering a wider range of uses and possibilities.

Responsibilities of more economically developed countries

The classification of whether a country is more economically developed (MEDC) or less economically developed (LEDC) is based on how well it matches a number of criteria based on socio-economic factors. These measures are normally related directly or indirectly to technology and, until quite recently, their specific manufacturing technologies.

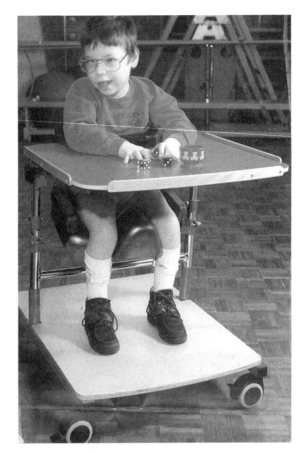

Figure 4.1.23 The SAM wheelchair

The terminology used suggests that all LEDCs have positive aspirations to one day becoming an MEDC. As a result of the requirements to meet the qualifying criteria to become an MEDC there is increased pressure to acquire these manufacturing technologies as a symbol of status.

In many ways LEDCs have already had a taste of the status as a result of the export/import markets and so they have increased accessibility of the products associated with success. The technology and the source of its origin is an important value in terms of its worth. However, the value of a technology must always be considered in relation to the product, the environment and social responsibilities. It would take a lot to convince anyone that companies are in business for no other reason than to make money for their owners or shareholders.

There are some very stringent conditions which have to be met as far as financial controls are concerned. However, there are not many controls to protect against environmental damage. Certain measures have been taken such as the banning of CFCs (chlorofluorocarbons) that damage the ozone layer. There is a move within some countries and industries to consider how they can meet the requirements and demands of the customer at the same time as conserving resources.

MEDCs have a responsibility to consider environmental impacts such as waste and by-products generated as a result of manufacturing processes, the disposal of a product when it reaches the end of its useful life or the pollution created as a result of the energy required in the manufacturing processes. Design and manufacture should take into account wherever possible the need for:

- more effective and efficient use of materials, and the recycling and **reuse of waste materials** and products
- reducing the impact on the environment from the products and processes used in their manufacture
- safer disposal of waste
- more efficient use of energy and natural resources
- increased use of **renewable resources** such as timber from managed forests.

It is the case with any design decision and solution, that an optimum is looked for and a balance drawn between the cost and benefit. Balancing the needs against the impact to the environment is becoming increasingly more difficult for manufacturers as they strive to develop new products and processes. **Life cycle assessment (LCA)** is a technique now widely used to assess and evaluate the impact of the product from its conception and realisation through to its disposal once it has reached its working end. Designers are now having to consider how parts can be reused, recycled or replaced and maintained so that the product can be serviced rather than simply thrown away.

Task

Consider the situation of the wheelchair company outlined on p195. List the advantages and disadvantages of the project from the point of view of:

- the customer
- the physiotherapist and
- the managing director of the company.

Exam preparation

Practice exam questions

Below are some specimen examination questions. They are similar to ones you will get in your examination.

1 Genetic modification (GM) is a subject constantly being discussed and debated in the media.
   a) Outline THREE ways in which GM can help in technology. (3)
   b) State one problem that is associated with GM and indicate what safeguard could be
      put in place to alleviate the problem. (2)

2 A chocolate manufacturer is developing a new range of packaging for Easter eggs.
   a) List THREE criteria, other than life costs, that should be considered when selecting
      the packaging material. (3)
   b) Life costs is an important criterion for material selection. Explain the meaning of
      life costs with reference to material selection. (2)

3 The Lotus super bike ridden in the 1992 Olympics games achieved the first British
   gold medal in cycle racing for more than 72 years. The design was revolutionary and
   made using carbon fibre moulded into hollow frame instead of the more conventional
   steel tubing.
   a) Give TWO advantages that a carbon fibre hollow frame has over tubular steel frame. (2)
   b) Explain how carbon fibre frame is produced. (3)

4 The use of recycled paper and card is becoming increasingly popular in society. However,
   recycling paper does have it's problems.
   a) State THREE limitations associated with the recycling of paper and card. (3)
   b) Give TWO factors that might make a manufacturer choose to recycle paper. (2)

5 a) Explain what is meant by the term 'mail merging'. (2)
   b) Explain how this process is used in administrative systems. (3)

Design and technology in society (G402)

Summary of expectations

1. What to expect
The content of this option builds on your AS study and enables you to develop further understanding of design and technology in society.

2. How will it be assessed?
The work that you do in this option will be externally assessed through Section B of Unit 4, Paper 2. There will be two compulsory long-answer questions, each worth 15 marks. You are advised to spend 45 minutes on Paper 2.

3. What will be assessed?
The following list identifies what you will learn in this option and subsequently what will be examined:

- Economics and production, including economic factors in the production of one-off, batch and mass-produced products.
- Consumer interests, including:
  - systems and organisations that provide guidance, discrimination and approval
  - the purpose of British, European and International Standards relating to quality, safety and testing
  - relevant legislation on the rights of the consumer when purchasing goods.
- Advertising and marketing, including:
  - advertising and the role of the design agency in communicating with manufacturers and consumers
  - the role of the media in marketing products
  - market research techniques
  - the basic principles of marketing and associated concepts.
- Conservation and resources, including:
  - environmental implications of the industrial age
  - management of waste, the disposal of products and pollution control.

4. How to be successful in this unit
You are expected to demonstrate a thorough understanding of what you have learned in the option. Examiners are looking for longer, more detailed answers that show a greater depth of knowledge and understanding at the Advanced GCE level.

You will be assessed on your ability to organise and present information, ideas, descriptions and arguments clearly and logically, taking into account the quality of written communication.

5. How much is it worth?
This unit, with the option, is worth 15 per cent of the full Advanced GCE.

Unit 4 + option	Weighting
A2 level (full GCE)	15 per cent

1. Economics and production

Figure 4.2.1 shows the UK's manufacturing sectors.

Economics and the production chain

The sequence of activities required to turn raw materials into finished products for the consumer is called the **production chain**. It is the aim of all manufacturing companies to undertake such activities in the most cost-effective manner. The purpose of this is to maximise profit.

The economic factors that combine in order to produce profit include:

- variable costs
  - the costs of production, such as materials, services, labour, energy and packaging
- fixed costs
  - related to design and marketing, administration, maintenance, management, rent and rates, storage, lighting and heating, transport costs, depreciation of plant and equipment.

In order to remain profitable, a manufacturing company must calculate accurately its total costs and set a suitable selling price. This calculation must allow for **variable costs**, **fixed costs** and a realistic profit.

The production chain includes the following:

- The **primary sector** is concerned with the extraction of natural resources such as mining and quarrying.
- The **secondary sector** is concerned with the processing of primary raw materials and the manufacture of products. Although this sector supplies a large proportion of exports from MEDCs (such as those of western Europe and the USA), it employs a decreasing proportion of the workforce, as changes in technology and the global economy occur.
- The **tertiary sector** industries provide services and include employment in education, retailing, advertising, marketing, banking and finance. This sector employs the most people in MEDCs.

In your design and technology course you are mostly concerned with the secondary and tertiary sectors. When you design and manufacture a coursework product, your main concern should be **product viability** in terms of its cost of manufacture and **market potential**. For any industry in the secondary sector, however, product viability relates to market potential and profit. Product viability is essential to the industry's existence and to the employment of its workforce.

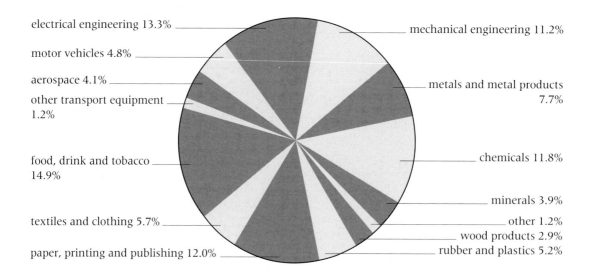

Figure 4.2.1 Breakdown of relative sizes of different manufacturing industries across the UK. Food, drink and tobacco, electrical engineering and paper, printing and publishing form the three largest sectors

Productivity and labour costs

A company's **productivity** is a measurement of the efficiency with which it turns raw materials (production inputs) into products (manufactured outputs). The most common measure of productivity is output per worker, which has a direct effect on labour costs per unit of production. The higher the productivity, the lower the labour costs per unit of production and therefore the higher the potential profit. Table 4.2.1 compares the weekly wage, output per worker and labour cost per unit of production for two companies. It shows how an efficient company with high output per worker can keep labour cost per unit of production low.

Table 4.2.1 *Comparison of productivity*

	Weekly wage	Output per worker	Labour cost per unit of production
Efficient company	£280	40.0	£7
Less efficient company	£250	25.0	£10

Labour costs are linked to the type and length of any production process. For example, a washing machine made from mild steel sheet, will have sharp edges where the steel has been cut to size. In order to produce a good quality product these sharp edges will need to be removed or covered. Introducing an additional process such as this will increase labour costs and therefore increase the cost of the product.

Scale of production

Within the economies of the MEDCs of the West and Asia, there is a responsibility for all the functions of a company, including manufacturing, to be profitable. For example, in many large companies:

- different internal departments are 'customers' to other internal departments within the same company
- each department has a budget and has to work within this budget or make a profit.

In manufacturing companies the level or scale of production is an extremely important factor for profitability, because it influences the process of manufacture, where the product is manufactured, the choice of products available on the market and their resulting selling price.

One-off products

One-off or custom-made products need to use materials, tools and equipment that are available or that can be resourced easily. They are often less efficient to manufacture because they may use different or 'specialised' materials and processes. One-off products are usually much more expensive to buy, since materials and labour costs are higher, but they do provide 'individual' or made-to-measure products, such as hand-made kitchen units.

Mass production and the development of new products

On the other hand, mass production has revolutionised the choice and availability of a range of relatively inexpensive products:

- It is increasingly difficult for a company to remain in profit without developing new products, even when existing ones are selling well. **Consumer demand** can be very fickle and sales of even a best-selling product tail off in time, as new or different competing models come on the market.
- The cost of developing new products is high. It includes initial setting up costs such as the factory and its layout and the cost of the workforce.
- New products may require changes in production, which are often difficult to achieve, due to the high levels of investment required.
- The development of new production methods needs long-tem planning and may need to overlap with existing production, in order to keep the company running and in profit.
- There is a constant need to reduce the time-to-market of new products. The most successful companies produce the right product at the right time, in the right quantity and at the right cost – one that the market will see as providing the right image as well as being 'value for money'.

Task

Select two similar products, such as a hand-made storage unit and a mass-produced flat pack. For the two products compare the following:
- the type of materials and components used
- the number and estimated length of assembly processes
- the selling price.

Sources, availability and costs of materials

The sources, availability and costs of materials depend on the type and quantity of materials required. As with all variations in manufacturing requirements, different scales of production often result in different levels of materials costing. Large-scale manufacturers generally carry more 'clout' with suppliers than do producers of batch or one-off products. For example, a motor vehicle manufacturer may negotiate long-term supply contracts for steel directly with steel makers. On the other hand, a manufacturer of custom-made kitchen furniture may source timber through a specialist merchant on a much smaller scale and at a comparatively higher cost.

Timber, paper and board

The major sources of the world's supply of commercially grown softwood are the coniferous forests of the northern hemisphere, stretching across North America and Europe. Careful management of forests enables the control of supply and demand of softwoods. Since conifers are relatively fast growing and produce straight trunks, they are economic to produce, with little waste. Softwoods are therefore relatively inexpensive. They are used extensively for building construction and joinery. Paper and card are made by blending cellulose fibres (obtained from trees) and water. Virgin fibre (as opposed to waste and recycled fibre) is obtained from softwood trees, grown especially for this purpose. The main supplier in western Europe is Scandinavia.

Hardwoods are slower growing, more durable and more expensive than softwoods. They come from broad-leaved trees which are mainly deciduous, growing in the temperate climates of Europe, Japan and New Zealand. Hardwoods grown in the tropical climates of Central and South America, Africa and Asia are mainly evergreen. These grow all year round and reach maturity earlier. Hardwoods are used for a range of furniture, kitchen utensils, flooring, toys etc.

Manufactured board

The UK imports almost 90 per cent of its timber needs, which are usually supplied in board form, ready for further processing. Importing timber in board form provides employment and income for the country of origin and avoids paying freight costs for raw materials that will produce waste. Another advantage of converting timber to board is that it can travel in freight containers, rather than bulk cargo ships, which are generally slower.

Plywood and blockboard are fairly expensive, because they need to be made from good quality timber:

- Plywood is made from birch, from northern Europe and North America and meraniti from South-east Asia.

- Blockboard uses birch for the facing surfaces but the core can be formed from small sections of pine, from North America and Europe.

Chipboard, particle board, medium density fibreboard (MDF) and hardboard are relatively inexpensive. They are made from small section timber or thinned timber and reconstituted wood-based materials.

Manufactured board is made in large quantities by specialist companies. It is extensively used in mass-produced, self-assembly kitchen and bedroom furniture, shop fitting and flooring.

Task

Investigate the sources of timber and wood-based products used in your school/college. Plot the sources on a map of the world. How many of these come from sustainable sources?

Metals

Metal ores form about a quarter of the weight of the Earth's crust, with the most common being aluminium, followed by iron. They have no particular pattern of distribution around the world, but some countries have larger deposits than others.

Iron and steel account for almost 95 per cent of the total tonnage of all metal production. Ships, trains, cars, trucks, bridges and buildings and thousands of other products depend on the strength, flexibility and toughness of steel.

Sources of ore

- The main sources of iron ore are Sweden, Ukraine, Australia and North America. The easy availability of the ore and its high metal content means that the price of steel is relatively low.
- Aluminium ore, known as bauxite, is mainly found in Australia, Guinea, Brazil and Jamaica. Bauxite is accessible, making it cheaper to make aluminium than steel.
- Copper ore is found in Chile, the USA and Canada. It is far more expensive than iron ore or bauxite because it is rare.
- Tungsten is found in China, Bolivia, North Korea and Portugal. It is also quite rare and therefore more expensive than iron ore or bauxite.

Metals in common use, such as iron and aluminium, are easily available and the supply problems associated with some hardwoods do not normally occur. Lower metal production costs are often achieved by smelting ore close to its source. This reduces transport costs and may make use of lower labour costs.

Figure 4.2.2 Offshore oilfields currently produce about a quarter of the world's crude oil. Great technological skill has been required to design platforms that are stable enough to allow drilling to take place and resilient enough to withstand the harsh conditions of wind and waves

Large users of metal, such as makers of tin plate for the canning industry, often buy direct from companies producing final sections. Smaller users, such as those involved in small-batch and one-off production, often buy from specialist metal stockholders.

The importance of oil

Crude oil is an important commodity because it supplies much of the world's energy needs. It is also the principle raw material for making plastics and polymers, without which modern society could not long continue to flourish.

Unfortunately, oil is rarely found where it is needed. The largest oil-producing countries are not themselves major consumers and they are therefore able to export much of their oil. These countries are part of the Organisation of Petroleum Exporting Countries (OPEC), a cartel which sets output quotas in order to control crude oil prices. The members of OPEC are the Middle East, South America, Africa and Asia, but not the USA, the Russian Federation or European oil producers such as the UK. In the 1970s, OPEC controlled 90 per cent of the world's supply of crude oil exports. The resulting high price of oil that OPEC maintained then allowed more expensive fields such as the North Sea and Alaska to be brought into production (see Figure 4.2.2). Oil costs continue to fluctuate, depending on the control of its supply.

This can result in higher petrol, energy and raw materials prices worldwide.

Plastics

Most thermoplastic and thermosetting plastics are derived from crude oil. They are easily available in sheet or rod form for processing into products, or as granules for injection moulding. Plastic resins are supplied in liquid form.

Large users of plastics often buy direct from manufacturers, but smaller users generally buy through plastics stockholders. In many industries traditional materials such as wood or metal have been replaced by cheaper plastics, which generally use fewer processes, as they are often made in one piece. For example, polyester (PET) bottles are safer, lighter and cheaper to produce than bottles made from glass.

Cost of materials

As we have seen, the cost of timber, manufactured board, metals and plastics is linked to their sources and availability. The following can be said about most materials:

- Raw materials that are in short supply cost more than those that are plentiful.
- Raw materials that are difficult or more expensive to process cost more.
- The transportation of raw materials, either in raw or finished form adds to their costs.

Figure 4.2.3 *One-off products are more expensive to produce, such as this cap and necklace, by Val Hunt. They are made from beer cans, which are annealed, woven, pleated, frilled or curled to produce exciting and creative products*

Advantages of economies of scale of production

Economies of scale are factors that cause average costs to be lower in high-volume production than in one-off production (see Figure 4.2.3). The unit price is lower because inputs can be utilised more efficiently. Economies of scale in high-volume production are brought about by:

- specialisation – the work processes are divided up between a workforce with specific skills that match the job
- the spread of fixed costs of equipment between more units of production
- bulk buying of raw materials at lower unit costs
- lower cost of capital charged by providers of finance
- the concentration of an industry in one area – attracting a pool of labour that can be trained to have specialist skills
- a large group of companies in one area – attracting a large network of suppliers whose costs are lower, because of their own economies of scale.

Task

Devise a checklist that will make your own production more cost-effective. Think about the range of materials, components and processes you might use.

The relationship between design, planning and production costs

SIGNPOST
Efficient manufacture and profit
Unit 3B1 page 124

In order to remain profitable, manufacturers must accurately calculate their total costs and set a suitable product selling price. Target production costs are established from the design stage and checked against existing or similar products. This is done so that the design team can make sound decisions early in the design task.

All costs in a manufacturing company are initiated in the design phase. It is in the manufacturing stage that the major costs are incurred. It is often said therefore that **designing for manufacture (DFM)** is directly related to designing for cost. The main aims of DFM are:

- to minimise component and assembly costs
- to minimise **product design cycles**
- to enable higher quality products to be made.

The cost of quality

SIGNPOST
Aesthetics, quality and value for money
Unit 3B1 page 124

Manufacturing a competitive product based on a balance between **quality** and cost is the aim of all companies. The costs of quality are no different from any other costs, because like the cost of design, production, marketing and maintenance, they can be budgeted, measured and analysed. There are three types of quality costs:

- the costs of getting it wrong
- the costs checking it is right
- the costs of making it **right first time**.

The costs of getting it wrong
There are two ways to get it wrong: **internal failure costs** and **external failure costs**.

Internal failure costs occur when products fail to reach the designed quality standards and are detected before being sold to the consumer. They include costs relating to:

- scrap products that cannot be repaired, used or sold
- reworking or correcting faults
- re-inspecting repaired or reworked products
- products that do not meet specifications but are sold as 'seconds'
- any activities caused by errors, poor organisation or the wrong materials.

External failure costs occur when products fail to reach the designed quality standards and are not detected until after being sold to the customer. They include costs relating to:

- repair and servicing
- replacing products under guarantee
- servicing customer complaints
- the investigation of rejected products
- product liability legislation and change of contract
- the impact on the company reputation and image – relating to future potential sales.

The costs of checking it is right

These costs are related to checking:

- materials, processes, products and services against specifications
- that the quality system is working well
- the accuracy of equipment.

The costs of making it right first time

There is one way to get it right: **prevention costs**. These are related to the design, implementation and maintenance of a quality system. Prevention costs are planned and incurred before production and include those relating to:

- setting quality requirements and developing specifications for materials, processes, finished products and services
- quality planning and checking against agreed specifications
- the creation of, and conformance to, a **quality assurance (QA)** system
- the design, development or purchase of equipment to aid quality checking
- developing training programmes for employees
- the management of quality.

Costing a product

The costing of a product involves much more than adding on a set percentage to give a profit once all manufacturing expenses have been taken into account (see Figure 4.2.4). Setting the selling price too high may reduce sales below a profitable margin, while setting it too low won't allow a profit even if vast numbers are sold. Checks against a competitor's product are often used to establish the potential price range of a new product because they can give an idea of what the market can stand.

Products are sometimes said to have a value, a price and a cost:

- Manufacturers want the income from selling the finished products, rather than keeping them in stock, so for them the product value is always lower than the selling price.
- On the other hand, consumers want the product more than the selling price, as they see the product value as being higher than the selling price.

The total cost of a product takes account of the following:

- *Variable costs* (**direct costs**) – the actual costs of making a product, i.e. materials, services, labour, energy and packaging. Variable costs vary with the number of products made. The more products made, the greater the variable costs of materials. Variable costs may account for around 50–65 per cent of the total product **selling price (SP)**.
- *Fixed costs* (**indirect costs** or overhead costs) – for example, design and marketing, administration, maintenance, management, rent and rates, storage, lighting and heating,

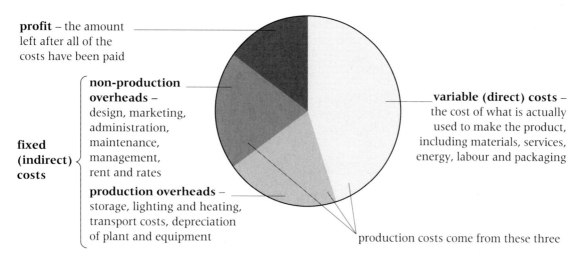

profit – the amount left after all of the costs have been paid

fixed (indirect) costs

non-production overheads – design, marketing, administration, maintenance, management, rent and rates

production overheads – storage, lighting and heating, transport costs, depreciation of plant and equipment

variable (direct) costs – the cost of what is actually used to make the product, including materials, services, energy, labour and packaging

production costs come from these three

Figure 4.2.4 *What's in a price?– The components of the product selling price*

transport costs, depreciation of plant and equipment. Fixed costs are not directly related to the number of products made, so they remain the same for one product or hundreds. A company's accountants will establish a way to divide up fixed costs between the various product lines made by a company, so that each product carries its share. Marketing and selling costs often account for 15–20 per cent of the total SP.

• *Profit* – the amount left of the SP after all costs have been paid. Profit is referred to as gross or net. The gross profit is calculated by deducting variable plus fixed costs from the revenue from sales. Net profit is gross profit minus tax. Net profit is used to pay dividends to shareholders, bonuses to employees and for reinvestment in new machinery or in new product development.

The break-even point

To cover the cost of manufacture, enough products need to be sold at a high enough price. Calculating this requirement is called 'break-even analysis'. It is a fundamental part of understanding the relationship between cost and price and is useful when working out a product costing. If a product selling price is £7, how many need to be sold to cover costs and break even? The answer can be expressed as:

$$\text{Break-even point} = \frac{\text{Fixed costs}}{\text{Selling price} - \text{variable costs}}.$$

Imagine that you work for a company that makes garden fountains. A new design has been introduced. The variable costs of making one fountain are £30 and it sells for £45. Fixed costs are £4 800. The formula above can be used to calculate the **break-even point**:

$$\text{Break-even point} = \frac{4\,800}{45 - 30} = 320.$$

So 320 products need to be sold to reach the break-even point.

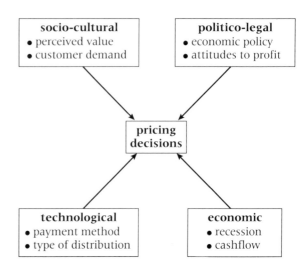

Figure 4.2.5 *The forces acting on pricing decisions*

The material and manufacturing potential for a given design solution

When manufacturers calculate the potential selling costs of a product, they have to take into account more than just manufacturing costs. Consumer expectations of the value of the product play an important part. This will depend on two things: what customers perceive as value for money and what the competition is offering.

The forces acting on pricing decisions are complex (see Figure 4.2.5). Perceived value is reflected in the likelihood of customers continuing to buy a product in the face of a price increase. This is the pattern for food items or services like heating. Unless they become prohibitively expensive, we continue buying them. For non-essential purchases such as a second car, demand can drop when price goes up. Other forces acting on price include the economic and political climate – how high are taxes, is there a recession, is there full employment?

Tasks

1 Estimate the cost of a product that you have made, including costs of materials, energy used in its production and your time for designing and making it. Produce a selling price that includes a profit.

2 Analyse your living costs for one week, assuming that you have to pay rent, energy costs and food. Calculate how many products you would have to make in a week to 'live' on your manufacturing profit.

Task

Some say that there is no direct connection between the product cost and the selling price. In the same way that materials in short supply cost more, the manufacturer of a popular product in limited supply can add a large mark-up to the price. Give examples of products that prove or disprove this theory.

2. Consumer interests

The relationship between manufacturers as producers and consumers as buyers is important. Manufacturers want their products to sell well and make a profit. Consumers want to buy high-quality, attractive products that are reliable, easy to maintain, safe to use and that provide value for money at a price they can afford. A manufacturer's success in producing goods that meet consumer requirements relies on keeping consumer needs and values at the heart of the business.

Systems and organisations that provide guidance, discrimination and approval

Most large companies use **market research** to establish consumer needs, values and tastes. They must also take into account consumers' statutory rights when buying products. These rights are enforced and regulated by a wide range of legislation relating to consumer protection and fair trading. There are many systems and organisations that provide guidance, discrimination and approval for consumers. These include:

- the Institute of Trading Standards Administration
- British, European and International Standards organisations
- consumer 'watchdog' organisations.

Consumer 'watchdog' organisations

These days consumers are much more aware of new products. One reason for this developing consumer awareness is the profusion of 'style' sections in newspapers and magazines. Products ranging from kettles to cameras are regularly featured and their desirability and value for money evaluated.

Consumer 'watchdog' organisations and specialist magazines also provide guidance, discrimination and approval for new products. These organisations are independent of product manufacturers and provide objective reviewing and testing of products. One such organisation is the Consumers Association, which publishes the magazine *Which?*. This provides reports about product testing and 'best buys'. It has a website, found at www.which.co.uk, which provides a range of information, such as an overview of its activities, 'headlines' for daily consumer news and links to electronic newspapers.

Task

Investigate the work of two different consumer organisations. Compare their roles and the range of products they evaluate. For each organisation explain how their product reviews:

- help or hinder product manufacturers
- guide consumer choice.

Other groups that provide consumer support include television and radio programmes. Many of these programmes give information and guidance on consumer rights. Motoring and government organisations, such as the Road Research Laboratory, also undertake objective reviewing and testing of vehicles. Product testing and reporting is also carried out by trade and professional organisations and journals.

British, European and International Standards relating to quality, safety and testing

SIGNPOST
British and International Standards
Unit 3B1 page 133

European and International Standards organisations set national and international **standards**, testing procedures and quality assurance processes to make sure that manufacturers make products that fulfil the safety and quality needs of their customers and the environment.

Most standards are set at the request of industry or to implement legislation. Manufacturers of upholstered furniture, for example, have to conform to established fire safety standards. The test procedures for checking fire safety have to comply with **British Standards Institute (BSI) guidelines** and must be carried out under controlled conditions.

Any product that meets a British Standard can apply for and be awarded a '**Kitemark**'. This shows potential customers that the product has met the required standard and that the manufacturer has a **quality control (QC)** system in place to ensure that every product is made to exactly the same standard.

There are a range of British Standards that apply specifically to packaging. They are classified under the following headings:

- Marking and labelling on product packaging – these refer to labels, signs, symbols and warnings, for example EN 71-6:1994; BS EN 71. Guidance is given on the appropriate use of the BS Kitemark, the CE mark and the 'e' mark, and special requirements are given for food and toy labelling (see Figure 4.2.6).
- Packaging materials – these deal with the performance requirements for packaging materials and appropriate tests, for example BS 6890. Guidance is given on tests to determine safety standards in terms of danger from asphyxiation, resistance to harmful or unwanted substances, odour contamination, etc.
- Glossary and checklist – these define and illustrate technical terms and conditions, for example BS 3130-2.

The relationship between standards, testing procedures, quality assurance, manufacturers and consumers

Common to all commercial product manufacture is the need to produce a quality product. For manufacturers, incorporating **quality management systems (QMS)** into the design and manufacture of products is therefore important. **ISO 9000** is an internationally agreed set of standards for the development and operation of a quality management system (QMS). ISO 9001 and 9002 are the mandatory parts of the ISO 9000 series. They specify the clauses manufacturers have to comply with in order to achieve registration with the standard. A QMS involves a structured approach to ensure that customers end up with a product or service that meets agreed standards.

Quality management systems

All industrial quality management systems use structured procedures to manage the quality of the designing and making process. The following designing and making procedures illustrate the kind of quality management process that industry adopts and uses:

- Explore the intended use of the product, identify and evaluate existing products, consider the needs of the client.
- Produce a design brief and specification.
- Use research, **questionnaires** and product analysis.

Figure 4.2.6 Quality control marks

- Produce a range of appropriate solutions.
- Refer back to the specification.
- Refer to existing products. Use models to test aspects of the design.
- Check with the client. Use models to check that the product meets the design brief and specification.
- Plan manufacture and understand the need for safe working practices.
- Manufacture the product to the specification.
- Critically evaluate the product in relation to the specification and the client. Undertake detailed product testing and reach conclusions. Produce proposals for further development, modifications or improvements.

You may be familiar with the procedures outlined above, because they are very similar to ones you use in your coursework.

Task
Compare the designing and making procedures listed above with the ones you used in your most recent project. Comment on any similarities and differences you find.

Applying standards

Risk assessment
In any product manufacturing situation, **hazards** must be controlled. A hazard is a source of potential harm or damage or a situation with potential harm or damage. Hazard control incorporates the manufacture of a product and its safe use by the consumer. As part of the Health and Safety at Work Act 1974 **risk assessment** is a legal requirement for all manufacturers in the UK. Its use is a fundamental part of any quality management system. The main purpose of the Health and Safety at Work Act is to control risks and to enable decisions to be made about the level of existing risk control. Figure 4.2.7 shows how a risk assessment may be planned.

A risk has two elements:

- the possibility that a hazard might occur
- the consequences of the hazard having taken place.

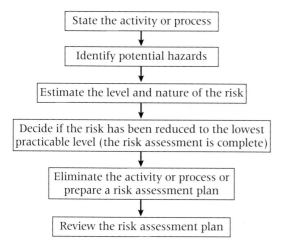

Figure 4.2.7 How a risk assessment is planned

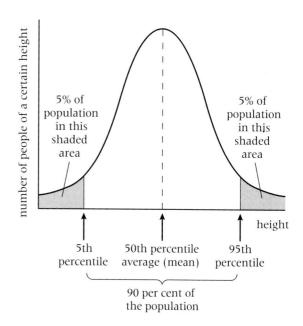

Figure 4.2.8 Anthropometric data uses measurements representative of 90 per cent of the population (the fifth to ninety-fifth percentiles)

Task

Using the risk assessment plan shown in Figure 4.2.7, draw up a risk assessment plan for your own product.

Ergonomics

Ergonomics is the application of scientific data to design problems. This means applying the characteristics of human users to the design of a product – in other words matching the product to the user. In order to do this, data about the size and shape of the human body is required – this branch of ergonomics is called anthropometrics.

Anthropometric data must take into account the greatest possible number of users. This data exists in the form of charts, which provide measurements for the 90 per cent of the population that fall between the fifth and the ninety-fifth percentiles (see Figure 4.2.8). Consumer products ranging from tools to storage units are generally easier, safer and more efficiently used if anthropometric data is used in their design.

Task

Anthropometric data is fine for the 90 per cent of people who fall between the fifth and ninety-fifth percentiles.
a) Explain what is meant by the fifth and ninety-fifth percentiles.
b) Discuss what problems the remaining 10 per cent of people might find in using products.
c) Explain how changes in the sizes and shapes of people might impact on the design of products.

SIGNPOST
Anthropometrics and ergonomics
Unit 3B1 page 131

Relevant legislation on the rights of the consumer when purchasing goods

Whenever consumers buy or hire goods and services, they are protected by a body of law called **statutory rights**. These set out what consumers should *reasonably* expect when buying products. The body of law includes legislation such as the Trade Descriptions Act 1968, the Sale of Goods Act 1979 and the Consumer Protection Act 1987.

Statutory rights protect consumers' 'reasonable expectations' when buying products in a shop, market, on the doorstep, by mail order, through direct response TV, leaflets, magazines and newspaper advertisements or on the Internet. In fact, it does not matter where or how goods are bought, consumers' statutory rights remain the same.

As a consumer you expect any product that you buy to look and perform satisfactorily. Even if you buy something in a sale, as a 'second' or as a 'shop soiled' item, the product must perform as you are told it does. In order to meet statutory rights products must satisfy three conditions:

- They must be 'of satisfactory quality', according to the sales description, cost and any other relevant factors. This rule covers the look, safety and durability of the product. Items should be free from defect, except when openly sold as substandard such as 'shop soiled'.
- They must be 'fit for purpose' and perform as described. If you are told a computer game can be used on a particular machine it must do so – or you have good reason to complain.
- They must be 'as described' on the packaging, the sales display or according to what the seller says. A 100 per cent cotton shirt must be 100 per cent cotton.

Limits to statutory rights

If you buy a product after being shown a sample, you should expect to get a product that is the same. Before making any complaint though, you should think about the circumstances under which you bought the product. Legally you cannot expect a product described as 'shop-soiled' to be as good as a new one and it may even be faulty. You have no legal grounds for complaint if you:

- were told about the fault
- examined the product and did not see the fault
- damaged the product yourself
- bought the item by mistake
- changed your mind about the product.

Statutory rights represent a consumer's minimum rights, but in order to build customer loyalty, many retailers often exchange products that are not faulty, as long as there is proof of purchase. Many retailers also exchange goods that consumers decide are the wrong size, fit or colour.

Second-hand goods

You have the same rights when buying second-hand goods as when buying new ones. You can claim your money back or the cost of repairs if goods sold to you are faulty provided:

- the faults are not due to the wear and tear to be expected when second-hand
- they were not pointed out to you at the time of sale
- they were not obvious when you agreed to buy the goods.

Sale goods

You have the same rights when buying sale goods as when buying new ones. Notices that say 'no refunds on sale goods' are illegal and local authorities can prosecute traders who

display them. Traders sometimes try to use these notices to limit their responsibilities.

If things go wrong

If there is something wrong with a product, a consumer should normally get a refund, provided the retailer is informed of the problem quickly. Keeping a faulty product beyond a reasonable time without complaining may lose any right to a refund because the product may have been accepted in a legal sense. The trouble is what is 'reasonable' is not fixed and depends on circumstances, but a consumer should normally be allowed to take a product home to 'try it out'.

- If it is not possible to return the product within a few days, a consumer should contact the retailer as soon as the fault is discovered, note the name of the person spoken to and keep a record of the conversation.
- A consumer should not hang on to faulty goods without complaining.
- The law says that sellers are responsible for the products they sell, so a consumer should not be fobbed off by a trader who says it's the manufacturer's fault.
- Losing the receipt doesn't mean a loss of statutory rights. There may be other evidence, such as credit card or bank statement.
- If, by chance, a faulty product should cause damage to another product, a consumer may be able to claim compensation.

If a product shows a fault after you have had it for a reasonable time, or you have used it for some time, then you cannot reject it. You are, however, able to claim compensation which could include compensation for the loss in value of the product or a repair or replacement.

Help in solving problems

Sometimes consumers need advice and guidance on everyday shopping problems. Local authority trading standards officers enforce and advise on a wide range of legislation relating to consumer protection and deal with problems and complaints. Details of local Trading Standards departments can be found on the Trading Standards website: www.tradingstandards.gov.uk.

> ## Task
> **Making a complaint**
> Sometimes a product does not perform as expected and a complaint must be made. Imagine you have recently bought a portable DVD player that fails to work properly. Describe the actions you would take to complain about the product:
>
> **a)** in person
> **b)** by telephone
> **c)** in writing.

3. Advertising and marketing

Advertising and the role of the design agency in communicating with manufacturers and consumers

Advertising is any type of **media** communication that is designed to inform and influence existing or potential customers. The cost of advertising is a major marketing expense and in spite of all the money that is spent on it no one really knows what makes it work!

Many large manufacturers employ advertising agencies to run their advertising campaigns. Successful campaigns are those that are effective in selling products and they are often the ones that we as consumers remember. Each advertising agency has its own distinctive approach which is different from its rivals in a highly competitive market place. There are said to be two approaches to advertising – the **hard sell** and the **soft sell**.

Hard sell

A hard sell advertisement has a simple and direct message, which projects a product's **unique selling proposition** (USP) – its unique features and advantages over a competitor's products. The advertising agency Saatchi and Saatchi was famous for hard-hitting advertisements, often on a grand scale. An example would be the 'Manhattan' advertisement for British Airways, which featured what appeared to be the whole New York landscape being carried across the Atlantic.

Soft sell

On the other hand, soft sell advertisements promote a product's personality or image, with which consumers can identify. This approach is often associated with brand advertising, which focuses on creating a positive product image. It takes for granted the benefits of the product and focuses on creating positive emotional and psychological associations with the product. Soft sell is often said to work best with expensive, high status products. Nowadays soft sell is successfully associated with even frequently bought items like detergent. Even this can be given an image that reflects its performance and quality.

> ## Task
> Find one hard sell and one soft sell advertisement in a newspaper or magazine.
>
> **a)** Devise a list of changes that would make the hard sell softer and the soft sell harder.
> **b)** Explain how your changes would make the advertisements still appeal to the same target market groups.

Successful advertising campaigns

Successful advertising campaigns often include elements of hard and soft sell. The advertising of Fairy Liquid, for example, promoted the benefits of kindness to hands and value for money, while giving consumers emotional images related to family life.

The success of a campaign can be measured in the effect on the sales of a product or service. One successful campaign was the BT 'It's Good to Talk' series, which featured the actor Bob Hoskins. This campaign was spread over 12 months from 1994 to 1995 and cost £44 million. It resulted in a large increase in calls and represented a 1.75 per cent return on the investment (work out 1.75 per cent of £44 million!).

Advertising standards

The **Advertising Standards Authority (ASA)** regulates all British advertising in non-broadcast media. The code has three basic points which state that advertisements should:

- be legal, decent, honest and truthful
- show responsibility to the consumer and to society
- follow business principles of 'fair' competition.

The role of the media in marketing products

The main aims of marketing through the media are to influence customers' buying decisions, to help sell more products and promote a good public image for a company. Paid-for media options include:

- the press – newspapers, magazines and direct mail
- broadcast media – television and radio
- cinema
- outdoor – posters and sports grounds
- electronic – direct email and the Internet.

An important source of information when choosing an advertising medium is a large-scale regular survey called the Target Group Index

Table 4.2.2 *Strengths and weaknesses of the major types of media*

Strengths	Weaknesses
Television (around 33% of UK advertising expenditure) • High audiences, but spread over channels • Excellent for showing product in use	• Short time span of commercials is limiting • High wastage – viewers not in target market
Newspapers, magazines (around 60% of UK advertising expenditure) • Can target the market with detailed info • Can get direct response (reply coupon)	• Can have a low impact on consumers • Timing may not match marketing campaign
Radio (around 2-3% of UK advertising expenditure) • Accurate geographical targeting • Low cost and speedy	• Low numbers compared to other media • Listen to it in the background to other tasks
Posters (around 4% of UK advertising expenditure) • More than 100,000 billboards available • Relatively cheap	• Seen as low impact/complicated to buy • Subject to damage and defacement

(TGI). This researches the consumption patterns of a representative sample of consumers – asking questions about the products they buy and use, and what they watch, read and listen to. Subscribing to this survey allows a marketing organisation to match its target market with the media it uses most.

Table 4.2.2 is a guide to the strengths and weaknesses of the major types of media.

Tasks

1 Collect a variety of advertisements from magazines and newspapers. For each advertisement explain:

- the image given about the product – is it luxury or bottom end of the market or something in between?
- the target market, relating to **demographics** and lifestyle
- any 'hidden ' messages about the product – the use of emotions, concern, fear, compassion, persuasion or politics
- any 'values' attached to the product
- if it is hard sell or soft sell
- if the product is advertised in other media or is part of wider campaign.

2 You need to promote your coursework product to the consumers in your target market group.

a) Describe the features and characteristics of your product.
b) Describe the **buying behaviour** and lifestyle of your target market group.
c) Choose two different media for an advertising campaign and explain why each is appropriate for your product.
d) Explain how an Internet marketing campaign could help you reach your target market.

Market research techniques

Questions such as 'Who are my customers?' and 'What are their needs?' must be answered before any decision is made about what to make, what to charge, how to promote it and how to distribute it. In order to find answers to such questions, it is necessary to undertake marketing research. This involves the use of market research techniques to identify:

- the nature, size and preferences of current and potential target market groups and sub-groups (called **market segments**)

- the buying behaviour of the **target market group**
- the competition – and its strengths and weaknesses (this includes pricing and marketing policies)
- the required characteristics of new products – matching these to target market needs (this information is also used to improve existing products and to identify 'gaps' in the market)
- the effect price changes might have on demand – how sales would be affected by a price increase, how the price compares to that of a competitor, the price to set for a new product
- trends in design, colour, demographics, employment, interest rates and inflation.

Market research comes from two types of sources:

- Primary sources provide original research, e.g. internal company data, questionnaires and surveys.
- Secondary sources provide published information, e.g trade publications, commercial reports, government statistics, computer databases. Other sources of secondary research include the media and the Internet.

When conducting primary research to find out, for example, about the size and buying behaviour of a target market group, two types of data can be collected: **quantitative** and **qualitative**.

Quantitative research often uses a survey to collect data about how many people hold similar views or display particular characteristics. Normally, the information is collected from a small proportion (a 'sample') of a target market group. The views of the whole target market group are then based on the responses from the sample.

Qualitative research involves collecting data about how people think and feel about issues or why they take certain decisions. Qualitative research explores consumer behaviour and is conducted among a few individuals. It is often used to plan further quantitative research to see if the views of a few are representative of the whole target market group.

Surveys

A survey is a way of collecting quantitative data, often about behaviour, attitudes and opinions of a sample in a target market group. The most effective way to conduct a survey is to follow a series of key activities (see Figure 4.2.9).

Questionnaires

The design of any questionnaire is critical in order to get the kind of answers that allow decisions to be made. Always test a questionnaire

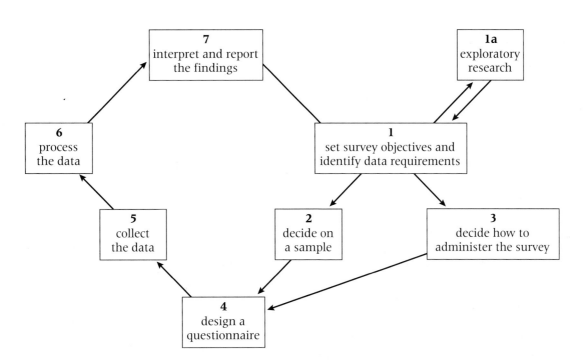

Figure 4.2.9 When conducting a survey it is essential to be clear about what needs to be found out

before using it. In general, shorter questionnaires are usually better than long ones. The wording of questions is important too. They should be:

- relevant – only include questions that target the information required
- clear – avoid long words, jargon and technical terms; questions must be easy to understand
- inoffensive – take care with questions about age, social class, salary, ethnicity, so as not to cause offence (use these types of question at the end)
- brief – short questions of less than 20 words
- precise – each question should tackle one topic at a time
- impartial – avoid 'leading' questions that influence the answer.

Questions can be of two types:

- Open-ended questions, such as 'What is your opinion of this new magazine?' These allow very wide and ambiguous answers that may not be useful.
- **Closed questions** provide a limited number of optional answers to choose from. They are easier to answer and are often used in surveys. Avoid offering two options such as yes/no because they are of limited use. Multiple-choice questions offer a range of different answers (see Figure 4.2.10).

Task
You are designing a folding stool for young people going camping. Draw up a questionnaire to find out their requirements.

What is your age? (tick one only)
Under 18 ☐ 18–64 ☐ 65 or over ☐

Which types of kettles do you prefer? (tick as many as apply)
Jug ☐ Stainless steel ☐ Patterned ☐

Where did you buy your kettle?
Supermarket ☐ DIY store ☐
Department store ☐ Mail order ☐
Don't know ☐
Other (please specify)

Figure 4.2.10 Multiple-choice questions from a questionnaire on kettles

Product analysis

> **SIGNPOST**
> Learning about manufacturing through product analysis
> Unit 1 page 27

As we have seen in Unit 1, product analysis is an important tool for the designer. It enables the analysis of a competitor's product and the development of specifications for new products.

Test marketing (test selling)
The purpose of **test marketing** is to find out if there are any potential problems with a new product before its full-scale marketing. It is done under real market conditions to find out customer and retailer reactions to the product – who are the buyers? how often do they buy? This enables forecasts to be made of future sales and profitability and enables production planning to take place.

Taking risks
The risk of not doing any research is high, because it can mean making the wrong decisions. Imagine that a company is thinking about adding a new product to its range, but is put off by the high cost of research. The company has the option of launching the product on the market and risking failure or funding research costs. A product failure could undermine the reputation of the company and its existing products. Should it decide to undertake the costly research or risk launching the product?

Market research case study: Madame Tussaud's
Madame Tussaud's waxworks is one of the UK's leading visitor attractions. In the 1980s, however, admissions were declining. To reverse this trend, and to help plan for long-term success, the company undertook an extensive programme of market research. To begin with, an extensive analysis of available data about levels of tourism, market segments and buying behaviour was undertaken using **secondary research** sources. While useful in indicating the potential within the market, this research revealed little about how potential customers perceived the 'product'.

The research team spoke to visitors before and after visiting the exhibition. They discovered that the enjoyment was not so much the quality of the wax figures, but the fantasy of being able to 'meet' famous people, or at least a life-like representation of them. As a result, the displays were reorganised to allow visitors to get close to the figures and make it easy for them to be photographed together. To encourage visitors to take photos, a deal was made with Kodak to sell film in the museum, and even to loan cameras. It was anticipated that this would lead to 'word of mouth' recommendation when visitors showed their holiday photos to friends and relatives on their return.

Having developed a new approach to their product, investment in a coordinated series of media communications was needed. The logo and marketing message were redesigned. The company's research had also revealed that the word 'waxworks' was a negative term, so the slogan 'Madame Tussaud's - where wax works' was adopted to suggest a more interactive experience. Press and TV advertisements carefully portrayed and projected messages about informality, atmosphere, humour and enjoyment. Specially commissioned photographs of people interacting with the wax figures were used for the leaflets distributed around London hotels, which form a key role in informing and attracting tourists.

Tussaud's market research also revealed the need to reconsider its entry prices. Instead of a single fixed entry price, the company introduced special deals for senior citizens, families and for early and late entry, when numbers are smaller.

By analysing the nature of the product being offered, the way it was being promoted and its pricing policy, Tussaud's successfully managed to halt the decline in admissions. Investing in market research, product development and promotion produced a dramatic increase in customers which the company has been able to sustain and continue to build on.

Since then, there have been many further developments, with the introduction of theme-park style rides, 'animatronic' figures, and new promotional strategies aimed at making Tussaud's a 'must see' attraction for visitors across the world.

The market research process

There are three key stages in the market research process. These are planning, implementation and interpretation.

Planning

Identify a clear reason and purpose for market research. This reason usually relates to a problem, an issue or opportunity for design. Once the reason is found (often in the form of a design brief), it is used as a starting point for research – what data need to be found out and how to collect them.

Implementation

There are many ways of collecting data, depending on the research plan and what needs to be found out. Surveys and questionnaires can be useful, but can sometimes be expensive. Always think about alternatives.

Interpretation

The information that is created when data are collected should be interpreted in relation to the design problem. All findings should be used to influence decision making.

Task

Read the case study and market research process above again, and then answer the following questions:

a) Identify a tourist attraction in your locality, or one you have recently visited, for example a small museum or heritage site.

b) Draw up an outline market research strategy, including information obtained from primary and secondary sources, to identify how it might increase its number of visitors.

c) Plan a market research questionnaire to learn more about the target market, including questions that will produce quantitative and qualitative data.

d) Develop outline ideas for a new logo and slogan, and plan an appropriate promotional campaign using a range of media.

The basic principles of marketing and associated concepts

SIGNPOST
Design development
Unit 1 page 16
Design and marketing
Unit 3B1 page 124

Marketing involves anticipating and satisfying consumer needs while ensuring a company remains profitable. The main objectives of marketing include:

- generating profit
- developing sales
- increasing market share
- diversifying into new markets.

In order to do this companies must create opportunities for meeting customer needs, for managing change within design and production and for promoting a **corporate image**. The main method of achieving this is through a marketing plan.

Marketing plan

A good sales and **marketing plan** involves developing a competitive edge through providing reliable, high-quality products at a price customers can afford, combined with the image they want the product to give them. This is sometimes called **lifestyle marketing**.

The basic structure of a product marketing plan includes:

- background and situation analysis, including **SWOT** (product strengths, weaknesses, opportunities and threats from competition) and **PEST** (political, economic, social and technological issues)
- information on markets, customers and competitors
- a plan for action and advertising strategies
- planning marketing costs
- time planning – the best time to market the product to an achievable timetable
- a plan for monitoring the marketing.

Target market groups

Companies supply their goods to customers. A market consists of all the customers of all the companies and organisations supplying a specific product, for example the car market or the domestic lighting market. As a result of undertaking marketing research, some companies decide that they cannot possibly supply all the potential customers in their markets – maybe the market is very large, geographically scattered, the competition is strong or customer needs too varied. In this case companies have to decide which types of customers to aim for and then to target their products at them – at a selected part of the market (known as the market segment). The process of identifying market segments and developing products for it is called **target marketing**.

Consumer demand and market pull

Customers in a market demand or 'pull' products and services to satisfy their needs. The job of the marketing department is to maximise the demand for its own company's products. Existing customers must be satisfied with these products and not be tempted to buy from other suppliers. At the same time the customers of other suppliers need to be persuaded to change their product **brand loyalty**. Further to this, potential new customers need to be attracted, in order to expand the company's market share.

> **Task**
>
> Identify three products which you think have been developed as a result of market demand. Identify the characteristics of these products.

Lifestyle marketing

Lifestyle is used as a basis for target marketing. Different lifestyles have been identified by investigating the geographic and demographic characteristics of the population. People with similar demographic characteristics often live in similar types of houses and have similar lifestyles. For example, young professional people living in towns and cities and who are either working or studying to improve their lifestyle form one group. They tend to be young, highly educated, live in high-status areas and have a high level of mobility. This group may then be broken down into further types, such as well-off town and city areas, singles and young working couples, furnished flats and bedsits, younger single people.

Lifestyle marketing is the targeting of these potential market groups and matching their needs with products. Market research identifies the buying behaviour, taste and lifestyle of these potential customers. This establishes the amount of money they have to spend, their age group and which products they like to buy. New products can then be developed to match their needs.

Brand loyalty

Branding is a key marketing tool for many manufacturers. A 'brand' is a marketing identity for a generic product that sets it apart from its competitors. The **brand name** protects and promotes the identity of the product, so that it cannot be copied by competitors. Typical brand names include 'Hoover' or 'Dyson'. A branded product usually has additional features or added value over and above other generic products – something that makes the product 'special' in the eyes of consumers.

Advantages of branding for the customer

For the customer, buying branded products provides an expected and reliable level of quality. Brand names also give consumers benchmarks when making their own purchasing decisions. For example, some people might use the AppleMac personal computer as a reference brand against which other computers

Figure 4.2.11 The distinctive Coca-Cola branding

Where do you want to go today?®

Figure 4.2.12

IBM

Figure 4.2.13

are compared. The status and image of a strong brand and brand loyalty can therefore be powerful influences over which products consumers buy. For consumers, the benefit of knowing which brands provide certain reliable or good quality products can also save time when deciding which product to buy.

The power of the brand
The most successful example of branding is probably Coca-Cola. It is also one of the oldest 'brands', being established in the late nineteenth century. Its distinctive colours and **typography**, and the unusual shape of the bottle it is supplied in are recognised the world over (see Figure 4.2.11). While the use of these elements is strictly applied and controlled, there are variations that associate the product more closely with the country it is being sold in. Most consumers will choose Coca-Cola instead of another brand, even if the other is cheaper, because Coca-Cola is the 'real thing'.

Other highly valuable global brand names include Microsoft and IBM. For Microsoft, its trademark question 'Where do you want to go today?' (see Figure 4.2.12) delivers the message that the user is in control and has the freedom to act as he or she chooses, also helping negate the fears of computers taking over the world and, indeed, of Microsoft's own commercial control over the operating and networking systems. IBM, which originally made calculating machines and typewriters before entering the computer business in 1953 (see Figure 4.2.13), is the world's largest information technology company with its 2001 revenue totalling US$85.9 billion. IBM's success is due to the fact that they have wrapped a set of products (hardware and software) and services into solutions. That is the drive of the IBM e-business message.

Branding is also widely and successfully used in the clothing industry. People don't buy jeans, they buy the expectation of a lifestyle offered by the label. In terms of advertising, Benetton is widely known. In aiming to associate itself with a global market, it adopted the phrase 'The united colours of Benetton', and used a series of controversial, disturbing shock-tactic images relating to issues of the day on its posters (see Figure 4.2.14). It is well known that there is 'no such thing as bad publicity', and the company has benefited from extensive press coverage of its campaign. The strategy eventually back-fired in the late 1990s when a series of posters featuring prisoners on death row in the USA caused Sears, the major department store in the USA, to refuse to sell Benetton products.

In contrast, Diesel has developed a brand image which mocks the classic advertising conventions, aiming to present itself as an 'honest' company that does not make outrageous claims about the abilities of its products to change the way we look, the things we do or the way the world works. The brand values of Diesel communicate being international, innovative and fun (see Figure 4.2.15).

Figure 4.2.14 Benetton often uses disturbing issues relating to issues of the day

Figure 4.2.15 Diesel presents itself as fun

Task

Working in pairs, study two different brands of either food, electronic or fashion products. Try to identify what each brand represents in the potential consumer's mind. Apart from the value of the product itself, what else is the advertising, promotion and label offering the consumer?

Competitive edge and product proliferation

If there are several companies producing a wide range of similar products for the same customers with similar needs, the only criterion upon which customers can base their buying decision is price. In theory the most expensive products would not sell, but this is rarely the case – since most customers are not aware of every product and price on the market.

To ensure that their products have a successful market share, many manufacturers try to make their products different from a competitor's. This involves creating unique features for the product in order to give it a **competitive edge**. For example, some manufacturers might offer 'special' features or different levels of quality. Different price levels can then be set – these are often based on the value that customers might put on these features or qualities.

Tasks

1 Identify a product which you or your family use, which you think has a competitive edge over other similar products. Give reasons for your choice.
2 Explain the benefits that this product brings to the customer, in relation to its price.

Price range and pricing strategy

Price is one of the most important aspects of marketing because a product's price affects profit, the volume of products manufactured, the share of the market and the image of the product. We've probably all heard the term 'cheap and nasty'!

The key to successful pricing is the attractiveness of the price to the customers. The concept of price, value and quality are difficult to separate. How much we are prepared to pay for a product depends on how much we value it. The justification for a higher price may depend on the following:

- a product's extra features, characteristics or innovative design
- the perceived quality of the product
- the increased reputation of the product brought about by advertising and promotion
- the possibility of paying by credit, which justifies a higher final price
- the guarantee of a specific delivery date and easy access to the product through mail order or the Internet.

Tasks

1 Market share and promotional gifts
   The aim of marketing departments and advertising companies is to increase the market share of brands and products. Research has shown that in advertising success breeds success. Increasing the market share of an already successful product or brand is easier than for a less successful or less well-known brand. Try to explain the reasons for this.
2 Direct marketing often includes mailshots with free 'gifts' or samples, often with a coupon that provides money off for the next purchase. Explain why you agree or disagree with this kind of marketing.

Distribution

The distribution of products covers a variety of operations. It makes the product available to the maximum numbers of target customers at the lowest cost. The way that technology and communications have changed in the past few years has had an enormous impact on the distribution of products and the way we shop.

Task

List three ways that consumers shop these days in comparison with shopping in the 1970s. Explain the main factors that have brought about these changes in shopping.

4. Conservation and resources

Environmental implications of the industrial age

One of the greatest problems relating to the industrial age is the consumption of non-renewable, finite resources which will eventually be exhausted. Recognition of this problem has led to an understanding of the need for conservation and the better management of resources. It also needs to lead to a better understanding of the meaning of 'design' and the purchase-attraction culture.

The future – designing where less is more

There are a number of questions relating to product design that need to be answered. For example:

- What will be the aims of product design in the future?
- How can we manage the changes necessary to our technological-industrial society to make life on Earth bearable in the future?

One of the greatest problems for product designers of the future will be how to design with the environment in mind. Previous sections of this unit discussed mass production and the marketing of products. We saw that the aim for most manufacturers is to sell as many products as they can in order to make a profit. Designers now and in the future will need to decide if this is an ethical way forward.

The key question is how design can influence the environmental safety of products and how it can contribute to a reduction in the number of products made.

At present we have what is called a 'purchase-attraction' culture, which results in the proliferation of products that we see in the market place today. In the future this may have to make way for a culture that supports long-term use and the conservation of resources. Changing the purchase-attraction culture could be achieved by:

- changing our attitude to products from purchase-attraction to their long-term use and usefulness
- developing products that would not be bought, but that remain the property of the manufacturer
- paying for the use of the product and its maintenance
- returning products to the manufacturer to be serviced, repaired, recycled and reused.

It is the task of all of us, including designers and technologists, to find starting points for changes to our purchase-attraction culture in order to ensure the future of the planet. Good design has an ethical and moral value – something that you have been asked to take account of, for example, when drawing up a **design specification**. Industrial production is becoming an increasing problem, with the production of more and more products. This is placing an enormous burden on the environment.

Tasks
1 Explain what is meant by the term a 'purchase-attraction' culture.
2 Put forward arguments for and against mass production and the marketing of products.

The use of non-renewable raw materials and fossil fuels during the manufacturing process

SIGNPOST
Forms of energy used by industry
Unit 1 page 29

Many modern products are made from non-renewable resources such as metals and plastics. The electrical energy used in their manufacture comes from coal, gas, oil or nuclear power. The management of these finite resources will increasingly become the responsibility of us all. Existing British, European and international legislation already places demands upon companies to design and manufacture products with the environment in mind. In this respect, product designers need to consider:

Figure 4.2.16 Glass is one of the most cost-effective materials to recycle

- reducing the amount of materials used in a product
- using efficient manufacturing processes that save energy and prevent waste
- reusing waste materials within the same manufacturing process
- recycling waste in a different manufacturing process (see Figure 4.2.16)
- designing for easy product maintenance, so that parts can be replaced, without the need to dispose of the whole product at the end of its useful life
- designing the product so that the whole or parts of it can be reused or recycled.

Sustainable resource management
Conservation and resource management
Conservation is concerned with the protection of the natural and the manufactured world for future use. In urban areas, for example, buildings may be protected because of their historic interest. In rural areas, plant species, animal habitats and landscapes may all be protected.

Conservation is also concerned with the sensible management of resources and a reduction in the rate of consumption of non-renewable resources, such as coal, oil, natural gas, ores and minerals. The aim of conservation is to achieve **sustainable development**. The 1987 Brundtland Report, *Our Common Future* (World Commission on Environment and Development), defined sustainable development as:

'development that meets the needs of the present, without compromising the ability of future generations to meet their own needs'.

Efficient management of resources includes:

- using less wasteful mining and quarrying methods

- making more efficient use of energy in manufacturing
- reducing fuel consumption in motor vehicles
- using cavity and roof insulation in buildings
- using low-energy light bulbs.

Renewable sources of energy, energy conservation and the use of efficient manufacturing processes

Renewable resources are those that flow naturally in nature or that are living things which can be regrown and used again (see Table 4.2.3). These include the wind, tides, waves, water power, solar energy, geothermal energy, biomass, ocean thermal energy and forests (see Figure 4.2.17).

Forests are renewable as long as they are not used faster than they can be replaced. In recent years, the indiscriminate destruction of the world's rainforests has led to a severe shortage of tropical hardwoods, such as Jelutong from South-east Asia. In order to conserve valuable renewable resources such as these, manufacturers are encouraged to use only those woods grown on plantations or in managed forests.

Most of what we consume is packaged in some way. The majority of bottles, cans, plastic containers and cardboard boxes are simply thrown away, wasting valuable resources. What is being done to minimise this waste?

Table 4.2.3 Renewable energy sources

Renewable energy source	Process	Advantages	Disadvantages
Wind	Power of wind turns turbines	Developed commercially Produces low-cost power	High set-up cost Contributes small proportion of total energy needs Wind farms sometimes seen as unsightly
Tides	Reversible turbine blades harness the tides in both directions	Occurs throughout the day on a regular basis Reliable and non-polluting Potential for large-scale energy production	Very high set-up cost Could restrict the passage of ships Could cause flooding of estuary borders, which might damage wildlife
Water	Running water turns turbines and generates hydro-electric power	Clean and 80–90% efficient	High set-up cost Suitable sites are generally remote from markets Contributes small proportion of total energy needs of an industrial society
Solar	Hot water and electricity generated via solar cells	Huge amounts of energy available. Could generate 50% of hot water for a typical house Relatively inexpensive to set up	High cost of solar cells Biggest demand in winter when heat from Sun is at its lowest
Geothermal	Deep holes in Earth's crust produce steam to generate electricity	Provides domestic power and hot water	Only really cost-effective where Earth's crust is thin, e.g. New Zealand, Iceland
Biomass	Burning of wood, plant matter and waste generates heat	Produces low-cost power	Environmental pollution Potential for deforestation

Figure 4.2.17 *Several large wind farms have been built in Europe and the USA, principally at windy coastal sites*

The majority of paper used in the UK comes from 'managed forests' in Scandinavia. This means that for each tree that is cut down, another is planted – indeed, there are now more trees in western Europe than there were 20 years ago. At the same time, new 'young' trees are most efficient in absorbing carbon dioxide and returning oxygen to the atmosphere, helping combat the 'greenhouse' effect.

Tasks
1 Explain your views on the use of renewable energy when compared to using coal, oil or gas.
2 Describe the benefits and disadvantages of renewable energy in relation to set-up costs, accessibility, production processes and environmental impact.

The use of efficient manufacturing processes
Even the most industrially advanced economies still depend on a continual supply of basic manufactured goods. This production will continue to require large amounts of raw materials and energy. Product manufacturers can contribute to sustainable development and to reducing costs by using more efficient manufacturing processes. This can often be achieved through redesigning an existing product.

In the early 1990s, BT used to print 24 million telephone directories each year, using some 80 000 trees. The company set out to try to reduce the size of the directories by redesigning the typeface and layout used.

First, a new typeface was chosen which allowed the spaces between letters and words to be slightly reduced. The exchange number code was used, replacing the longer name. The repetitive use of the subscriber surname was replaced by just using the initials.

These changes significantly increased the number of entries that could be contained on one line while at the same time making the information easier to read. Increasing the number of columns from three to four produced further savings of space. These changes brought about a 10 per cent reduction in the amount of paper used, with further cost savings in terms of the time taken to do the print run and reduced transportation costs. Since then, further savings are being made by encouraging users to access online directory services.

Tasks
1 Identify a product that you think is badly designed. Give reasons for your choice.
2 Using annotated diagrams, explain how the product could be redesigned to improve the efficiency of its manufacture.

New technology and environmentally friendly manufacturing processes

SIGNPOST
New materials, processes and technology
Unit 3B1 page 122

Redesigning a product is one way of achieving efficiency in manufacturing. For many companies this also improves their environmental performance and helps them to increase profits.

Since 1994 one organisation that has been helping UK companies to do just this, is the Environmental Technology Best Practice Programme (ETBPP) recently renamed Envirowise (see Figure 4.2.18). This is a joint initiative of the Department for Trade and Industry (DTI) and the Department of the Environment (DoE). The ETBPP aims to help manufacturing companies improve their environmental performance and increase their competitiveness.

Figure 4.2.18 *The logo of the Environmental Technology Best Practice Programme, recently renamed Envirowise*

The main themes of the ETBPP programme are **waste minimisation** and cost-effective **cleaner technology**:

- Waste minimisation often generates significant cost savings, through the use of simple no-cost or low-cost measures.
- Cleaner technology is the use of equipment or techniques that produce less waste or emissions than conventional methods. It reduces the consumption of raw materials, water and energy and lowers costs for waste treatment and disposal.

Two important aspects of cleaner technology are **cleaner design** and **life cycle assessment** (LCA):

- Cleaner design is aimed at reducing the overall environmental impact of a product from 'cradle to grave'.
- Life cycle assessment evaluates the materials, energy and waste used in or produced by a product through design, manufacture, distribution, use and end-of-life, which could be disposal, reuse or recycling.

These two aspects of cleaner technology enable areas to be identified in design and manufacture where companies can make changes that result in environmental benefits and cost savings.

The importance of using sustainable technology

In the above sections you have been introduced to different aspects of manufacturing that relate to the environment: minimising waste and using cleaner technology. These are both aspects of what is known as 'sustainable development'. This concept puts forward the idea that the environment should be seen as an asset, a stock of available wealth. If each generation spends this wealth without investing in the future, then the world will one day run out of resources. The concept of sustainable development includes a number of key concepts:

- that priority should be given to the essential needs of the world's poor
- meeting essential needs for jobs, energy, water and sanitation
- ensuring a sustainable level of population
- conserving and enhancing the resource base
- bringing together environment and economics in decision making

Sustainable development is a problem for the whole world and many countries are involved in trying to develop policies which support it. In the UK, for example, many government programmes are involved, such as the ETBPP. Bio-Wise is another government initiative that supports and advises companies and organisations on developing sustainable practices that make use of biotechnology.

For more information on both initiatives, visit the ETBPP website on www.etsu.com/etbpp/ and the Bio-Wise website on www.dti.gov.uk/biowise.

Information about biotechnology can be found in the next section.

Task

Life cycle assessment

Identify two different items of packaging. Draw up a list of questions that you could use to assess the overall environmental impact of these products from cradle to grave. Some of the questions are done for you:

- Are the raw materials renewable?
- What types of processes are required to manufacture the products?
- Do the manufacturing processes cause risk to people or the environment?

Once you have added to the above list of questions, use them to evaluate the environmental impact of your chosen products. You could evaluate your coursework project in a similar way.

Tasks

1 Discuss the importance of using sustainable practices when manufacturing products. Include references to raw materials and manufacturing processes.
2 Explain how you could adapt sustainable technology to your own manufacturing.
3 Research further examples of good practice in manufacturing, using the Internet, CD-ROMs or libraries.

Management of waste, the disposal of products and pollution control

SIGNPOST
Life cycle assessment and environmental issues
Unit 1 page 29

As we saw from the Environmental Technology Best Practice Programme, waste minimisation often generates important cost savings, through the use of simple changes to manufacturing processes. There are three key approaches to reducing waste:

* reduce the amount of materials used in manufacture
* reuse materials in the same manufacturing process where possible
* recycle materials in a different manufacturing process if possible.

Reducing materials use

The possibility of reducing the amount of materials used to manufacture products is often found in processes that involve cutting and stamping shapes from sheet materials. For example, in can manufacture careful calculations must be made to limit the amount of aluminium used for making the circular tops of the cans. There are two ways of arranging the can top on a rectangular sheet, as shown in Figure 4.2.19. In (a) the tops are in a square formation, each sitting in its own square of aluminium. In (b) the tops are in a triangular formation, which is the closest that they can be packed together. When the scrap for each is worked out (a) is found to produce a staggering 21.4 per cent scrap, while (b) works out at only 9.3 per cent scrap. Clearly the placement of the can tops on the aluminium sheet will have an enormous impact on the amount of aluminium required to manufacture the cans.

The disposal of products and pollution control

The disposal of products when they have reached the end of their useful life is a major problem. Around 90 per cent of household rubbish in the UK is buried in landfill sites, 5 per cent is incinerated and only 5 per cent is recycled. One of the best options for products is, firstly, to create the need for fewer of them to be made and, secondly, to design for recycling.

For the disposal of industrial waste a simple solution is 'skip and tip' into either landfill sites or sewers. Disposal by landfill is currently inexpensive and popular, but legislation is

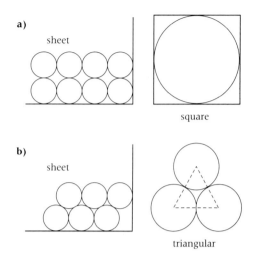

Figure 4.2.19 *Reducing the amount of aluminium required to make aluminium can tops depends on how they are arranged on the aluminium sheet*

expected to enforce change. Pollution control is the responsibility of a variety of agencies which enforce the 1990 Environmental Protection Act (EPA). This Act introduced wide-ranging legislation with tight controls on the discharge of waste into air, water and land in the UK. It also reinforced the policy of '**polluter pays**'. The aim of this policy is to restrict potentially harmful materials from entering the environment and to place greater responsibility on those generating, handling and treating wastes. Legislation is strictly enforced so that any company or organisation that causes pollution can be fined huge sums.

Task

Discuss the impact of the policy 'polluter pays' on product manufacture. Relate your discussion to one product type, such as PET bottles. Explain how product pricing might be affected.

Impact of biotechnology on manufacture

Biotechnology is the use of biological processes which make use of living proteins, called **enzymes**, to create industrial products and processes. These enzymes are the same kind that help us digest food, compost garden rubbish and clean clothes. The use of yeast in making bread and wine was one of the first examples of the use of biotechnology. Technologists also developed enzymes that could be added to detergents to improve their cleaning properties, resulting in 'biological' washing powders.

Biotechnology is being used in an increasing number of industries to provide efficient and ecologically sound solutions to environmental problems. Current applications of environmental biotechnology include: the treatment of industrial air emissions; aerobic and anaerobic treatment and reed beds to treat industrial effluent; composting to treat domestic, industrial and agricultural organic waste; and 'bioremediation' techniques to clean up contaminated land. Not only are many companies using biotechnology to increase their competitiveness, but they are using it to meet strict new environmental legislation and to improve their competitive edge.

Compared with more traditional methods, biotechnology can often produce better, faster, cleaner, cheaper and more efficient ways of doing things. One example, in the engineering industry, is the treatment of waste cutting fluids.

Treating waste cutting fluids

In the engineering industry, many companies use cutting fluids to cool and lubricate components while they are machined. After being used for a time, the fluids become contaminated with bacteria. This makes them go 'off' and become useless. Every year around half a million tonnes of used cutting fluids and oily wastes are disposed of – mainly to landfill sites, sewers and incinerators.

Biotechnology is being used to solve this problem (see Figure 4.2.20). The cutting fluids are passed through an anaerobic treatment tank to digest the oil. Any remaining fluids are passed through an ultrafiltration system to extract the remaining contaminants. The result is purified water, which is reused. This biotechnology process gives the following benefits:

- All types of cutting fluids can be treated.
- Water and effluent disposal costs are reduced.
- Site health and safety is improved.
- Costs related to handling and disposal of contaminated wastes are removed.

Biochips

In the electronics industry, developments in integrated circuits (ICs) have almost reached the limit of that technology. The number of circuits which can be placed on a chip is limited by:

- the width of the circuit tracks
- the occurrence of short circuits
- the heat produced by large numbers of components.

Biochips have been developed in which the silicon used in ICs is replaced by inserting semiconducting molecules into a protein framework. The proteins can assemble themselves into complex and predetermined 3D structures, so they can grow to take up the shape required for an electronic circuit. The manufacture of biochips will lead to the further **miniaturisation** of electronic products. This may one day allow a personal computer to be as small as a watch!

The advantages and disadvantages of recycling materials

The advantages for the environment of recycling are numerous. Recycling conserves non-renewable resources, reduces energy consumption and greenhouse gas emissions, reduces pollution and the dependency on raw materials.

For manufacturers the advantages of recycling are cost related, with metal, glass and paper being the most cost-effective materials to recycle:

- Scrap non-ferrous metals are sorted into different grades for recycling, due to their relatively high commercial value.
- Steel and cast iron are graded by size, due to their lower commercial value.
- Items made of glass, rubber and plastics are more difficult to sort and have a much lower value than metals.
- Plastics, paper and glass recycling are growing industries.

Scrap materials from the manufacturing process are the most valuable because their material content is known and they are easily available. The disadvantage of recycling scrap

Figure 4.2.20 Biotechnology can be used in the treatment of cutting fluids used to cool and lubricate components while they are machined

from old products is that the chemical or physical make-up of their materials is often very complex, so for some products recycling is too expensive or impossible. The result is that 'old' scrap from products has a lower commercial value.

Design for recycling

Under proposed European regulations, manufacturers may have to recycle electrical and electronic parts rather than disposing of them in landfill sites or incinerating them. By 2006 customers may be able to return electrical goods to stores for recycling. This will have a major impact on the way that products are designed.

With this in mind UK engineers have developed a mobile phone that falls apart when heated, so that the component parts can be recycled. The phone is made from **shape memory polymers**, plastics that revert to their original form when heated. Different components of the phone will change shape at different temperatures, allowing them to fall off at different times, when they pass along a conveyor belt. The reusable parts can then be recycled. The phones will only fall apart under extreme temperatures – so they will be safe if left in the car in the Sun!

Recycling packaging

Nowadays most packages are designed specifically with recycling in mind. First paper, plastic, metal and glass materials need to be separated as they are treated differently. They then need to be further sorted into the different types of paper, plastic, metal and glass (see Figure 4.2.21).

Bales of waste paper are shredded and dropped into a large tank of hot water, called a hydrapulper. This chops and stirs the paper until it becomes pulp. The water is drained out and

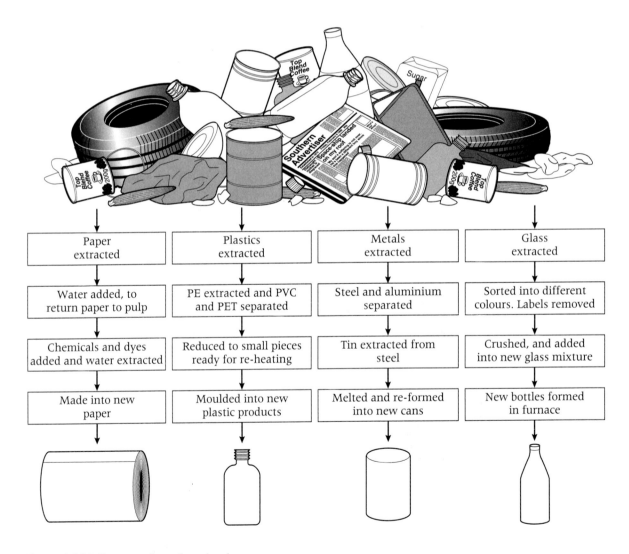

Figure 4.2.21 The recycling of packaging

various chemicals and dyes are added to give the paper its final colour. The pulp is then poured on to a mesh which acts like a sieve, through which any remaining water can be squeezed out. The paper is dried using heavy rollers and steam-heated cylinders before being wound up on to big paper reels.

Most plastics used in packaging are made out of one of three main types:

- PE (polyethylene) – used to make things like bin liners, road signs and flower pots
- PVC (polyvinyl chloride) – commonly used for pipes
- PET (polyethylene terepthalate) – used as stuffing in cushions, sofas, etc.

To aid the recycling process these plastics are increasingly identified by the numbers 1, 2 or 3 on the base of many plastic products. There are various methods of separating unnumbered plastic containers. One method is first to drop them into a large tank of water: PE floats on the surface and can be skimmed off, while PVC and PET sink. PVC and PET can then be separated using sensors that can detect the different chemical make-ups of the plastics. The separated plastics are chopped into cornflake-size pieces and are thoroughly washed. After drying they can be moulded into new products.

Aluminium and steel cans are first crushed together into blocks. They are then fed through a shredding machine. A magnet is used to extract the steel, from which any tin needs to be extracted. The remaining steel can be melted back into an ingot, ready to be re-rolled to make new cans. Meanwhile the aluminium is melted down and thinly stretched so that new cans can be made from it. Recycling a tonne of steel saves 3.6 barrels of oil.

Glass needs to be sorted into different colours, and this is initially done at the bottle bank by providing different containers. Any extraneous, non-glass materials, such as bottle tops and labels need to be removed. The bottles are then crushed into tiny fragments and mixed with sand, limestone and soda ash, ready to be made into new glass. Most bottles include about 20 to 30 per cent of recycled glass, although some recycling plants can produce recycled bottles that include up to 75 per cent of recycled glass.

Exam preparation

In order to revise successfully for this unit, you need to make sure that you have a good grounding in all of the topics included in the unit, so that you can apply your knowledge to the context of an exam question.

Many questions will ask you to make specific reference to one or more products and give answers related to that product.

It is useful to cut out and keep news items in newspapers and magazines that relate to design and technology, since these will keep you up to date with product information. The Internet is also a good way to research products. You can gain more information if you get together with other students in your group and share your information.

Make brief notes about each topic in the unit. Try to summarise key points on one or two sheets of paper. Learn these key points.

Practice exam questions

1 There are a number of economic factors which affect product manufacture.
   a) Identify two such factors and explain how they affect the manufacture of mass produced products. (6)
   b) When costing a product, it is important to get it 'right'. For a mass-produced product of your choice, discuss the importance of accurate costing. (9)

2 Discuss the main objectives of marketing and the importance of developing a competitive edge. Relate your answer to product examples. (15)

3 a) Explain the role of British Standards in ensuring that manufacturers make products that fulfil the needs of their customers and the environment. (9)
   b) Discuss the importance of incorporating a quality management system into the design and manufacture of products. Make reference to any relevant International Standards. (6)

4 Describe and explain the implications for the environment of using non-renewable resources and the benefits of recycling. (15)

4 B2 CAD/CAM (G403)

Summary of expectations

1. What to expect

The content of this option builds on your AS study and enables you to develop further understanding of CAD/CAM.

2. How will it be assessed?

The work that you do in this option will be externally assessed through Unit 4, Paper 3. There will be two compulsory long-answer questions, each worth 15 marks. You are advised to spend 45 minutes on Paper 3.

3. What will be assessed?

The following list identifies what you will learn in this option and subsequently what will be examined:

- Computer-Aided Design, Manufacture and Testing (CADMAT), including:
  - Computer-Integrated Manufacture (CIM)
  - Flexible Manufacturing Systems (FMS).
- Robotics, including:
  - the industrial application of robotics/control technology and the development of automated processes
  - complex automated systems using artificial intelligence (AI) and new technology
  - the use of block flow diagrams and flow process diagrams for representing simple and complex production systems

  - the advantages and disadvantages of automation.
- Uses of ICT in the manufacture of products, including:
  - the impact and advantages/disadvantages of ICT within the total manufacturing process.

How to be successful in this unit

You are expected to demonstrate a thorough understanding of what you have learned in the option. Examiners are looking for longer, more detailed answers that show a greater depth of knowledge and understanding at the Advanced GCE level.

You will be assessed on your ability to organise and present information, ideas, descriptions and arguments clearly and logically, taking into account the quality of written communication.

There may be an opportunity to demonstrate your knowledge and understanding of the content of this option in your coursework. However, simply because you are studying this option, you do not have to integrate this type of technology into your coursework project.

5. How much is it worth?

This unit, with the option, is worth 15 per cent of the full Advanced GCE.

Unit 4 + option	Weighting
A2 level (full GCE)	15%

1. Computer-aided design, manufacture and testing (CADMAT)

CADMAT

Systems that fully integrate the use of computers at every level and stage in the manufacturing process can be described as CADMAT systems. Unit 3 looked at the benefits of CAD and the advantages that CAM brings and you were introduced the wider role that computers now play in all sectors of graphic design and manufacturing. In addition to their application in CAD/CAM, computers are also used extensively for decision making within CADMAT in a variety of ways because of the operational flexibility they provide. The graphic design and product manufacturing industries use computers for activities like:

- information control through the gathering, storage, retrieval, and organisation of data, information and knowledge
- simulations in which computer 'models' are used to help to provide answers to 'What if …?' questions; for example, 'What would be the consequences if we changed to a flexible manufacturing system compared to our present sequential system of manufacturing?'
- narrowing the field of design choices available using number-based and other analytical methods; for example, graphs can be used for analytical and mathematical purposes rather than simply as visual aids to explain data

- communicating and discussing product designs with clients
- managing the full range of manufacturing data including the tracking of components and stock
- controlling equipment and production processes to include the routine scheduling of maintenance or minimising the effects of machine **downtime** when production tools have to be replaced
- monitoring quality and safety to the appropriate national standards.

As with all systems, the efficiency of a CADMAT system is determined by the effectiveness of many interrelated sub-systems. A failure in any of the input, processing or output sub-systems leads to production lines that malfunction or operate below capacity. Work in progress can be delayed in many ways, especially when the required materials and components are unavailable, which means that suppliers need to be part of the production planning process. The complex relationships in the design and manufacturing process can be described in a systems diagram (see Figure 4.3.1). Maximising profit and reducing costs is important for all manufacturers but the global economy that has been made possible by the use

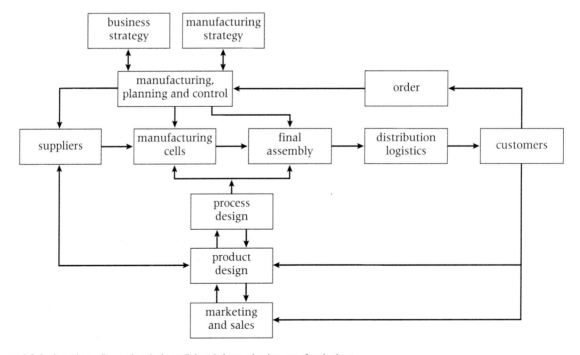

Figure 4.3.1 A system flow chart describing integrated manufacturing

of computers has created many new pressures. Increasingly, manufacturers have to compete with low-cost imports, satisfy customers who require shorter production runs and improved delivery times. In order for companies to compete for business more effectively they need to manage data in 'real time' by using customised computer programmes to manage orders and control production times.

Product data management (PDM)

Product data management (PDM) is a complete (holistic) data management system that aims to integrate all aspects of manufacturing from product modelling to the management of the business. All the design and production processing data is generated once and stored electronically on a secure database that can be accessed at different levels according to a specific need. PDM enables instant and rapid communications between departments, manufacturers and retailers across the world so that a product that has been designed in one country can be manufactured to the design specification in another.

Online order tracking is another example of the potential power of PDM because of the advances in information and communications technology (ICT) and the growth of the Internet as a way of selling goods and services across the world. **Telematics** is an ICT-based technology that allows a product to be managed from the receipt of a customer's order. The product will be tracked from development into manufacturing, on through delivery and finally to after-sales support via real-time feedback from telematic reporting systems. An everyday example of real-time feedback is the customer service section on an Internet website that sells CDs and DVDs. On this type of site it is possible for the customer, who has an unique access code, to track the progress of his or her order, when it was or will be dispatched and what the final costs of the transaction will be. If the customer requires further information, then an email facility is provided that allows an instant and low-cost dialogue with the company, especially if it is based in another country.

The advantages of using PDM include:

• increased productivity because of fewer bottlenecks
• improved quality control
• more accurate product costing
• more effective production control
• a reduction of work in progress and a smaller inventory of components and materials
• increased flexibility
• higher profits.

Production data from a manufacturing plant or a volume production line such as a printing press can be simultaneously merged and analysed along with other data, including product demand and financial figures. This means that raw materials and components for the production line are bought in as required. As a result, financial efficiency improves as the size of the stock inventory and its storage space reduces. Reduced business overheads equal increased profitability. This type of approach is known as 'just-in-time'.

Just-in-time (JIT)

Just-in-time (JIT) is a management philosophy which is applied in businesses around the globe. It was first developed in Japan in the 1960s by Taiichi Ohno within Toyota's car manufacturing plants as a 'quick response' method of meeting consumer orders with the minimum of delay, to the required level of quality and in the right quantity. It has also come to mean producing with minimum waste. Waste here is taken in a general sense to include such things as time and resources as well as raw materials (see JIT manufacturing below). JIT is achieved by a strategy of computer-aided inventory management in which raw materials, components or **sub-assemblies** are delivered from the 'supplier' just before they are needed in each stage of the manufacturing process. The advantages of JIT include reduced stocks of raw materials leading to a reduction in the area, volume and cost of storage space required. It also reduces the levels of finished goods kept in stock, waiting to be sold.

JIT 'suppliers' and 'customers'

A 'supplier' can either be the organisation that provides the original raw materials, components or stock needed for the product or it can be the previous stage of the manufacturing process (downstream). A 'customer' can either be the purchaser of the final product or it can be another stage of the manufacturing process further along the production line (upstream).

JIT manufacturing

This is a systems approach to developing and operating all the processes in a manufacturing system. JIT has been found to be so effective that it increases productivity, work performance and product quality, while still saving costs. Information flow through the system is vital to its success, and advances in computer-based applications, such as PDM, have further enhanced its effectiveness. Before discussing the specific elements of JIT, it is useful to identify the underlying principles:

- JIT is a continuous operation, not a one-off event, and workers are responsible for the quality of their individual output.
- There has to be synchronisation (matching) or balance between operations so that all production occurs at a common rate that reduces 'bottlenecks'.
- Simple is considered to be better, so that a continuous effort is made to improve (**Kaizen**) by operating with fewer resources and less waste in terms of time, personnel, materials and equipment.
- Work is carried out in less-complicated ways through employing foolproof (**poka-yoke**) tools, jigs and fixtures to prevent mistakes.
- It is necessary to concentrate on fundamentals and remove anything that does not add value to the product.
- Factory layouts should be product-led so that less time is spent moving materials, components and parts reducing 'waiting time' to a minimum.
- It looks for opportunities to use machines that are self-regulating (**Jidoka**-autonomation) and that make production decisions based on the data they are processing rather than relying on human intervention.

The key features of JIT workplace organisation

The control systems for JIT are the most visible manifestation of the JIT approach but there are other important features:

- Operational set-up times are reduced, increasing flexibility and the capacity to produce smaller batches more cost effectively.
- There is a multi-skilled workforce that is capable of operating multiple processes leading to greater productivity, flexibility and increased job satisfaction.
- Production rates are either levelled or varied, as appropriate, to smooth the flow of products through the factory.
- **Kanban** is a control technique that is used to 'pull' products and components through the manufacturing process. The Kanban controls and schedules production rates including the flow of materials by using computer applications.

Task

Research what role computers played in manufacturing operations in the 1960s compared to now. Write a short report explaining what it was like then and how advances in ICT since then have supported the further development of modern manufacturing systems.

Computer-integrated manufacture

In Unit 3 we saw that to achieve full **computer-integrated manufacture (CIM)**, all aspects of a company's operations must be integrated so that they can share the same information and communicate with one another. A CIM system uses computers to integrate the processing of production and business information with manufacturing operations in order to create cooperative and smoothly running production lines. CIM ensures that all existing systems can talk to one another within both local and wide area networks. The tasks performed within CIM will include:

- the design of the graphic product to be produced
- planning the most cost-effective workflow
- controlling the operations of the machines or equipment needed to make the product
- performing the business functions such as ordering of stock, materials and customer invoicing.

Digital print solutions

The print industry has been revolutionised by the application of CIM. Digitalisation means change, and change brings new business. The demand for short, personalised print runs and faster job turnaround has quickly created a new, dynamic market for printers. Producing short colour runs – with extremely fast delivery times, last-minute changes and uncompromising print quality – is now a must for one-to-one marketing.

On-demand printing – quickly supplying exactly the right number of copies to meet each customer's needs – provides a competitive edge by providing greater flexibility and less need for expensive stockpiling. State-of-the-art digital technology means that images are scanned or transmitted electronically into memory, to be stored for further use. Data only needs to be input once, reducing much of the set-up time involved in traditional printing and copying methods. In addition, less set-up time means lower costs for the customer. Printing with variable data enables print products to be easily personalised as required.

Some printing companies allow Internet access with online ordering and they also provide a fast delivery service, so that customers can get the job printed and delivered back to them without ever leaving their office. Digital ICT can now be employed effectively in all three stages of the printing process – pre-press, press and post press.

Pre-press solutions

The software a company has used to produce its document or artwork defines the speed of production and the printing hardware has to keep up. Digital technology and media grow more important with the increasing data volumes involved. The goal of large volume printers is to minimise production times with completely digitised production workflows and database-supported networks. Digital solutions provide a full range of image-handling capabilities, which enables greater flexibility in DTP, colour management, proofing and workflow systems. Image-handling capabilities of digital technology include: image clean-up, reducing, enlarging, photo screening, cropping, masking, rotating, inline collating, moving of text and graphics to enable printing on unusual paper sizes and other media.

Press solutions

Printing costs reduce with **digital printing** machines that can operate at up to 14,400 pages per hour. Digital printing is also well suited to the production of images in low-volume or 'short runs', as it does not require the making of **plates**, as does offset printing. Digital technology means that images are scanned or transmitted electronically into memory, to be stored for further use. When higher print volumes are required, other digital methods are used.

Computer to plate

In conventional print making, images are converted into film, which is used to create the printing plate, which is then used to print the final image. The introduction of computer to plate (CTP) has increased workflow times as images can now be sent direct from a computer to a digital plate-making machine without the need to create a film first.

Post-press solutions

Post-press is about the finishing of print or graphic products which are increasingly complex. It adds value and leads to increases in profit. Customers' wishes are rapidly driving technological innovations so their ideas can be produced as originally envisioned by the designer. Whether folding, gluing, cutting, perforating, die cutting, wire-stitching or perfect binding is involved, maximum precision is necessary, and so is a standard of productivity that lets finishing machines keep up with the output of the latest digital printing presses. Digital finishing equipment can be more easily programmed and provides greater accuracy when compared to traditional mechanical labour-intensive equipment.

The development of CIM in the packaging industry

CAD/CAM technologies were first developed for the aerospace and automotive industries. For the companies that invested in these technologies they provided a competitive advantage by allowing the faster development of new ideas. Once the financial advantages they provide were realised they quickly spread to many other manufacturing sectors, especially those where there is an increased emphasis on quality, accuracy and consistency. One of the fastest growing applications for computers is in the graphics area and particularly in packaging.

Designers in all manufacturing disciplines are under increasing pressure to develop distinctive design concepts in order to create a strong corporate or brand image. Recognisable product identities can establish a secure and viable market share for their products. Graphic designers, particularly in the packaging industry, must ensure that their clients' products stand out clearly on the shelf. In many cases, this means that products are becoming more difficult to design and manufacture by conventional techniques.

CAD systems, no matter how sophisticated, are not a substitute for the creative abilities and visual skills of a graphic designer, especially during concept development when ideas are first being generated. The computer provides real time savings in all the subsequent operations that need to take place to convert an idea into reality. As an example, photo realistic .visualisations rather than costly one-off models can be presented to the client electronically allowing unsuitable designs to be eliminated more quickly. On the functional side of packaging the computer's mathematical capability can be used to calculate the 'volumes' that can be contained by different designs more quickly and accurately. In the case of manufacturing technologies, such as injection or blow moulding used to produce plastic containers, reliable manufacturing data can be generated from the design drawings to ensure accurate and effective dies and moulds.

For graphic product designers the power of modern CAD systems is their ability to allow designs to be visualised quickly in different material and colour combinations within different settings. For example, the designer can combine the digital images of an actual supermarket display with the images for proposed product design that he or she has created. In effect, a shampoo or cosmetic manufacturer could be given a visualisation of what the pack would look like on a supermarket display when compared with its various competitors.

Many printing operations including those in the packaging sector are now employing CNC machines in the key production areas of pre-press (artwork and printing plate preparation), press and binding (assembly of the finished article). Traditionally, films are used to transfer images from original artwork to the printing plates. Computer to Plate (CTP) processing equipment means that digital artwork can be used to output a set of plates directly from the desktop. The quality of reproduction is greatly improved because printing plates can be imaged directly by laser and workflow for plate production is greatly improved.

Other CIM developments include computer-controlled digital presses that mean four-colour printing in relatively small quantities is more efficient and cost-effective. These types of press can produce images on a wide variety of materials such as plastic sheets and laminates. This means that personalised and short-run printing such as promotional mouse mats or printed products with an individual customer's name on can now be produced competitively.

Task
Use a digital camera to take pictures of a product display of your choice. Transfer the images into a CAD system. Create a new package design for a product of your choice and produce a visualisation of what your product might look like on the digital images of the real display that you captured.

Flexible Manufacturing Systems (FMS)

In the late 1960s, all the leading Japanese manufacturers moved towards the flexible factory as a new way of gaining a competitive advantage. Since that time there has been increasing global competition and the creation of more open markets. Additionally, consumers are more sophisticated and are demanding an increasingly diverse range of products with regular updating beyond simple cosmetic changes. This, in turn, is influencing manufacturing to invest in more flexible plant and equipment. A piece of flexible equipment is one that has the ability to perform multiple processing tasks on a wide range of products, for example CNC machining centres for machining a variety of parts, or robots for material handling.

The introduction of machinery that can operate flexibly enables manufacturing to explore various processing sequences that might be available through alternative configurations of plant and equipment. The following are characteristics of most flexible manufacturing solutions:

- A system that responds quickly to changes from whatever source on the 'supply' or the 'demand' side.
- A range of techniques to increase operational flexibility are used and their effectiveness is constantly monitored and evaluated.
- The lead times from design to manufacture are significantly reduced and future changes in design can be made quickly.
- Levels of stock are kept to a minimum.
- Vertical partnerships, with effective two-way, ICT-based communication, exist between suppliers, product manufacturers and retailers.
- Increased sales and stock turnover for all partners in the enterprise.
- Computer-based management tools such as manufacturing resource planning (MRP) are used to collect real-time processing data as well as to track work accurately through the production cycle and to re-plan production in response to changing demand.

Creative and technical design: the role of computer-aided engineering

CAD is at the core of the graphics industry. **Computer-aided engineering (CAE)** can be part of a CAD application or it can form a 'stand-alone' system that can analyse the effectiveness of a design by creating simulations in a variety of conditions to see if it actually achieves what the designer intended. The success of modern CAD/CAM and CAE systems has been the dramatic reduction in design and development time. This allows a company to develop products in quick response to the needs of the market.

Modelling and testing ideas

With the rapid developments in mobile phone technology over the last few years the outward appearance of the phone as a visual or graphic product has changed dramatically from the early models that were the size and appearance of a small brick. The changes in technology have allowed product designers more freedom to change the visual appearance of the phone without compromising its technical efficiency. Consumers can also customise their phones by downloading different ring tones or adding covers with different graphic designs on them.

CAD modelling techniques allow the graphic

designer to try out different shapes and styles on screen. They can experiment with button layouts and the placing of graphic images or logos. CAE enables the production designers to simulate the manufacturing process, such as plastic injection moulding, to see if the design can be manufactured. Software is used that converts the solid graphic representation of the design to be converted into the thin shelled case that is needed to protect the operational circuitry and control devices on the phone. The 'engineered model' can be looked at from any angle and all the strengthening ribs and flanges required in the moulding process can be put in place. Once the design dependencies are determined the software can be used to generate the data needed for the computer-aided manufacture of the product.

Task

A more recent development in computer-aided modelling is **Rapid Prototyping (RPT).** This process was first developed in the 1990s as a means of using a computer to generate a 3D model of an object such as a perfume bottle or similar 3D objects that had been drawn by CAD software. Use the Internet to investigate the emerging technology of RPT and compile a short illustrated report showing its application in CAD/CAM.

Virtual Reality Modelling Language (VRML)

3D virtual product modelling is an emerging technique, looked at in Unit 3. **Virtual Reality Modelling Language (VRML)** further extends the power of product modelling as a design tool. It is a specification for displaying 3D objects on the Internet. It is the 3D equivalent of HTML. Files written in VRML have a *wrl* extension (short for world); HTML files have the extension *html*. The VRML script produces a virtual world or 'hyperspace' on the computer display screen. The viewer 'moves' through the world or around an object by pressing computer keys or using another input device to turn left, right, up or down, or go forwards or backwards. This technology is still in the early stages of development; the first VRML standards were only set in 1995. It is set to become a powerful tool for the computer-based modelling of products (see Figure 4.3.2).

Task

Use the Internet and other sources of information such as printed media to collect information on how **virtual reality** is being used increasingly in the 'e-marketing' of products on the world wide web. Present your findings in a suitable graphic format.

Production planning and control

Production planning and control are areas where computers are used to good effect. Earlier in this unit we learned how modern manufacturers regard 'time' as a weapon that gives them an edge over their competitors. Scheduling is part of the planning process and its key features are to specify:

- the scope and detail of the work to be done
- the date when production has to start
- the latest date that production can be completed by
- any specialist machinery or manufacturing processes that are required
- what labour capacity is available.

The range of scheduling software applications that is available is evidence of the many different planning approaches used across different manufacturing sectors. The term finite capacity scheduling describes a processing schedule that is based on the overall manufacturing capacity available. By contrast, infinite capacity schedules use the customer's order due date as an end stop. The aim of these schedulers is to complete the order by working back from the due date using the available capacity.

Figure 4.3.2 *3D virtual reality model of a CIM system*

Modern manufacturers regard time and responsiveness as strategic weapons; to them time is the equivalent of money or increased productivity. They focus on reducing 'non-value-adding' time rather than trying to make people or machines work harder or faster on value-adding activities. The ways that leading companies use computers to manage time – in production, in new product development, in sales and in distribution – represent the most powerful new sources of competitive advantage. Today's new-generation companies compete with flexible manufacturing and rapid response systems, expanding variety and increasing innovation.

A company that has a time-based strategy is a more powerful competitor than one with a traditional strategy based on low wages, lower scales of production or a narrow product focus. These older, cost-based strategies require managers to do whatever is necessary to drive down costs. This could involve moving production centres to a low-wage country, building new facilities by consolidating, in effect closing down, old plants to gain economies of scale, or focusing operations down to the most economic activities. These tactics reduce costs but at the expense of responsiveness.

In contrast, strategies based on the principles of flexible manufacturing and rapid response systems need factories that are close to the customers. They serve to provide fast responses rather than low costs and control. The whole process is coordinated via computer databases and software applications that provide a full range of scheduling functions.

Master production schedule

The Master Production Schedule (MPS) is a top-down scheduling system that sets the quantity of each product to be completed in each week or **time bucket** over a short-range **planning horizon**. It is derived from known demand, forecasts and the amount of product to be made for stock. The planning assumption is that there is always sufficient manufacturing capacity available. For this reason, this method is sometimes called infinite capacity scheduling.

Other computer-aided scheduling functions

Resource scheduling is a finite scheduling function that concentrates on the resources that are required for converting raw materials into finished goods. Other scheduling functions may include the entering and invoicing of sales orders (sales order processing), stock recording and cost accounting. These functions combine to provide a powerful integrated database for the company. The only real problem with these applications is related to the accuracy of the data that is input into the database – inaccurate data causes cumulative problems as products move through the processing stages.

Types of computer-aided scheduling

- *Electronic scheduling board*. The simplest scheduler is the electronic scheduling board, which imitates the old-fashioned card-based loading boards that were used to sequence machining operations. The advantage of this computer-based system is that it calculates processing times automatically and warns the production staff of any attempt to load two jobs on to the same machine.
- *Order-based scheduling*. In order-based scheduling the sequencing and delivery of individual component parts and resources is determined by the overall priority of the order for which the parts are destined. It is a distinct improvement on infinite capacity schedulers but its biggest drawback is that it allows gaps to appear in resources. Some schedulers allow the process to be repeated to try to reduce gaps before it is put into practice in order to reduce the time through the system but this can be a very time consuming process.
- *Constraint-based schedulers*. With these the aim is to locate potential bottlenecks in a production line and ensure that they are always well provided by synchronising the flow through the MPS.
- *Discrete event simulation*. This computer simulation loads all the required resources at a chosen point on the production line. When all processing problems and queues are resolved at that point, it moves on to the next point on the line. Because the simulation moves from one set of processing events to the next, there are far fewer gaps in the schedules and consequently production lines are far more stable.

Task
Describe the costs and benefits of using computer systems to aid production planning.

Control of equipment, processes, quality and safety

Different types of systems are used in modern manufacturing centres to control production tools, to monitor and evaluate quality and ensure safe working environments. These systems can be electrical, electronic, mechanical pneumatic, computer or microprocessor controlled. On modern machines such as printing presses or plastic moulding equipment these systems are often used in combination and their effective operation is dependent on the feedback of information on the state of each part of the system. Feedback makes control systems more efficient. Control systems improve efficiency, accuracy, reliability, safety and reduce waste. They can be found in:

- materials handling, such as moving resources so they are in the right place at the right time
- processing of materials, for example the automated manufacture of plastic-based packaging, using temperature control
- joining materials, such as the bonding and heat-sealing of plastics in a point-of-sale display using electronic or computer control
- monitoring quality by using feedback from colour sensors to regulate the feed rate on a digital printing press to ensure more even inking of the paper or card
- ensuring safety using feedback from electronic sensors to stop machines such as printing presses if hands or other obstructions are in the way
- graphical displays in a central control room that often show the floor-plan of the factory, the manufacturing cells or machines in use, the position of materials and components and a visual and audible alert system for when there are problems.

Total Quality (TQ)

One of the characteristics of a 'world class business' is that a 'quality culture' exists throughout all levels of the organisation. There is a strong and clear commitment to getting things right first time, every time. Total Quality links together quality assurance and quality control into a coherent improvement strategy. **Total Quality Control (TQC)** is the system that Japan has developed to implement Kaizen or continuous improvement for the complete life cycle of a product. It operates both within the function of the product and within the system to develop, support and retire the product. Total Quality Management (TQM) is the US equivalent. What both these systems have in common is that they are 'purpose driven', they involve comprehensive change and both are long-term processes rather than short-term fixes for problems that arise in design and manufacturing operations.

Implementation of TQ systems

Computer-based technologies are used to ensure TQ throughout design and manufacturing organisations. The use of computers ranges from supporting employee training by providing interactive learning methods to computer-aided statistical tools and methods for checking quality. For example, a product can be assigned a unique

bar code that can be read by laser readers at different points in the manufacturing process. The data that is collected is fed into the central PDM system and that information can be used in a variety of ways. The quality control team can use the data collected to trace a faulty product back through the manufacturing process to pinpoint where the particular problem occurred. They can then take the necessary remedial action to remove the cause of the problem so the manufacturing process is further improved.

There are three stages in the development of TQ in an organisation:

1 Awareness raising – to recognise the need for TQ and learning its basic principles.
2 Empowerment – learning the methods of TQ and developing skills in practising them.
3 Alignment – harmonising the business and TQ goals with the manufacturing practices of the company.

TQC and TQM tools

Thinking tools are equally as important as the physical tools we use to make things. Thinking or intellectual tools enhance the way we plan to do things at a strategic, tactical or operational level. At a strategic level, such as business planning, thinking can be highly symbolic or abstract and this relies on accurate qualitative information. At a tactical level, such as production scheduling, the information required is a mixture of qualitative and quantitative data. The importance of qualitative information decreases as you move from the strategic management level down to the operational level of factory coordination where achievement is judged against the planned outcomes (see Figure 4.3.3).

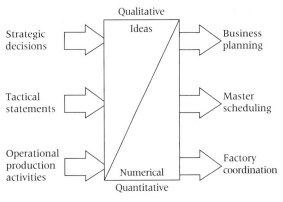

Figure 4.3.3 *Thinking tools link to planned outcomes*

In the context of TQC or TQM, computer-aided systems contribute to the collection, dissemination and analysis of the qualitative and quantitative data that is needed to bring about improvement. In general, computer-based thinking tools can be classed under the following functions:

- management
- product planning
- quality functions
- statistical process control (SPC).

Using computers in SPC

There are several established graphically based methods for representing statistics and analysing processes. Computers now play an important role in representing and analysing the causes and effects of different actions. For example, Pareto charts are used as a decision making tool. The Pareto Principle suggests that most effects come from relatively few causes. In quality control this is known as the 80–20 rule – 80 per cent of faults come from only 20 per cent of the causes. If it is possible to identify the 20 per cent accurately, you can eliminate 80 per cent of your faults (see Figure 4.3.4). Pareto charts can be used to compare before and after situations. This allows managers to decide where to apply the minimum of time for maximum effect. Other areas where computers can be applied effectively to improve quality include the following:

- Flow charts are useful for modelling processes, feedback and **critical control points (CCPs)**. They use symbols, text and arrows to show direction of information or data flow (see Figure 4.3.5).

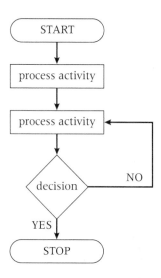

Figure 4.3.5 *A flow chart – used to show the movement of people, materials, paperwork or information and processes*

- Cause and effect, fishbone or Ishikawa diagrams describe a process or operation as a sequence. They provide a method for analysing processes to establish a cause and its effects (see Figure 4.3.6).
- Bar graphs or histograms are used to analyse variations in data in graphic format. This makes it easier to 'see' graphical variations than it would be to read a table of numbers (see Figure 4.3.7).
- Check sheets can be used to collect quantitative or qualitative data during production. They are used in high-volume manufacturing where there is greater repetition than you would find in shorter production runs.
- Checklists are an important management tool used in specific operational situations to ensure quality by establishing consistency and reliability. They list all the important steps or actions that must be correctly sequenced to achieve the best possible quality outcome.

Monitoring and inspecting quality

Despite the progress in designing for quality and incorporating quality into manufacturing processes, inspection still remains a necessary component of many quality assurance systems. In all areas of manufacturing, whatever the product, the cost of defects escalates as materials and components move through the supply chain to the customer. The manual or 'human' monitoring of any continuous process, such as in the printing industry, becomes less efficient as

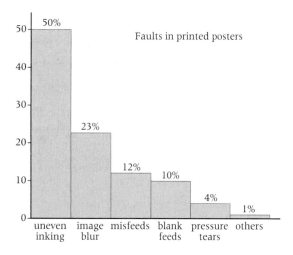

Figure 4.3.4 *A Pareto chart identifies the production areas where there are faults*

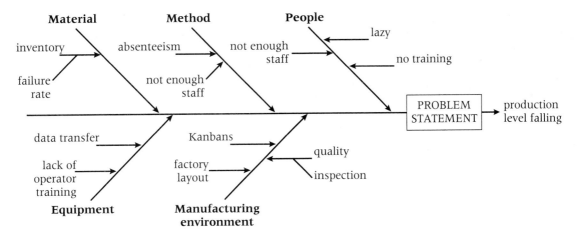

Figure 4.3.6 *A cause and effect diagram*

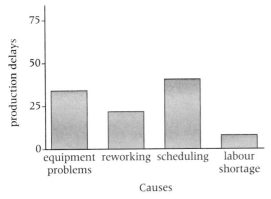

Figure 4.3.7 *A bar graph – an analytical tool to compare outputs, percentages or activities*

the time spent on it increases. In many industries, this effect is minimised by the setting up of teams that work together in a manufacturing 'cell'. Each cell is responsible for the product from when it is received by the cell until it moves on to the next stage of production. As a team the cell has to meet specified **quality indicators**. Each member of the team is responsible for maintaining quality level so that the burden of monitoring and inspection is shared. This leads to the improved detection of faults.

In high-volume, continuous production industries such as packaging and printing 'artificial vision' is now used to good effect in the monitoring and inspection of the processing operations. In a print room, for instance, the optical quality of a colour print can be more accurately judged over a longer period of time by using digital image-processing technology rather

than relying on an experienced operator who might be subject to many physiological defects such as tiredness. The benefits of optical quality control are:

- high-speed, real-time fault detection, recording and reporting throughout a processing operation from receipt of raw materials and components through to the delivery of the product
- reduced labour costs
- low running costs after initial installation
- specific quality specifications can be programmed into a computer and checked using the digital data collected during the full range of processing operations.

Manufacturing to tolerances

Most mass-produced 3D products are assembled from components that are manufactured or machined to specific or 'nominal' measurements so that they can all be fitted together. However, it would be too expensive and wasteful to produce each part to the exact dimension every time so dimensions on a drawing often have a **'tolerance'** that indicates what the acceptable maximum and minimum dimensions are. Providing the dimensions of the component or part lie within this range it is acceptable, as the product will still function effectively. There are two types of dimensional measurement that can be applied:

- On-process measurement where a part is measured while under production. This is used where manufacturing cycles for a component or part are long or high-cost materials are being processed. The early detection of error removes the high cost of failure.

- Post-process measurement involves making measurements of critical operational dimensions after the component or part has been produced. It is a method that is often used in high-volume, low-cost manufacture or in specialist one-off manufacturing such as the production of promotional gifts using a CNC machine. The measuring methods used are mostly mechanical in nature such as vernier **calipers**, height gauges and profile projectors. Increasingly, computers and lasers are being used because of their increased accuracy, as they do not rely on the visual interpretation of a scale or rule.

Controlling complex manufacturing processes

A modern integrated system of manufacturing is heavily reliant on the use of computers throughout the process of design and make. Computers are also used to determine the optimum factory layout of plant and equipment, the effective deployment of labour and the scheduling of processing operations. Computers can also be used to monitor and control 'workflow' and the dissemination of information about work in progress, especially in high-volume manufacturing situations such as printing and packaging.

Managing workflow

Production plans and control systems manage the input of materials and components to the different processing departments within a manufacturing operation. They also monitor output and work in progress. This information informs and controls the workflow through the factory to ensure that orders are met on time and to cost. It is particularly important in batch and short-run manufacturing operations such as those involved in the production of promotional gifts or sign making.

Monitoring workflow

The role of artificial vision and the use of lasers were described earlier in this unit and they are a vital component in computer-controlled parts recognition systems. These visual recognition systems rely on **data communication tags** (bar codes) as a way of monitoring products through a high-volume manufacturing system. They can be attached to pallets of stock that are to be processed or to individual assemblies. A sensor or bar code reader attached to an individual workstation can read these tags. Laser scanners are used to read bar codes because they are well suited to applications requiring high reading performance, small size and low cost. A digital pre-processor receives and decodes the signal into data that can be read and analysed by a computer. This data is then transmitted to the 'supervising' computer controlling the production line. The development of these systems has improved the operational effectiveness of manufacturing systems based on the concurrent model.

Controlling workflow

The term workflow describes the tasks required to produce a final product. Project management software allows a manufacturer to define different workflows for different types of jobs and coordinate production cells in an overall MPS. For example, in a design environment a CAD file might be automatically routed from the designer to a production engineer to purchasing for comment or action. At each stage in the workflow, one individual or group is responsible for a specific task such as ensuring the right materials are in the right place when required. Once the task is complete the workflow software ensures that the individuals responsible for the next task are notified and receive the information they need to execute their stage of the process.

Sequential or concurrent manufacturing?

The design of an effective and marketable product that will be sold at a profit is dependent on the input from a range of specialists and the efficiency of the manufacturing or processing system that is chosen. The scale of the manufacturing operation is a key factor in deciding on a system of design and make.

Sequential manufacturing

In a design studio, where a small group of graphic designers is working together, each person may take responsibility for a particular project or graphic outcome. The designer works in a similar way to you when you are producing your practical coursework for this examination. In this linear or sequential approach to design and making an idea or a product will pass through a series of discrete, self-contained stages and at each stage outcomes are evaluated before moving on to the next. If there is a fault, it is passed back through the stages until the fault is found and corrected when it then begins its journey through the process again. This linear or 'function-based' approach is slow to respond to change or demand. It has longer lead times and is often characterised by low product quality because of the separation of the design and manufacturing functions and the costly design and redesign loops.

Concurrent manufacturing

Concurrent manufacture is widely regarded as an effective system for scales of production ranging from batch to mass and high volume. The key feature of concurrent or simultaneous manufacturing is a team-based approach to project management, so that the right people get together at the right time to identify and resolve design problems. The underpinning philosophy of this approach is that quality decisions have to be made at every stage to ensure the intended outcome of a quality product that makes a profit. This means setting appropriate specifications and 'quality indicators' to evaluate both the design and intended manufacturing processes. The concurrent approach also forces manufacturers to consider all elements of the product life cycle from conception to disposal.

Many companies also involve their suppliers and retailers at an early stage in the product development cycle so that they create a product development team (see Figure 4.3.8). They work closely in a vertical partnership known as a 'value chain'. For this partnership to work effectively, information has to flow quickly between the partners. Increasingly, this takes place electronically through **electronic data interchange (EDI)** systems, which we shall look at in more detail later in this unit.

Computers are becoming increasingly useful when companies decide to adopt a concurrent approach to design for manufacture (DFM). The importance of this is clear when you consider that 70 per cent of the manufacturing costs of a product, materials, processing and assembly, are determined by design decisions. Production decisions such as process planning or machine selection are only responsible for 20 per cent.

Companies can now create their own 'expert systems'. These are databases of established good design practice and specialist knowledge that are available on a network of computers within the company known as an **intranet**. This internal computer network enables fast, efficient communication between all the members of the product development team.

The development of Internet technologies means that this information can also be accessed via the world wide web from anywhere in the world. Different access codes or passwords can control access to commercially sensitive areas of development. Concurrent systems enable the use of just-in-time (JIT) and Quick Response Manufacturing, which reduces the time from idea to market. The times to market for sequential and concurrent manufacturing are compared in Figure 4.3.9.

In concurrent systems, to ensure that the time to market is further reduced, different members of a team will take responsibility for ensuring specific production deadlines or 'milestones' are met. The team agrees these milestones and they are recorded in a **Gantt chart**, a simplified form of which is shown in Figure 4.3.10. This chart shows

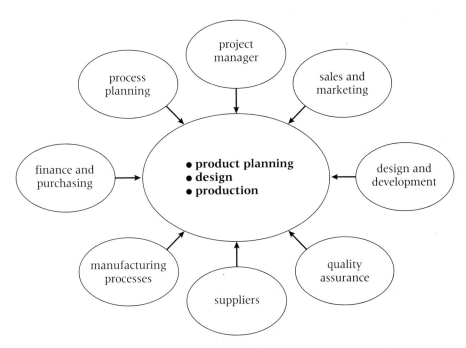

Figure 4.3.8 *Concurrent manufacturing – a team approach to product development*

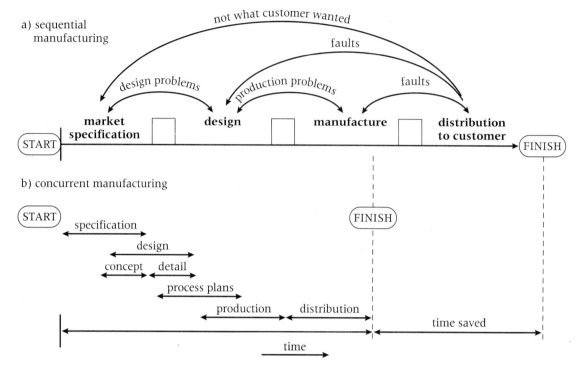

Figure 4.3.9 *A comparison of sequential and concurrent manufacturing. The latter reduces time to market*

where operations are to start and finish and also those activities that can be run concurrently because they are not dependent on other operations being completed first. You will see, for instance, that the design and production of advertising, promotional and sales literature that would include printed and electronic media starts before the first prototype has been produced. You will also see that the final list of suppliers is only completed once the milestone of a successful prototype has been reached.

Tasks

1 Explain why the use of ICT is such an important feature in a concurrent manufacturing system.
2 For a product of your choice, explain the benefits of using a concurrent manufacturing system in comparison with a sequential manufacturing system.
3 Explain how **milestone planning** reduces the 'lead in' time for a new product.

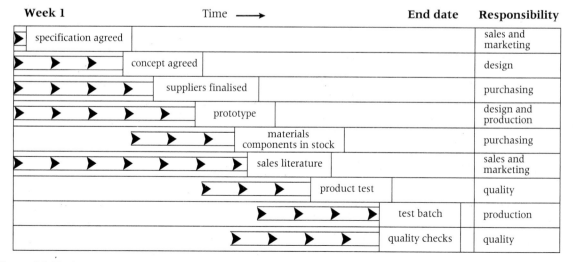

Figure 4.3.10 *A milestone plan for a typical manufactured product*

2. Robotics

Automation is a term describing the automatic operation and *self-correcting* control of machinery or production processes by devices that make decisions and take action without the interference of a human operator. Robotics is a specific field of automation concerned with the design and construction of self-controlling machines or robots. More recently, mechatronics, a Japanese term, is an exciting development in the field of automation. Mechatronic devices integrate mechanical, electronic, optical and computer engineering to provide mechanical devices and control systems that have greater precision and flexibility.

The first operator-guided robotic manipulating devices were developed in the 1940s for use in radioactive conditions encountered in the development of the atomic and hydrogen bombs during World War II. The first industrial robots developed in the late 1950s and early 1960s included sensors and other simple feedback devices. The use of robots is growing with the development and wider application of CIM and FMS. Figure 4.3.11 shows a bank of robots working in a manufacturing cell in a car plant. The next generation of manufacturing robots will have enhanced sensory feedback systems and possess a degree of **artificial intelligence** (AI).

The robots we see in the movies, such as C-3PO in *Star Wars* and in the *Terminator* series, are portrayed as fantastic, intelligent, even dangerous forms of artificial life. However, the robots of today are not exactly like these walking and talking intelligent machines. Most of the world's robots are working for people in factories,

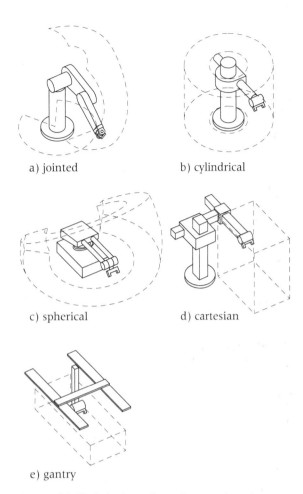

a) jointed   b) cylindrical

c) spherical   d) cartesian

e) gantry

Figure 4.3.12 *Robot configurations and work envelope*

warehouses, and laboratories. Specialist suppliers now offer a large range of robot sizes/payloads/reach (work envelope) and joint capabilities with differing degrees of precision and repeatability. Figure 4.3.12 shows different robot configurations and their work envelope.

Robots have been applied to assembly, painting, palletising, packing, welding, dispensing, cutting, laser processing, material handling and other emerging applications in the advertising, promotional, leisure and entertainment industries such as animatronics. Animatronics, which has a lot in common with robotics and mechatronics, encompasses automata and is concerned with bringing inanimate sculptures to life. These devices known as animatrons (see Figure 4.3.13) require many design and manufacturing skills drawn from mechanical engineering, pneumatics, sculpture and animation.

Figure 4.3.11 *A manufacturing robot*

Figure 4.3.13 *An animatric iguana puppet hams it up during the taping of a children's show for National Geographic television*

Basics of robot design

Most robots are designed to be a helping hand. They are in effect robotic arms. Figure 4.3.14 shows a schematic view of a typical jointed-arm robot. They help people with tasks that would be difficult, unsafe or boring for a real person to do alone. At its simplest, a robot is a machine that can be programmed to perform a variety of jobs, which usually involve moving or handling objects.

For a machine to be classified as a robot, it usually has five parts:

* *Controller*. Every robot is connected to a computer, which keeps the pieces of the arm working together. This computer is known as the controller. The controller functions as the 'brain' of the robot. The controller also allows the robot to be networked to other systems, so that it may work together with other machines, processes or robots.

    Robots today have controllers that are run by programs – sets of instructions written in code. Almost all modern robots are entirely pre-programmed by people; they can do only what they are programmed to do at the time, and nothing else. In the future, controllers with artificial intelligence (AI) could allow robots to think on their own, even program themselves. This could make robots more self-reliant and independent.

* *Arm*. Robot arms come in all shapes and sizes. The arm is the part of the robot that positions the end-effector and sensors to do their pre-programmed business. Many (but not all) resemble human arms, and have shoulders, elbows, wrists, even fingers. This gives the robot a lot of ways to position itself in its environment. Each joint is said to give the robot 'one degree of freedom'. So, a simple robot arm with three degrees of freedom could move in three ways: up and down, left and right, forward and backward. Most working robots today have six degrees of freedom (see Figure 4.3.16).

* *Drive*. The drive is the 'engine' that drives the links, the sections between the joints, into their desired position. Air, water pressure or electricity powers most drives.

* *End-effector* (end of arm tooling). The end-effector is the 'hand' connected to the robot's arm. It is often different from a human hand – it could be a tool such as a gripper, a vacuum pump, tweezers, scalpel, blowtorch or heat

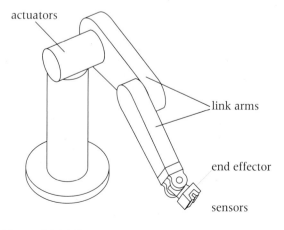

Figure 4.3.14 *The parts of a robot arm*

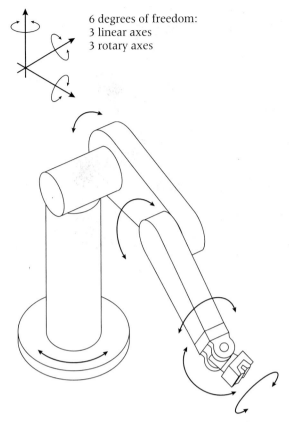

6 degrees of freedom:
3 linear axes
3 rotary axes

Figure 4.3.16 *Six degrees of freedom on a robotic arm*

sealing gun, just about anything that helps it do its job. Some robots can change end-effectors and be reprogrammed for a different set of tasks. If the robot has more than one arm, there can be more than one end-effector on the same robot each suited for a specific task.

- *Sensor.* Most robots have limited awareness of the world around them. Sensors can provide some feedback to the robot so it can do its job. Compared to the senses and abilities of even the simplest living things, robots have a very long way to go. The sensor sends information in the form of electronic signals back to the controller. Sensors also give the robot controller information about its surroundings and let it know the exact position of the arm or the state of the world around it. Sight, sound, touch, taste and smell are the kinds of information we get from our world. Robots can be designed and programmed to get specific information that is beyond what our five senses can tell us. For instance, a robot sensor might 'see' in the dark, detect tiny amounts of invisible radiation or measure movement that is too small or fast for the human eye to see.

The industrial application of robotics/control technology and the development of automated processes

Ninety per cent of all robots used today are found in factories. These robots are referred to as industrial robots. Although many kinds of robots can be found in manufacturing, jointed arm robots are particularly useful and common. Ten years ago, car manufacturers were buying nine out of ten robots – now, car manufacturers buy only 50 per cent of robots made. Robots are slowly finding their way into warehouses for automatic stock control, laboratories, research and exploration sites, energy plants, hospitals, even hostile environments like outer space (see Figure 4.3.17).

Japan is a world leader in robotics. It has 400 000 robots working in factories – ten times as many as the United States. Robots are used in dozens of other countries. As robots become more common, and less expensive to make, they will continue to increase their numbers in the workforce.

The manufacturing industry, including those sectors involving graphics, is in a non-stop state of change and is evolving rapidly because of the availability of a whole range of improved computer-based control technologies and the recent emergence of the global marketing of

Figure 4.3.17 *The loading bay on the Space Shuttle showing the Canadian Robot Arm in use by astronauts*

products on the Internet. These new digital technologies are supporting rapid changes in working practice from the shop floor through to business processes such as the web-based advertising and selling of products. The application of robotic technologies has brought about significant changes in those sectors of industry concerned with batch and mass production.

Robots in batch production

Robotic devices have supported the batch and short-run production of items such as CD-ROMs that contain entertainment or business software. Figure 4.3.18 shows a CD printing machine that is loaded using a robotic application. In mass-production they are used in large, highly automated operations such as printing and packaging.

The advantage of these 'robotic' systems is that they support continuous production and they allow many processes to run concurrently, sequentially or in combination. The numbers of human operators that work are minimal and those that do are working in safe and secure operating conditions. This is because electronic, computer or microprocessor controlled systems have many sensing devices that monitor events in real time and are designed to shut down to a 'fail-safe' state in the event of a failure or accident. These robotic technologies are **enabling technologies** that allow innovative solutions to complex processing operations.

Figure 4.3.18 *CD printing machines use a robotic application*

Control systems

The common factor in all automated processes, including the use of robots, is that they all require a computer or microprocessor-based control system. Control systems regulate, check, verify or restrain actions. Automatic or robotic manufacturing systems are able to sense and control how processes are operating by combining sub-systems of electrical, electronic, mechanical and pneumatic devices (see Figure 4.3.19). All these devices produce performance data that may need to be monitored by instrumentation systems.

Instrumentation systems provide data in the form of visual displays or electronic signals that indicate how effectively the manufacturing sub-systems are operating, both individually and in relation to one another. Because of their ability to store, select, record and present data, computers and microprocessors such as programmable **logic** controllers (PLCs) are widely used to direct and control actions and process operations.

In many manufacturing and assembly applications robots are internally programmed to perform a programmed cycle of operations when given a simple start signal. However, there are many robots which combine an internal computer program with compliance to commands from outside the program. These can be classified as sensor or remotely controlled robots. Many robot tasks cannot be programmed exactly in advance because real-world operating conditions cannot be predicted exactly and they rely on sensors to provide the information required by the robot. Below are some examples.

Automatic storage and retrieval systems (ASRS)

Automatic storage and retrieval robots that are used in automated warehouses are commanded from outside either to transfer an object from a designated pickup point to a designated position in storage or vice versa. The required motions are internally programmed into the device. Such robots are made in sizes to handle 'objects' from tape cassettes, DVDs and other data storage devices to pallets carrying cardboard in a carton-making factory. The commands may come from a computer, which controls a larger operation, in which case the robot computer and the computer that commands it are said to form a control 'hierarchy'.

Mobile robots or automated guided vehicles (AGVs) are typically used in component or pallet transfer. The AGV is an unmanned vehicle that carries its 'load' along a pre-programmed path. AGVs use different navigational systems and travel around under some combination of automatic control and remote control. Other

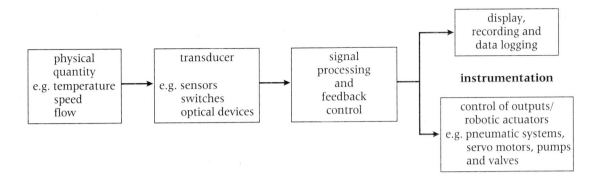

Figure 4.3.19 *A simple systems model can represent a manufacturing control system*

uses of AGVs are in mail delivery, surveillance and police tasks, material transportation in factories, military tasks such as bomb disposal, and underwater tasks like inspecting pipelines and ship hulls and recovering torpedoes.

Robots for advertising and promotional purposes

There are other uses for automated or mobile robots. 'Fun' robots are innovative graphic products that are used widely in the United States, and their use is increasing in the UK. Fun robots are extremely effective in situations where there is a need to grab people's attention or where a company wants to 'stop traffic' by engaging potential customers. When robots talk, people listen!

These robots make a lasting visual impression at trade shows, parades, theme parks and museums such as NASA Houston. They are also used to publicise special educational programmes such as anti-drugs campaigns and fire safety. The types of robots found most effective for advertising promotions are mobile robots, talking signs and animated figures, especially musicians.

Mobile robots, which are powered by rechargeable batteries, can seize anyone's attention by driving up to or chasing him or her and starting an interactive 'conversation'. The robots can 'talk' via wireless microphone systems using a transmitter and microphone that allows the operator to speak through the character. In some products such as futuristic or 'alien' robots, a voice modifier is included to electronically modify the operator's voice to sound like an alien, robot, monster or normal, simply by changing a four-position switch. The character's mouth movement is automatically synchronised with the operator's voice and there are other movements such as mouth lights that are automatically synchronised with the 'voice'.

Robots can be 'customised' by painting and the application of vinyl decals designed to convey corporate images in the form of logos and specific colour combinations. Some robots are equipped with a digital messaging system that allows a customer to record messages up to three minutes long on a digital chip and activate message playback either automatically by infra-red sensor or manually by push buttons. In noisy situations such as a trade show or shopping centre the robots can be equipped with a remotely controlled siren or back-up alarm, or a combination of these attention-getting noisemakers.

A remotely controlled futuristic mobile robot known as 'Z-Tron' features a remote camera that allows moving graphic images to be relayed between the operator and the person that is engaged. It has full 360-degree mobility and a full 'talk-back' system that simulates hearing and allows the operator to respond to questions the robot is asked.

Figure 4.3.20 *Robots, like this one entertaining children at a fair, are often used to promote companies and their products*

Talking signs are stationary robots that are used for promotional and advertising campaigns (see Figure 4.3.20). They 'talk' by a wireless microphone system or by a recorded message on a cassette tape or digital chip. The message can be activated automatically by infra-red sensor that senses when a person is nearby or manually by push buttons or by a combination of both. Their mouth movement is automatic, in synchronisation with the sound, and they can also turn their heads automatically. They are excellent as 'greeters' in restaurants, retail stores or anywhere you want to capture customers' attention. Talking signs can be strategically placed in a store to talk about sale items, special promotions, directions, instructions and forthcoming events.

New uses for robots in manufacturing

The majority of the plastics industry is based on a limited number of production processes. Technical advances have been made in recent years by applying six-axis robot work cells to the primary and secondary processes of injection moulding, blow moulding, extrusion and sheet fabrication. For example, within a robotic work cell for the trimming of plastic components, a six-axis robot would manipulate a trimming device around a plastic part to accomplish drilling, routering, blow moulded bubble removal, flash removal, de-gating or any other type plastic removal application. It has applications in the packaging and display areas as well in the production of other plastic components such as packaging and point-of-sale displays. The key benefits of this robotic cell are:

- consistent cutting quality
- safer work environment
- reduction of labour costs.

Other applications of robots in the plastics industry include:

- insert loading systems in which the robot will load metal threaded inserts, other plastic components and appliqués into the moulds to keep a continuous cycle of production
- part removal systems in which the robot will remove parts from the moulding machines to maintain consistent production
- infra-red plastic welding systems in which the robot presents two halves of an assembly to an infra-red heat source to melt the weld areas and then assembles the two halves together
- laser cutting systems in which the robot will manipulate a laser around the plastic part or manipulate the part around a stationary laser to remove unwanted plastic.

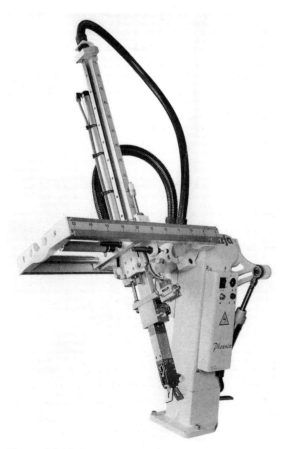

Figure 4.3.21 *A swing arm robot*

Complex automated systems using artificial intelligence and new technology

Artificial intelligence (AI) was a term defined by John McCarthy at the Massachusetts Institute of Technology (MIT) in 1956. It describes the branch of computer science concerned with developing computers that think and act like humans. AI is a broad area of development covering a wide range of different fields, including engineering, ranging from 'machine vision' to 'expert' systems. The common thread that links research in this area is the creation of devices that can 'think'. Computers are not yet able to fully simulate human behaviour in manufacturing environments but this is a major research area in manufacturing and it is only a matter of time because there have already been significant developments in the use of AI in other areas. In 1997, an IBM super-computer called 'Deep Blue' defeated the world chess champion.

What is a thinking machine?

In order to 'think', it is necessary to possess intelligence and knowledge. Intelligent behaviours may consist of solving complex problems or making generalisations and constructing relationships. To do these things requires perception and understanding of what has been perceived. Intelligent systems should be able to consider large amounts of information simultaneously and process them faster in order to make rational, logical or expert judgements. Perhaps the best way to gauge the intelligence of a machine is British computer scientist Alan Turing's test. He stated that a computer would deserve to be called intelligent if it could deceive a human into believing that it was human.

Knowledge-based or expert systems

A knowledge base stores the knowledge related to a particular area or domain. Expert systems in which computers are programmed to make decisions in real-life situations already exist to help human experts in several domains including engineering, but they are very expensive to produce and are helpful only in special situations or in hostile working environments like working in space. Expert systems are designed by knowledge engineers who study how expert designers and others make decisions. They identify the 'rules' that the expert has used and translate them into terms that a computer can understand.

Application of AI in design and manufacture

In Unit 3 we saw that CAD systems already provide a quick and efficient means of representing the technical form of a product. Present systems currently provide limited design information or advice to inform the decision making part of the design process. CAD/CAM software readily generates sets of manufacturing instructions but a designer receives limited information from the software to decide whether the designed part is capable of being economically manufactured. At present, these types of decision are reached by combining and applying the experience and expertise of the whole product development team.

Applying design or production rules

It is possible to represent some knowledge in the form of a set of linguistic facts or logical rules. In electronics, for instance, there are the IF and THEN statements which can be combined with **logic gate** truth tables. When designing an electronic system, IF a set of conditions is true, THEN a conclusion can be made or an action can be taken. In a warning system on a printer, for example, if two inputs are true – the machine is on but there is no paper – then an audible alarm will sound or an on-screen warning is displayed to alert the operator. If two inputs are both true and an audible warning is required, then logic dictates that you would use an AND gate.

Future uses for new technologies

Enabling technologies such as vision recognition and AI are being used to improve production planning such as the optical analysis of visual images on the production line for quality, safety and process control. Vision systems are used in product distribution and bar coding systems. These computerised warehouses can make electronic links (see below for a detailed explanation of electronic data interchange) between suppliers and customers. These intelligent 'vision' systems can 'see', 'make decisions', then 'communicate' those findings to other 'smart' factory devices, all in a fraction of a second. Modern digital vision technology means that little to no extra external requirements such as special lighting are required. With a central computer hooked up, it can update the pictures it takes very quickly with high resolution making the system more responsive and easier to program.

Developing artificial intelligence

Because of the pressure to develop more responsive manufacturing systems considerable research and development time is focusing on developing manufacturing systems that fully integrate the use of **CAD modelling**, artificial intelligence, ICT and knowledge-based databases. These areas include:

- neural networks
- voice recognition systems
- natural language processing (NLP).

Neural networks

A neural network is a computer system modelled on how the human brain and nervous system operate. Whereas a computer manipulates data in zeros and ones, a neural network reproduces the types of processing connections (neurones) that occur in the human brain. Neural networks are particularly effective for predicting events, when they have a large database of examples to draw on. They are proving successful in systems used for voice recognition and natural language processing (NLP).

Voice recognition systems

These are computer systems that can recognise the spoken word and currently they can take dictation but cannot understand what is being said to them. Since such systems are high cost and have operating limitations, they have only been used as an alternative to a computer keyboard. They are used when working in hostile environments such as space, or when the use of a keyboard is impracticable because the operator is disabled. In the future, an operator will be able to talk directly to an expert system for guidance or instruction.

Natural language processing (NLP)
If successfully developed, it is hoped NLP will enable computers to understand human languages. This would allow people to interact with computers without the need for any specialised knowledge. You could simply walk up to a computer and talk to it. Unfortunately, programming computers to understand natural languages has proved to be more difficult than originally thought. Some rudimentary translation systems that translate from one human language to another are in existence, but they are not nearly as good as human translators.

Tasks

1 Explain and give examples of what is meant by an enabling technology.
2 Choose three enabling technologies and describe how they support the development of Flexible Manufacturing Systems.

The use of block flow diagrams and flow process diagrams for representing simple and complex production systems

Graphical system diagrams
Systems thinking and design are evident all around us in the manufactured world. For example, an audio system can contain many different components and devices that can be configured to operate in a variety of ways. We can make sense of the natural world by applying systems thinking, as when we talk about a weather system. Block flow diagrams have been developed to explain how the parts or activities in a system are organised and related to one another. They explain the processes that change an input into an output (see Figure 4.3.22). There is a set of drawing or graphic conventions for representing what the blocks that make up a block flow diagram do (see Figure 4.3.23).

Complex processing within a manufacturing system can be modelled by breaking down the system into a collection of sub-systems each with their own input and output. Together they describe the flow of information or actions through a process. These are sometimes called flow diagrams. Figure 4.3.24 shows a simplified flow diagram of a plastic injection moulding process.

Open- and closed-loop control systems
A system operating **open-loop control** has no feedback information on the state of the output. It will continue without interference from the system even when the output changes. This is a major disadvantage in an automated process, as we shall see later when we consider 'lag'. A system operating **closed-loop control** can have either positive or negative feedback. In these systems, information about the state of an output is fed back into the processor where it is combined with the input signal in order to control or change the size of the output.

Positive and negative feedback
In an automated process positive feedback would result in the system becoming unstable because an increase in the output leads to an increase in the input which creates an increase in the output and so the loop starts again. Negative feedback is

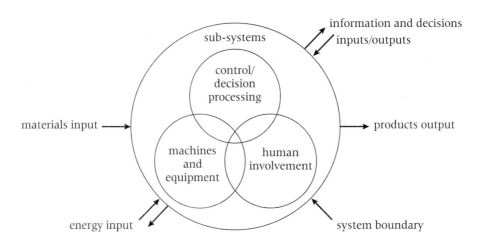

Figure 4.3.22 *A system has a boundary that defines the limits of the system*

often used in automated control processes because it works to change the input in such a way that the output is decreased to provide stable operating conditions. Kanbans are used in some manufacturing systems to control the flow of work on a production line. Without negative feedback they would not operate effectively.

Error signals

In any system with feedback, the difference between the input signal and the feedback signal is called the error signal. The size of the error signal determines how much the system output will need to be changed. When the system is in a near stable state, the error signal will be nearly zero. In a high-volume industry such as printing, for instance, an error signal will be generated when there is a difference between the projected printing production and the actual production. If production is higher than required, this will generate a positive error and signal the need to decrease production levels. If production falls, this generates a negative error and production is increased.

Lag

In any large-scale system or automated process, it will take time for the system as a whole to respond to the feedback signals it is getting. If a faulty batch of products has ended up in the distribution chain, it will not be noticed until customers or the outlets selling the product notify the company of the problem. This time delay before the system is able to respond is known as lag and it is a common feature in closed-loop control systems. The ways in which manufacturers and others are using ICT to improve the speed of communications at all levels of business processes will be discussed later. The measures being taken to reduce lag include electronic data interchange (EDI) and improved 'real-time' sales data from electronic point of sale (EPOS) information systems. These inform the manufacturer of the need to adjust production to correct the 'fault'.

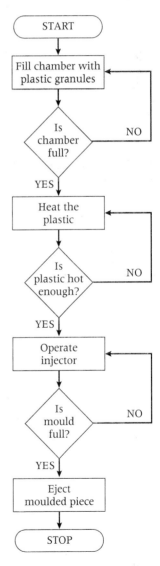

Figure 4.3.24 *A simplified material processing system and its sub-systems of injection moulding*

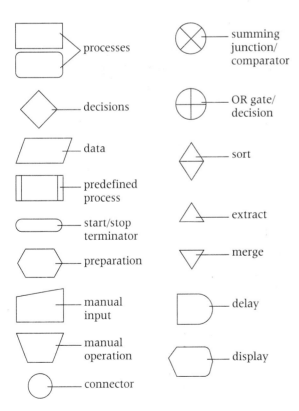

Figure 4.3.23 *Block diagram symbols*

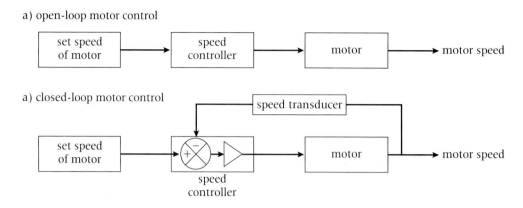

a) open-loop motor control

a) closed-loop motor control

Figure 4.3.25 *Open- and closed-loop control of a motor*

Automated systems using closed-loop control systems

Open-loop control systems have a major disadvantage when it comes to controlling devices such as the motors on a conveyor belt. If the motor is put under too great a load its speed decreases and it may stop completely with disastrous effect such as overheating. The alternative is a closed-loop system in which negative feedback is used to provide proportional control of the motor speed. Any difference between the required motor speed and its actual speed produces an error signal that is fed back to the speed controller in order to stabilise the conveyor system by adjusting the speed of the motor. Figure 4.3.25 shows open- and closed- loop control of an electric motor.

Sequential control
Robotic and automated processes often use sequential control programs in which a series of actions take place one after another. For instance, on the automated CD loading system discussed on page 246 each action depends on the previous one having been carried out. If the system has not 'sensed' that the CD to be loaded is in place, the screen printing process will not operate. It will sit and wait until the sensor sends it the required information signal that the CD is in place.

Logical control of automated and robotic systems
Combinational logic is used in situations where a series of conditions have to be met before an operation can take place. This is also known as 'multiple variable' control. The conventional logic gates that you may be familiar with are the AND/OR/NAND gates. **PLCs** can also perform similar operations. Figure 4.3.26 is a simplified diagram of how multiple inputs are controlled in the operation of a CNC machine.

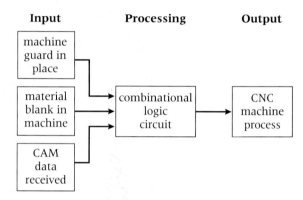

Figure 4.3.26 *All three inputs have to be 'on' for the CNC machine to operate*

Fuzzy logic
Fuzzy logic was developed in the 1960s and it recognises more than the simple true and false values such as those used in digital electronic systems that are based on zeros and ones. An input or output is either on or off. With fuzzy logic, propositions can be represented with degrees of truthfulness and falsehood. For example, the statement 'Today is sunny' might be 100 per cent true if there are no clouds, 80 per cent true if there are a few clouds, and 50 per cent true if it is hazy and 0 per cent true if it rains all day. Fuzzy logic allows conditions such as rather warm or pretty cold to be formulated mathematically and processed by computers in an attempt to apply a more human-like way of thinking. Fuzzy logic has proved to be particularly useful in expert systems, other artificial intelligence applications, database retrieval and engineering.

Fuzzy control
Fuzzy logic controllers (FLC) are the most important applications of the recently developed

fuzzy logic theory. They work rather differently compared to conventional logic controllers employed in process control operations involving electrical, hydraulic and pneumatic devices. Expert knowledge is used instead of algebraic equations or linguistic conventions to describe a system. This knowledge can be expressed as fuzzy sets in a very natural way using linguistic variables similar to the way that humans express their ideas. A fuzzy set is a collection of objects or entities without clear boundaries.

Many people think that until computers can think like humans their wider application will be limited. Fuzzy control is useful for very complex non-linear processes when there is no simple mathematical model or if the processing of (linguistically formulated) expert knowledge is to be performed. Fuzzy control is less useful if the conventional single or multiple variable control theory based on mathematical models described earlier provides an effective result. This is a rapidly developing area and application is found in several areas ranging from electronics design automation (PCB design and manufacture) and product analysis, sampling or testing to e-commerce applications such as document management systems, data warehousing and marketing.

The advantages and disadvantages of automation and its impact on employment, both local and global

The pressure for automated manufacturing

In many large manufacturing organisations the main focus of research and development is on improving the ability of all their production operations to respond to unpredictable disturbances and increasing change on a global scale. Global competition forces companies to compete on all fronts in terms of cost, quality, delivery, flexibility, innovation and service.

Traditionally, the performance of all production systems, including those that are automated, has been assessed in terms of their output under steady-state operating conditions. However, greater product variety, smaller batch sizes and frequent new product introductions, coupled with tighter delivery requirements are becoming the norm. This means that manufacturers have to devise and develop automated operations that are capable of performing consistently under continually changing or disturbed conditions. Disturbances may be external to a production process (for example sudden changes in demand for the product, variations in raw material supply) or internal (for example machine breakdowns).

Automated production will play an increasing part in production responsive strategies. We saw in Unit 3 that a characteristic of these approaches is the negative impact it has on local and global employment patterns.

Advantages of automated manufacturing systems

- They increase 'value-added' by reduced labour costs including compensation costs for physical problems associated with repetitive tasks and back injuries.
- Precision and high speed improves cycle time, reliability and reduces downtime.
- They offer a faster time to market .
- Multi-axes robots are capable of servicing multiple machines/stations/operations in addition to having the ability to re-orient parts between operations without expensive options or use of complicated jigs and fixtures.
- They enable increased production rates not only with the accuracy and speed of a robot but also through the elimination of load-out inaccuracies and downtime required to change from product to product.
- They improve machine tool uptime productivity as much as 30 per cent by eliminating 'door open' time.
- Several sensing, motion, process and system options allow for greater control, consistency and quality output in less time with less chance of scrap parts.
- Typically, production rates are improved and do not vary by more than 3 per cent.
- There is no indirect labour training of potentially large numbers of operators.
- There is a shorter pay-back time on the capital invested in the machinery and equipment for an automated production line compared to machines with human operators.

Disadvantages of automated manufacturing systems

As we have seen, most types of robotic or automated devices are designed for manufacturing assembly or for materials handling, retrieval and storage. They cannot be used in all manufacturing situations – the greater the manufacturing complexity, the greater the complexity of calculations that need to be done before instructions can be given by the computer control system to the robot. Too complex a manufacturing process slows down a robot's speed of action and therefore increases manufacturing time. This could be a no-win situation, resulting in no manufacturing advantage – a human workforce could be more cost-effective. There are other significant cost factors:

- the cost of buying and installing and commissioning new technology so that it is effective
- the cost of recruiting and training operators with the necessary skills to enable them to use the new technology
- the cost of keeping up with new technological advances that enables the company to maintain its competitive edge.

Sensing, control and AI technologies are continually evolving and there are expected to be significant increases in the processing power and operational capability of computers in the future. The range and complexity of tasks that can be performed by manufacturing robots will therefore increase.

Tasks

1 Explain the difference between automation and robotics, giving examples of their application in your chosen materials area.
2 What are the advantages and disadvantages of flexible automation using robots in comparison with employing direct labour?

The impact of automation on employment

The most obvious impact of automation is on the numbers of people that are required to service the manufacturing process. Those people that are working in an automated manufacturing environment have to be able to operate and think in a more flexible way than ever before. This requires education, training and the development of ICT skills such as the collection and analysis of production data in order to measure their effectiveness. In the UK there is a chronic shortage of people with technical skills who are willing to take up a career in manufacturing. The people who want to work in this sector of industry need a wider range of basic skills including literacy and numeracy than previously and they must be able to transfer their skills and knowledge into new situations because the speed of change is so great. 'A job for life' is no longer the option for millions of workers across the globe. Workers in an automated working environment have to be capable of multi-tasking so that they are capable of responding to different machines within a production cell.

3. Uses of ICT in the manufacture of products

Since the 1990s some of the market forces driving manufacturing improvements have included an increased emphasis on design, innovation, operational flexibility, quick response to changing demand and the need for improved product quality. ICT is the key to enabling companies, large and small, to respond to new challenges effectively. Computers talk to one another across networks such as the Internet, which is a global network of computers. The recent development of **Integrated Service Data Networks (ISDN)** and broadband technology now means that huge amounts of information and data can be transferred across computer networks at far greater speeds than ever before. This is having a profound effect on the range and scope of electronic communications that are now possible in modern manufacturing.

Electronic communications
E-manufacturing

The term electronic manufacturing or e-manufacturing reflects the impact that ICT has made on the way that manufacturing is organised and managed. ICT is used from the boardroom, throughout the business and out into the supply and distribution (**logistics**) chains. Figure 4.3.27 shows the complex organisation of an e-business that is only cost-effective because of readily available electronic information and communication.

Beyond electronic mail (email)

ICT has improved the range and reach of electronic communication. Electronic mail or email is the simplest form of electronic communication and it has a comparatively low level of reach and range when it is used for messaging or sending files to an individual or a work group. Reach refers to the level of communication that is possible with other users across a communications network and range refers to the types of data transfer that can take place. When business or manufacturing data can be shared by anyone, anywhere, irrespective of a computer's operating system, then the building blocks for an integrated design and manufacturing system with extended range and reach is in place. Electronic data exchange (EDI) makes the system work.

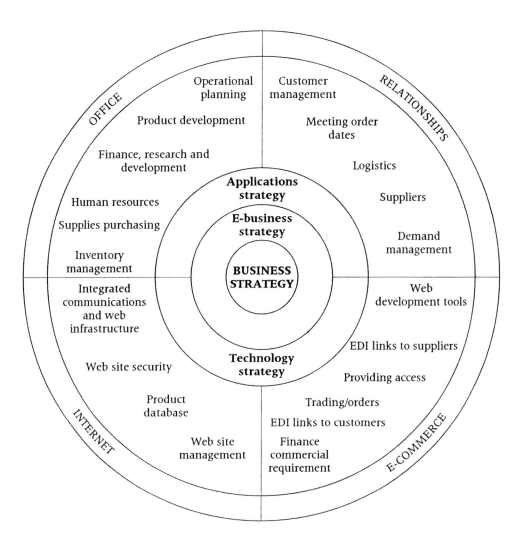

Figure 4.3.27 *The organisation of an e-business*

Electronic data interchange

ICT is the tool that integrates the use of computer systems and electronic links to create paperless trading. Figure 4.3.28 shows how this can happen in the manufacturing business so that all the participants or trading partners can communicate with each other electronically. This cost-effective and efficient technology known as electronic data interchange (EDI) is gradually being adopted by many industries as an essential tool to increase competitiveness. EDI has been a core element for many companies who have adopted the quick response and just in time as manufacturing strategies discussed earlier.

Electronic data exchange

The development of EDI into a means of exchanging technological data about all aspects of a product is well under way. CAD/CAM data interchange (CDI) is the process of exchanging design and manufacturing data. The system by which EDI and CDI are combined to provide automated transfer of data over a computer network is called electronic data exchange (EDE). There are various networks available for implementing EDE systems. The key to their usefulness in the field of graphics is their connection speed and the rate at which data can be transferred (throughput).

The modem was a big breakthrough in computer-based electronic communications. It allowed computers to communicate by converting their digital information into an **analogue signal** to travel through the public telephone network. The amount of information an analogue telephone can transfer is limited to about 56 kilobytes per second (kb/s). Commonly available internal or external modems have a maximum speed of 56 kb/s, but the actual speed

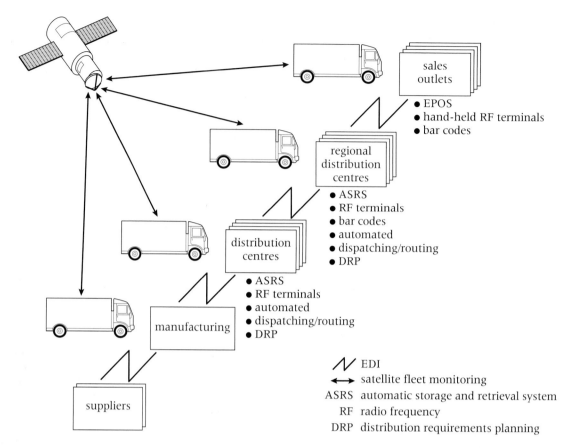

Figure 4.3.28 *The future uses of EDI in the total manufacturing process*

of data transfer is limited by the quality of the analogue connection. Computer modems routinely transfer data at about 35–40 kb/s. CAD files and other large files such as workflow data take a vast amount of time to be transferred from computer to another. This is ineffective and costly for all business, especially those involve in the graphics sector.

Integrated Services Data Network (ISDN)

Integrated Services Data Network (ISDN) allows multiple digital channels to be operated simultaneously through the same regular telephone wiring used for analogue lines. This permits a much higher data transfer rate than analogue lines. In addition, the amount of time it takes for a communication to begin on an ISDN line is typically about half that of an analogue line. This improves response for interactive applications with high graphic content and decision making such a games or conferencing.

Multiple devices

Previously, it was necessary to have a separate telephone line for each device you wished to use simultaneously. For example, one line each was required for a telephone, fax, computer, or live videoconference. To transfer a CAD file to someone while talking on the telephone or seeing his or her live picture on a video screen requires several potentially expensive telephone lines. ISDN allows multiple devices to share a single line. It is possible to combine many different digital data sources and have the information routed to the proper destination. Since the line is digital, it is easier to keep the electronic 'noise' or interference out while combining these signals.

ISDN interfaces

The ISDN Basic Rate **Interface** (or BRI) is ideal for home, small business and remote working, also known as teleworking, since it gives fast access to most of the services available in an office including telephone, fax, email, Internet. ISDN BRI offers two simultaneous connections (any mix of fax, voice and data). When used as a data connection, ISDN BRI can offer two independent data channels of 64 kb/s each or 128 kb/s when combined into one connection compared to 30–40 kb/s for a modem. The fast dial-up speed and high throughput (when compared to a conventional modem) offers a more effective remote working environment for individual graphic designers or computer artists contracted to provide graphic services such as desktop publishing or website design to larger companies or end users

The ISDN Primary Rate Interface (PRI) is intended for high-volume telephone users such as large businesses as it is well suited for central-site tasks where many concurrent connections or calls will be handled in one place. ISDN PRI provides 30 channels of 64 kb/s each, giving 1920 kb/s. As with BRI, each channel can be connected to a different destination, or they can be combined to give a larger **bandwidth**. These channels, known as bearer or B channels, are at the heart of the flexibility of ISDN. A server-based connection to ISDN can act as a gateway offering telephone services to users on the **local area network (LAN)** in the office. For example, a server connected to ISDN can accept incoming data such as CAD files, and route them to individual users on the LAN.

Broadband

Most recently, ISDN services have largely been displaced by broadband Internet service, such as cable modem services. These services are faster, less expensive and easier to set up and maintain than ISDN. Still, ISDN has its place, as backup to dedicated lines, and in locations where broadband service is not yet available. Broadband technologies offer the possibility of extending a product visualisation right into the customer's home through the electronic equivalents of mail order or catalogue shopping. This kind of selling is often referred to as e-marketing and it is commonly found on the Internet and digital television channels.

Local area networks

Local area networks (LAN)are closed networks that are used in a variety of ways to meet individual needs. They are used for email and the exchange of other electronic data within an organisation. For example, a LAN can be used to communicate manufacturing data from the design office directly to CNC manufacturing equipment within the same building.

Wide area networks

A wide area network (WAN) allows data to be transferred to processing or business centres around the world using existing digital telephone systems. To ensure compatibility between different systems WANs follow agreed standards and **protocols**. This can be so costly and ineffective, as they require dedicated equipment and software, that many companies are now using the global network provided by the Internet as it is easier to access through standard **Internet service providers (ISPs)**. There are other types of network in use.

Intranets and extranets

An intranet is an Internet-based system used for communications and data exchange with web-type pages but there is a '**firewall**' that prevents unauthorised access to the site. Companies use intranets in the same way as the Internet is used, to share information and expertise, but they are much less expensive to set up compared with a WAN.

An extranet is an increasingly popular way for a company to share its data with its business partners though usernames and secure password-controlled access. The 'identity' of individual users will determine which parts of the network they can enter. These networks are also used for subscription services on the Internet to provide access to the 'expert' knowledge systems that were discussed earlier. Some companies now combine the ease and flexibility of access provided by an Internet website with web pages where access is password controlled. For instance, a company providing graphics equipment might have a series of pages that describe its range of products and retail pricing. The secure section accessed by retailers will include details of local suppliers, wholesale costs and conditions of resale to a customer.

Global networks

The world wide web is a global network of Internet **servers** that process and communicate data via cable, radio and satellite. **Web browsers** and search engines with which you will be familiar, such as Netscape Navigator, Microsoft's Internet Explorer and Ask Jeeves, make it easy to access information sources. Each website has a unique address called a **URL (universal resource locator)**. Web pages are specially formatted documents that are written in a language called **HTML (Hypertext Mark-up Language)** that supports links to other web pages, graphics, audio and video files. **Hyperlinks**, the specially marked 'hot spots' sometimes underlined, sometimes in a different colour or shown as a graphic 'button', allow the user to jump quickly from one website to another.

Advantages of using the Internet and the web

The Internet is becoming an invaluable tool for designers, manufacturers and suppliers. It provides:

- an easily accessible means of sharing ideas at a relatively low cost

- a vast and ever-growing body of knowledge and information (but remember it might not always be correct as there is no control on information that can be put into the web)
- a medium for communicating with current and potential customers and for seeing what the other designers or manufacturers are producing
- a readily accessible online source of reference product information and design data ranging from product or material specifications and catalogues of parts and components, to marketing trends and other commercial data.

Disadvantages of using the Internet

Industrial espionage is a real problem in a paper-rich environment but there is no guarantee that the Internet is any more secure without sophisticated data encryption software. The problem is that the more secure a system, the more difficult it becomes to use and this is said to be putting off potential customers of commercial websites. The growth of e-business on the Internet can be further restricted by scare stories of computer viruses destroying computers after accessing emails via the Internet. Again, virus software that can be updated via the Internet will eventually ease the fears of users.

Tasks

1 What are the differences between a LAN, WAN and a global network, and explain where they are used.
2 Describe the benefits to designers and manufacturers of using the Internet.
3 It is easy to waste a lot of time when trying to find information on the world wide web, so it is important to be clear about what you want to find out. Using a web browser or search engine of your choice, produce a short list of websites that provide information about CAD, robotics or any other area studied on this course. To share the information you have found present your links as a reference source for others to access. This might be in the form of a web page produced using web-publishing software or a presentation put together on a software package like PowerPoint.

Videoconferencing

Computers, electronic communications and video technologies have revolutionised the way people live and work. **Videoconferencing (VC)** is a rapidly growing segment of the ICT sector that integrates these three technologies to enhance communications and speeds up decision making by eliminating the need for time-consuming travel to meetings, which may be across the other side of the world. The use of ISDN and broadband technologies with improved rates of data transfer means that problems such as video pictures that are jerky or 'choppy', sound and picture out of synchronisation and poor quality or fuzzy images are a thing of the past. Two main types of VC organisation are used today:

- Desktop videoconferencing (DTVC) works like a video telephone between two people. Each person has a video camera, microphone and speakers mounted on to a desktop computer, equipped with a sound card. This means that each person can hear and see the other talk in a small window on his or her computer. Lower hardware costs have meant that this a readily available system that has even reached into the home computing market.
- Multi-point VC enables three or more people to sit in a 'virtual' conference room and communicate as if they were sitting next to each other.

Benefits of videoconferencing for manufacturers

We have seen how ISDN lines allow simultaneous file transfers. After a design meeting ends, the design information which has just been discussed needs to be sent to a regional office. This electronic file can be transferred either after the VC meeting has ended or, if a two-channel ISDN line is in use, it can be sent in the background on one channel while sight and sound information is still being exchanged on the other channel. VC can be used in other applications:

- Marketing presentations. In an era of global companies, marketers and manufacturers may not operate in the same country. When a new product is to be launched at a trade show or directly to the public, it is important to know how the product will be presented to persuade people to buy it. VC provides the opportunity to see the presentations as they develop and enables instant opinions on their effectiveness.
- Corporate training. If there is something employees all over the world need to know, it may be much faster and cheaper to train them using videoconferencing.
- Remote diagnostics. Experts in a particular process may work in one office but the problem requiring their immediate expertise

might occur halfway around the world. VC offers the experts the opportunity to solve the problem without travelling from the office. Production downtime is reduced and the expert and the other employees can get back to whatever it was they were originally doing without wasting time.

Remote manufacturing of components using VC
ICT provides the opportunity for expertise and expensive equipment to be made available to a number of schools, colleges and universities from one central location. Denford Ltd has pioneered the educational use of CAD/CAM via video conferencing in the UK, enabling staff and students to manufacture parts at a distance in real time using industry standard equipment and software (see Figure 4.3.29 and Figure 4.3.30).

Task

Describe the benefits that videoconferencing might bring to the design, manufacture and sale of a graphic product.

Figure 4.3.29 *Denford has identified the potential of videoconferencing as an educational tool by incorporating the vital dimension of data sharing to allow staff and students to open notebooks, documents, drawing files and work together discussing problems and jointly solving them*

The steps involved in remote manufacturing through video conferencing

(see Figure 4.3.30)

Step 1. The student creates a design on CAD/CAM software on a PC (stand-alone or networked).

Step 2. The student and teacher participate in a live VC with the Remote Manufacturing Centre or the Denford on-demand facility.

Step 3. The student's design is downloaded to the Remote Manufacturing Centre where it is discussed and amendments made, where necessary.

Step 4. The Remote Manufacturing Centre creates the CNC file containing the student's design.

Step 5. The student's design is manufactured on a CNC machine while he or she views via the remote video camera.

Step 6. The finished component is evaluated, shown to the student on the VC and then posted back to the school/college.

Source: Denford Ltd

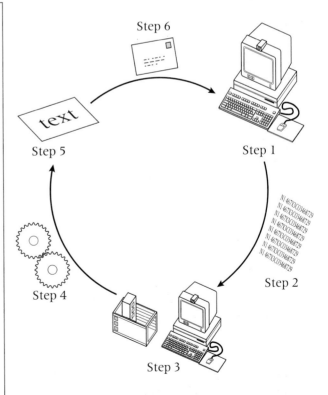

Figure 4.3.30 *The steps involved in remote manufacturing through videoconferencing*

The use of ICT for graphic communication
Electronic whiteboards
Sometimes called a 'smart board', this recently developed communication technology integrates the simplicity of a whiteboard with the power of a computer to make a tool for graphic communication. It is an interactive, flexible visual device for use in presentations, videoconferences, training sessions and for recording data (see Figure 4.3.31). Simply touching the board allows access to:

- a sensitive writing surface with a scanner and a thermal printer for producing hard copy
- computer software such as Microsoft Word, PowerPoint, spreadsheets and databases, for file manipulation and storage, printing or emailing
- data and video images, either stored on a computer or in the form of a live stream from a camera, video recorder or the Internet
- an automatic way of recording a video conference, which is available for playback and reference at a later date.

Information centres or PC kiosks
These interactive kiosks process, communicate and display specific graphic information and data stored on a computer (see Figure 4.3.32). Designed to automate information distribution, they are easily accessed by hotel guests, council tenants, tourists, etc. The kiosks can operate 24 hours a day and often include maps, display forthcoming events, employ graphic symbols and important contact information.

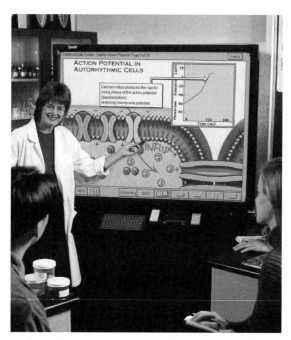

Figure 4.3.31 *Electronic whiteboard*

Figure 4.3.32 *A modern PC information kiosk*

Automated information appears in various forms:

- multimedia presentations of local information such as the location of restaurants, shopping centres, movies that are showing, hospitals, schools, addresses, businesses or local attractions, all accessed from a touch-screen, keyboard or mouse-driven interface
- information broadcasts via a customised mini-Internet broadcasting system using web-cams for video teleconferencing
- interactive services at a range of trade and product shows that allow potential customers immediate access to critical marketing information and product reviews. This type of interactive system can be used to create newsgroups, chat-rooms and portals to a vast range of community information in organisations such as local and regional authorities.

Kiosks have also been designed for museum displays, public Internet access and information booths in public spaces. Changes in the size and power of personal computers have revolutionised the design of these information centres. The large and bulky kiosks of yesterday have been transformed into the compact, slim-line designs of today. The PC kiosk with back-lit advertising signage shown in Figure 4.3.32 has a state of the art Pentium processor with 128 Mb RAM at its heart, combined with a 15" LCD 'touch-screen', touch-pad and standard keypad. Extra functionality and security is provided by the

Figure 4.3.33 *An information kiosk for a public place*

inclusion of a credit card reader, camera and fully linked 'windows' type information screens. Figure 4.3.33 shows a PC kiosk of the future.

Task

Discuss how electronic communications have impacted on manufacturing and business practice worldwide.

Electronic information handling

Agile manufacturing

An agile manufacturer recognises the uncertainty that change brings and puts flexible manufacturing systems (FMS) and mechanisms in place such as **quick response manufacturing (QRM)** to deal with it. The organisation moves from being production-driven to customer-driven and it realises that customers will not pay a premium for product quality: quality is always assumed. Agile manufacturers work in partnerships with customers and suppliers, and understand that the so-called 'soft' business information processes are important to the entire manufacturing process. In an agile manufacturing environment, information is the primary enabling resource and ICT is the enabling technology. In order to find out what the customer needs and wants manufacturers set up a range of computer-based systems for collecting market information.

Computer-aided market analysis (CAMA)

Market research is a term that describes the collection and analysis of data about consumers, **niche markets** and the effectiveness of marketing programmes. Market analysis focuses on the collection, analysis and the application of research data to predict the future of a particular market (trends).

The analytical process includes the examination and evaluation of relevant information (data) in order to select the best course of action from the possible business options. The analysis can be undertaken in-house by the manufacturers themselves, through specialist market research agencies, or consultancy firms who will provide tailor-made or customised data sets related to specific companies, customers or markets.

The analysis of business data and market trends generates a vast flow of information and data that is now most efficiently managed in a computerised relational database. Such a database can be interrogated in various ways depending on the type of analysis required:

- A qualitative analysis will tell a company who is buying the products and why, or what customers like or dislike about a product and its after-sales support.
- A quantitative analysis will provide facts and figures such as where the products are being bought and when, or a comparative analysis of the company's financial ratios over time.
- A trend analysis will tell a company what is happening in a particular sector of a market and put the company's performance into a local, regional, pan-European or global context.
- **Market timing** is about attempting to predict future market directions, usually by examining recent product volume or economic data, and investing resources based on those predictions.
- In situations of high product volumes, existing customer information can be profiled against lifestyle surveys and demographic data to give a detailed picture of the company's ideal customer.

Benefits of CAMA

In a modern e-business environment, a manufacturer needs access to up-to-date research, in-depth product and market analysis and industry-specific expertise to make the best ICT-led decisions in relation to the core business goals. Using advanced computer-based marketing tools such as CAMA will help a manufacturer to:

- convert data into actionable information for sound marketing and planning decisions
- calculate demand for products and services more accurately and set the right sales targets
- identify markets and find out where potential customers are shopping
- employ a marketing technique – **market segmentation** – that targets a group of customers with specific characteristics
- launch new products with focused strategies such as regional or mini product launches rather than whole country product 'roll outs'.

Computer-aided specification development

An effective product design that satisfies functional requirements and can be manufactured easily requires vast amounts of 'expert' knowledge. Integrated ICT-based systems already exist where design features can be generated by CAD software and then checked for ease of manufacture (design for manufacture – DFM) and for ease of assembly (design for assembly – DFA) by a knowledge-based expert system. With the increased volume of design information generated by computer-based systems, designers now have to analyse and evaluate large data sets of design constraints to find a design specification that satisfies those constraints.

Computer-aided specification of products (CASP)

A design specification is a document which explores and explains the internal design of a given component or sub-system. It should be sufficiently detailed to permit manufacturing to proceed without significant reworking resulting from flawed or missing design details. It will often draw on previous specifications or other design information. The computer-aided specification of products saves time and costs through basing new product specifications on those already held in a product data management (PDM) system which is, in effect, an intelligent design system.

There are three classifications of information or knowledge held within an intelligent design system: CAD data, a design catalogue and a knowledge database. The CAD data contains specific information about the physical characteristics of each component part being designed. The design catalogue is a reference for data such as the cost, weight and strength characteristics of the standard materials, or parts and fasteners that are available to the manufacturer. The knowledge database contains 'rules' about design and manufacturing methods. These automated and intelligent design environments when fully developed will enable a designer to perform and manage complex automated design tasks including decision making.

> **Task**
>
> During this course you have developed different products to meet different demands. In other words, you have created your own 'expert' database. Using this knowledge, list six design and six manufacturing 'rules' that you have applied. Write your rules as an if/then statement. For example 'if the logo has been designed on a CAD package, then I can use a CNC plotter to manufacture the vinyl product.'

Automated stock control

Earlier in the unit, we looked at the benefits that arise out of applying computers to implement the just-in-time (JIT) philosophy where materials and components are ordered just in time for production. A characteristic of JIT is that product variety increases as the range of processing systems decreases and waste is reduced by a variety of means including automated stock control systems. These ensure that the size of the **inventory** is optimised but available on demand when and where it is needed. These automated stock control systems support the move from batch to continuous flow production. They are also an advantage in the process of line balancing, which is a scheduling technique that reduces waiting times caused by unbalanced production times.

Production scheduling and production logistics

Computer-based scheduling and logistics systems ensure that a production plan is implemented and that production is 'smoothed' so that small variations in supply and demand are managed without causing problems. This is achieved by spreading the product mix and the quantities of each produced evenly over each day in a month. Different manufacturers have different **planning horizons** that determine when the detailed production plan is produced; typically, it is one month in advance. The advantage of these computer-based scheduling and logistics applications is that:

- they are flexible and easily adapted if the product mix or quantity required is changed at short notice
- they minimise work in progress and reduce the inventory
- they maintain balance between the stations on a production line
- they raise productivity levels.

Flexible Manufacturing Systems

Quick Response Manufacturing (QRM)

QRM focuses on reducing product lead times and the production of small batches such as advertising or promotional gifts. QRM provides a better 'time to market'. Real-time reprogramming of manufacturing is a production management tool that offers high-volume, rapid turnover manufacturers the potential for enormous savings in time to market and increased business efficiency.

The ability to re-programme automatically both manufacturing and business processes in response to market pressures is on the horizon. Stock levels will be constantly re-evaluated as demand patterns change. The demand patterns also affect capacity planning and so the relevant individual within the business would be automatically alerted. The impact that a change to the business or a manufacturing process may have will be immediately available for review from anywhere in the world. The change to real-time re-programming also presents the possibility for disastrous levels of confusion unless the information flow is carefully managed by an effective PDM system using highly integrated knowledge bases.

Production control

Quality monitoring

Mechanical methods of inspection involve direct physical contact between a probe and the component or product being evaluated. The probe sends data back to a computer or microprocessor for checking against the product specification. These systems are concerned with checking the dimensional accuracy of a product such as the injection mould for producing a plastic package such as a shampoo bottle.

The computer-aided optical monitoring of quality involves the use of electronic sensors that use scanning and optical devices, digital cameras and vision systems. They are used in a range of situations such as checking the inking pattern on a poster or other printed product. They can be used for ensuring that colour mixes and inks are consistent throughout a high-volume printing process. They can be used for comparing the quality of a visual product against the original artwork or design.

Using digital cameras for monitoring quality

Inspection has always been part of the manufacturing process. However, this often involved the inspection of a random sample of finished components; if any errors are detected at this late stage, the main processing operations have already been carried out and putting the fault right is costly and may not be possible. This results in high scrap rates and wasted processing time. To help alleviate this problem, 100 per cent piece-by-piece inspection of work in progress can be carried out using low-cost digital cameras at critical points on the production line. The cameras are connected to a computer or to a dedicated microprocessor containing the original specification. Differences between what has been tested and the design specification are automatically evaluated and an audible or visual signal identifies when there is a problem. This allows rapid real-time quality control. The other main advantages of optical methods of monitoring quality are that:

- there is no direct mechanical contact with the product so the quality of the product, especially if it is a graphic product, is not compromised
- the distance between the optical sensor and the piece being measured can be large
- the response time is fast because of the electronics built into the system.

Product marketing, distribution and retailing

Using ICT for the business processes in manufacturing

Throughout this unit we have considered how ICT is being employed to improve profitability and product quality on the production side of manufacturing. CAD/CAM technologies and the use of robotics lead to new ways of organising manufacture. As we have seen, ICT can also be used to speed up business processes such as

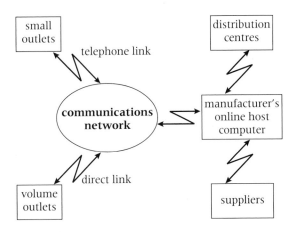

Figure 4.3.34 *Manufacturer's communication links with distribution centres, suppliers and retail outlets*

invoicing and stock control but this on its own will not necessarily increase productivity. One of the key benefits that increased use of ICT brings is in better business integration. This means that manufacturers not only have to change their manufacturing processes but also the way that they do business outside the company. Refer back to Figure 4.3.27 on page 255 to review the complex organisation of a large e-business.

Electronic point of sale (EPOS)

Information from many sources, including EDI, is at the centre of any business organisation. EPOS is a form of EDI that allows companies to monitor product sales in order to respond by supplying and delivering their products and services to their retailers or customers faster (see Figure 4.3.34). For retailers it allows them to keep their costs down by reducing the stock that has to be held in reserve. Almost all products sold in retail outlets have a unique bar code that is scanned by fixed or hand-held laser-operated bar code readers. The bar code contains a range of electronic information about the product which is collected automatically or at scheduled times during the day. In addition, the two-way information flow can include financial data, emails and price updates from the supplier to the product retailer. EPOS and the associated management software provides manufacturers with:

- a full and immediate account of the financial transactions involving the company's products
- the data needed to undertake a sales/profit margin analysis, as it can be exported into financial management software such as spreadsheets
- the means to monitor the performance and popularity of all product lines, which is particularly important in high-volume situations as it allows the company to react quickly to demand fluctuations and be able to deal with surges in demand by being notified of them immediately
- accurate information for identifying buying trends when making marketing decisions rather than relying on guesswork
- a full and responsive stock control system by providing real-time stock updates for product sales, transfers between retail outlets, goods received, etc.
- a system to ensure that sufficient stock is available to meet customer needs without overstocking and tying up capital. This is known as a continuous product replenishment system (CPR) which is a feature of high-volume merchandising.

Internet marketing

Entering the Internet market is a major strategic decision for any manufacturer. It generally means an almost total restructuring of the internal and external business processes as well as supply chains and relationships with customers. For many manufacturers this redesign of the business process presents a bigger challenge than the change to an automated production system or other ICT-based manufacturing process.

For many companies the rush to the Internet and the development of websites was ill conceived and badly thought out. Many businesses invested heavily in terms of time and money merely to put their existing product catalogues out on the world wide web to little commercial effect. There was no underpinning business strategy and customers were able to do little more than browse the catalogue electronically. If they wanted to contact the company, it had to be by conventional means such as phone or fax. Some companies added email addresses in recognition of a need to communicate interactively. Using the Internet to market a business is more than that. Some large companies are looking to small and medium-sized graphic design consultancies companies who specialise in the design of websites or the development and use of 3D images.

The move towards using interactive 3D virtual products in e-marketing on the web is a natural extension of modern marketing and sales strategies. Manufacturers are faced with ever-shortening product life cycles, increasing product complexity and greater global competition. They must ensure that existing and potential customers understand their products better. They might even have to provide a means by which individual customers are able to configure a product to meet their own specific needs as we saw when we looked at digital print solutions.

The benefits of Internet marketing include:

- instant global reach, access to new markets and an increased customer base
- faster processing of orders and transactions, resulting in efficiency savings and reduced overhead costs
- detailed knowledge of user preferences, leading to improved customer service through better product and after-sales support
- reduced time to market
- an increased company profile on a worldwide basis
- for small and medium-sized companies, especially those offering bureau or other specialist graphic services, it creates new

markets because it is often easier and cheaper to create an Internet presence than it is to promote their products by conventional printed or visual means.

The future use of ICT in the graphics field

We have seen that advances in ICT are already providing manufacturing with benefits throughout its business. The impact of ICT is also powerful in the field of graphic art.

Interactive graphic design

Games playing and simulations are examples of interactive graphic design but there is a growing and exciting range of interactive and visual experiences to be found. Digital media is influencing the arts and is giving form to new artistic languages that make bridges between the real world and imaginary virtual environments. Electronic visualisation is the art and science of creating images on electronic screens and on virtual reality display devices. Electronic visualisation uses the tools of advanced computer graphics, computer animation, interactive graphics, video and virtual reality (VR) to create a more intensely aesthetic interpretation of imagery and visual design in a virtual reality environment.

The role of the printed book as the prime information source has also been transformed by the arrival of digital inputs that create new information pathways that are interactive, versatile and aesthetically challenging in addition to engaging the intellect. Interdisciplinary groups are being set up around the world to develop the creative potential that evolves from the integration of technology, communication, art and design to expand the boundaries of traditional artistic media and of visitor participation in artwork. For example, one group has created an emotional, virtual reality experience for the public of being in different centuries and in different worlds all in one unique container, the Multi-Mega Book in a virtual environment. The range and scope of this exciting digital fusion is beyond the scope of this book but you can gain an insight into the future of digital media by visiting the following web sites www.evl.uic.edu in the USA and www.fabricator.com in Italy. These will lead you to other areas where science technology and art come together to create informative and educational interactive graphic applications.

Exam preparation

You should now have completed the work and covered all the sections in this unit option. Much of the work covered will possibly have been done in a theoretical context, with some practical or investigational work using CAD/CAM resources and equipment. You will have developed your basic skills in using information sources including the Internet. You will probably have answered many of the questions or tasks set in the text throughout the unit. These will have enabled you to apply what you have learned in that particular section.

There are some additional questions for you to do below. They are similar in format and style to those that you will expect to find in the exam paper.

Here are some tips to help you prepare for your exam:

- Plan your programme of revision well in advance – and stick to it.
- Do not leave it until the night before.
- Little and often is better than a lot all at once.
- Try condensing all you know about a specific topic on to one side of A4. Then make notes under key headings or by using bullet points.
- There are a number of flow diagrams, organisational charts and other visual means of communicating information that are used in this unit. You will find it helpful to learn these and also to develop diagrams of your own for this exam. A clear diagram or chart will demonstrate what you know and understand effectively. It also shows your ability to organise and present information clearly and logically.

Practice exam questions

1 Consumers in the industrialised world have differing and ever-changing needs. When this is combined with the advances in technology and reduced waiting times, it increases the pressure on manufacturers to produce new, improved or innovative products. Using specific examples of products in your chosen materials area, explain how manufacturers are using ICT to develop quick response manufacturing (QRM) systems to meet market demands. (15)

2 a) The organisation and management of manufacturing has been completely changed by the use of computer systems. Explain the following terms related to computers used in manufacturing:
   i) computer-aided production planning (CAPP) (2)
   ii) computer-integrated manufacturing systems (CIM). (2)
   b) Describe how computer technology provides support for monitoring and inspecting the quality of products. (5)
   c) Using examples from your chosen materials area, evaluate the use of computers in a flexible manufacturing system (FMS). (6)

3 Discuss the impact of the use of CNC machines in the design and manufacture of products within your chosen materials area. (15)

UNIT 5

Product development II (G5)

Summary of expectations

1. What to expect

You are required to submit one coursework project at A2. This A2 project is described as being **'synoptic'** to the AS project because it enables you to draw together and apply all the knowledge and understanding related to designing and making that you learned in Unit 2. The A2 project should, therefore, show clear progression from AS to A2.

> **SIGNPOST**
> For more information about 'synoptic assessment', see Unit 6 page 292

The A2 project should comprise a product and a coursework project folder. It is important to undertake a coursework project that is of a manageable size, so that you are able to finish it in the time available.

2. How your project will be assessed

The A2 coursework project covers skills related to designing and making. It is assessed using the same assessment criteria as the AS project (see Table 5.1)

Table 5.1 AS coursework project assessment criteria

Assessment criteria	Marks
A Exploring problems and clarifying tasks	10
B Generating ideas	15
C Developing and communicating design proposals	15
D Planning manufacture	10
E Product manufacture	40
F Testing and evaluating	10
Total marks	100

At A2 the assessment criteria demand a different level of response, requiring you to demonstrate a greater level of sophistication and more in-depth knowledge and understanding. You are also expected to take more responsibility for your own project management.

Your A2 project will be marked by your teacher and the coursework project folder will be sent to Edexcel for the Moderator to assess the level at which you are working. It may be that after moderation your marks will go up or down. The Moderator also gives you a grade for your coursework project.

3. Choosing a suitable project

At A2 you are expected to take a more commercial approach to designing and making a prototype product, which could be batch or mass produced for a range of users in a target market group. The project may be developed in collaboration with potential users, or in collaboration with a client (such as a local business or organisation). Collaboration could take the form of consulting with a client or users to develop a design brief and specification and to consult with them as the project progresses.

There are two possible approaches to project work for the Advanced GCE:

- Two different starting points.
- The progressive project.

The most likely approach you will take is two different starting points. This means starting from one design context, topic or theme for the AS coursework and then a different starting point for the A2 coursework. This should result in two completely different coursework project folders and two completely different end products, both of which must satisfy the assessment criteria.

If you take the progressive project approach, the evaluation of your AS project is used as the starting point for your A2 project. This should result in two completely different coursework project folders and two different products, each of which must satisfy ALL the assessment criteria at AS and A2. Think about the following before deciding to take a progressive project approach:

- The development from the AS project needs to be sufficiently interesting to develop into a full project at A2.
- The A2 project must meet ALL the assessment criteria shown in Table 5.1.
- Two years is a long time to spend on a project that has the same or similar design context, even if the A2 project develops in a different direction.

4. The coursework project folder

This should be similar in content to the AS folder, comprising around 20–26 pages of A3/A2 paper.

- Include clear, concise information that is relevant to the project.
- Be very selective about what to include, so that you target the available marks.
- Include a contents page and a numbering system to help its organisation.

- The title page, contents page and bibliography are extra pages, on top of the suggested 20–26 pages.

5. How much is it worth?

The coursework project is worth 20 per cent of the full Advanced GCE.

Unit 5	Weighting
A2 level (full GCE)	20%

A Exploring problems and clarifying tasks (10 marks)

1. Identify, explore and analyse a wide range of problems and user needs

Look at Figure 5.1 and consider the question in the caption.

SIGNPOST
Identify, explore and analyse a wide
range of problems and user needs
Unit 2 page 41

Developing a project at A2

The A2 project is 'synoptic' to the AS one because it enables you to draw together and apply all the knowledge and understanding related to designing and making that you learned in Unit 2. This experience should provide you with a clear understanding of the assessment requirements and give you a solid basis for developing a new project at A2.

It may be helpful, at this stage, to read the AS and A2 exemplar projects in the Edexcel Coursework Guide, which demonstrate the standard at which you are required to work during the A2 project. You will be required to work more independently, which may involve using a wider range of people to support you in your work. Your most difficult decision now is to decide what to design and make that will enable you to demonstrate your knowledge, understanding and skills. Your choice of A2 project must enable you to produce:

- a folder (in A3 or A2 format) that summarises the development of 2D/3D coursework elements
- a 2D element, developed from traditional and modern graphics media – the 2D element should be linked to and support the 3D outcome
- a 3D model or prototype product, constructed from modelling materials *and* resistant materials (which could include some wood and/or metal and/or plastic). The 3D outcome should be semi-functioning.

New ideas or problems may have come to mind during your work on Unit 2, or when you discussed your project with others. In order to develop your A2 project, you will need to decide on a design context and a target market group.

Developing a commercial approach

At A2 you are expected to take a more commercial approach. This means designing and making 2D and 3D elements that may have the

Figure 5.1 *The Nike logo has been used on sports shoes since the 1970s and is one of the most successful logos of all time. Explain why you think that this is so.*

potential for batch or high-volume production. In industry a prototype product is made prior to manufacture to test every aspect of the design before putting it into production. This commercial prototype product is as close as possible to the 'real' end product. Your prototype product also needs to be as close as possible to the 'real' thing, so you should make it to the highest possible quality.

One way of developing a commercial approach to your project is to collaborate with potential users or with a 'client', which could be a local business or organisation. If you do have the opportunity to collaborate on a project, you will be working in a similar way to a professional designer, who has to work to a client brief, meet the needs of others, work to a budget and to a deadline.

There are a number of ways to develop a commercial approach to project work and you could consider some of the following:

- Use work-related materials produced by a business to inspire a project, for example using an information pack about a company to develop a product that projects a brand image that is appropriate to that company or retailer. In this case, you would need to carry out market research to find out about the kind of customers that the company targets.
- Use the expertise of a visitor from business to inspire a project, for example after discussing a company's marketing strategies and the type of products they make. In this case, you may be lucky enough to benefit from advice and support from the company throughout the project.
- Make an off-site visit to identify a specific problem that will meet the needs of the local community, for example visiting a local community centre, primary school or a

workplace, where different needs can be investigated.
- Use work experience as a context for designing and making a product, for example using a part-time job or work experience to spark off ideas about developing a product to meet specific needs.

Identifying and analysing a realistic need or problem and exploring the needs of users

SIGNPOST
Identifying and analysing a realistic need
Unit 2 page 42

Once you have decided your approach for your A2 project, you can start the actual development process. It is always difficult at the start of a project to know exactly where and how to start, and what the end product will be – this is part of the excitement of product design. The best way forward is to undertake two tasks that are interrelated:

- identify problems that may lead to product development
- identify the needs of potential users and a product that may fulfil these needs.

You can explore the needs of potential users and look for product ideas by undertaking market research. In many industries, market research is carried out to identify the taste, lifestyle and buying behaviour of potential customers. This establishes the profile of a potential target market group.

Task

Using market research techniques

Collect information about your target market group and investigate products using market research techniques:

a) Produce a product report through window shopping, going into stores, visiting galleries or museums. Identify product type, price ranges, market trends and new ideas.

b) Use questionnaires/surveys to identify user needs and values. Identify age groups, available spending money, favourite product types and brand loyalty.

c) Use product analysis to find a 'gap in the market'. Identify design styles, construction processes, quality of design and manufacture and value for money.

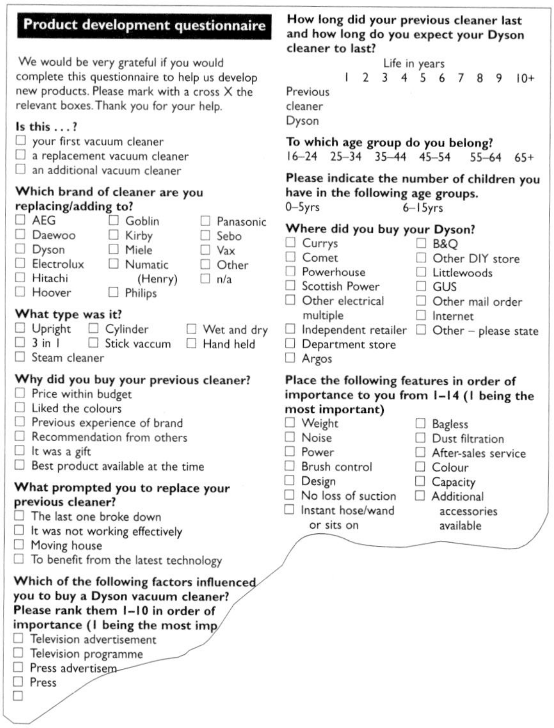

Product development questionnaire

We would be very grateful if you would complete this questionnaire to help us develop new products. Please mark with a cross X the relevant boxes. Thank you for your help.

Is this . . . ?
☐ your first vacuum cleaner
☐ a replacement vacuum cleaner
☐ an additional vacuum cleaner

Which brand of cleaner are you replacing/adding to?
☐ AEG ☐ Goblin ☐ Panasonic
☐ Daewoo ☐ Kirby ☐ Sebo
☐ Dyson ☐ Miele ☐ Vax
☐ Electrolux ☐ Numatic (Henry) ☐ Other
☐ Hitachi ☐ n/a
☐ Hoover ☐ Philips

What type was it?
☐ Upright ☐ Cylinder ☐ Wet and dry
☐ 3 in 1 ☐ Stick vaccum ☐ Hand held
☐ Steam cleaner

Why did you buy your previous cleaner?
☐ Price within budget
☐ Liked the colours
☐ Previous experience of brand
☐ Recommendation from others
☐ It was a gift
☐ Best product available at the time

What prompted you to replace your previous cleaner?
☐ The last one broke down
☐ It was not working effectively
☐ Moving house
☐ To benefit from the latest technology

Which of the following factors influenced you to buy a Dyson vacuum cleaner? Please rank them 1–10 in order of importance (1 being the most imp
☐ Television advertisement
☐ Television programme
☐ Press advertisem
☐ Press
☐

How long did your previous cleaner last and how long do you expect your Dyson cleaner to last?

	Life in years									
	1	2	3	4	5	6	7	8	9	10+
Previous cleaner										
Dyson										

To which age group do you belong?
16–24 25–34 35–44 45–54 55–64 65+

Please indicate the number of children you have in the following age groups.
0–5yrs 6–15yrs

Where did you buy your Dyson?
☐ Currys ☐ B&Q
☐ Comet ☐ Other DIY store
☐ Powerhouse ☐ Littlewoods
☐ Scottish Power ☐ GUS
☐ Other electrical multiple ☐ Other mail order
☐ Internet
☐ Independent retailer ☐ Other – please state
☐ Department store
☐ Argos

Place the following features in order of importance to you from 1–14 (1 being the most important)
☐ Weight ☐ Bagless
☐ Noise ☐ Dust filtration
☐ Power ☐ After-sales service
☐ Brush control ☐ Colour
☐ Design ☐ Capacity
☐ No loss of suction ☐ Additional accessories available
☐ Instant hose/wand or sits on

Figure 5.2 *Part of a Dyson company 'Product development questionnaire'*

For manufacturers the 'customer' plays a key role in the product development process. Without customers there is no need to make products, so their views are vital if a product is to sell in the market place. Manufacturers need to ensure that their products meet customer requirements, at the right quality, at the right price and at the right time. For example, it is no good trying to market a product that is made from purple plastic if the customers' concept is that purple is last year's colour! Feedback from customers is, therefore, crucially important if their requirements are to be met. Figure 5.2 shows how one manufacturer sought the views of potential customers.

Clarifying the task

SIGNPOST
Clarifying the task
Unit 2 page 42

Clarifying the task means deciding on the purpose for your prototype product and how it will benefit users – including the image it needs to project in order to promote its market potential. Market research about buying behaviour may help here. You should also consider the factors that affect customer choice when buying products. These may include considerations of:

- price and value for money
- aesthetic factors, like the product's appearance and the image it provides the users
- marketing factors, such as brand image
- functional requirements of packaging materials or point-of-sale display, such as ease and safety in use.

> **To be successful you will:**
> - Clearly identify a realistic need or problem.
> - Focus the problem through analysis that covers relevant factors in depth.

2. Develop a design brief

> **SIGNPOST**
> Develop a design brief
> Unit 2 page 42

Whichever approach you take in developing your A2 project, you should find that your early exploration of problems and user needs should help you develop a design brief. Take into account:

- the views of potential users
- information about similar products
- the market potential of the product
- its benefit to the users.

Your brief should enable you to plan what, how and where you need to research in order to develop a product that meets the needs of your potential users.

> **To be successful you will:**
> - Write a clear design brief.

3. Carry out imaginative research and demonstrate a high degree of selectivity of information

> **SIGNPOST**
> Carry out imaginative research and demonstrate a high degree of selectivity of information
> Unit 2 page 43

At Advanced level you should aim to demonstrate clear progression from your AS work, a greater level of sophistication and more in-depth knowledge and understanding. What this means is that you should demonstrate evidence of a higher level of design thinking by your ability to do the following:

- Undertake research that targets the design brief more closely. Do not waste time doing unnecessary research.
- Select and use relevant research information. Only include in your coursework folder information that is directly connected to your project. For example, do not include everything you know about hardwood if you are not using wood to make your product. Even if you are using some wood, you still do not need to include everything you know about it, only information relating to its appropriateness for making your product.
- Make closer connections between your research, your product design specification and your design ideas. Making connections is what this unit is all about (see Figure 5.3).

If you are collaborating with users or a client on your project, it may be a little easier to target your research, because you will have specific requirements and users in mind. Include a range of primary and secondary research. This may include using questionnaires, analysing existing products, using statistical data and identifying market trends.

At Advanced level you should be thinking about a range of issues related to the design and manufacture of your 3D prototype product, as well as issues related to the 2D elements you develop. These may relate to the ease and cost of manufacture; how to ensure the performance, quality and safe use of your product; and if there are any environmental issues related to your chosen materials.

In industry, product design teams do not generally undertake market research themselves, but rely on market information produced by the marketing department. Many manufacturers use product data management (PDM) software to organise, manage and communicate accurate, up-to-date product information in a database. Everyone in the product design team has access to this information, making it easier and faster to get the product to market. Since you are not in this happy position, you have to do your own research, so bear in mind that you have to meet deadlines. If you have access to the Internet, you may be able to find useful information about existing products, for example about materials properties, product prices and styling. Many companies also provide information through

Research

As far as I am aware, this is the first time that a project combines all aspects of a modern office. As there are no previous designs to look at and improve upon, I am going to look at the design of each individual aspect of the modern office. Areas that I am most interested in are the desk/storage area, seating and equipment.

While the office has become a target for design influenced equipment, the desk/storage area has not benefited much from design in recent years. Few pieces of office furniture can be considered aesthetically pleasing as opposed to purely practical. The Dyson dual cyclone vacuum cleaner, however, does achieve both. Like the Dyson, a desk/storage solution that combines both good design and practicality would be a success.

While seating is one area of office design that has been developing rapidly, I feel that having the seat attached to the unit would provide areas of design to be exploited that have not been available to the designers of seats before. I feel that seating is an area in which it will be hard to innovate, as it has reached a point where nearly all practical solutions have already been met. I will also need to research the equipment that a well equipped office would have.

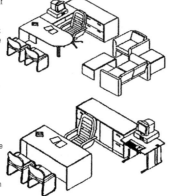

Here are two designs of recommended office layouts by 'office by design.com', an Internet-based design company. Guest seating and desk space are considered in both layouts. Guest seating is one area I have yet to investigate. There are separate areas of desk space on both designs. The seat designed for the employee appears to be comfortable and of a high quality. This is likely because comfortable surroundings induce more productivity and higher standards of work. Another aspect that features in all the designs is the computer – an integral part of any modern office. It cannot be left out of any realistic design.

To the left you can see a drawing showing the ideal posture for someone sitting at a desk. This will be useful as I want the unit I design to be as comfortable as possible to increase productivity.

Here you can see a desk that has been designed specifically for a modern office environment. It is minimalist and uses bright colours along with natural curves to achieve a ergonomic look and feel. However, it does not afford the functionality that I want to include in my design.

To the left, you can see another design for the modern office desk. It is smaller than the previous example and I think that this option is unattractive due to its bitty look The second surface looks like an afterthought and is not at all to my taste. To the right, there is a retro modern desk It is supposed to look old, but uses modern construction techniques and design This is pointless, in my opinion, as there seems no reason to dwell on old fashioned design.

Figure 5.3 *Research sheet*

company newsletters and case studies about their products. However, before you start any Internet research you need to know exactly what you are looking for and how to find it. Find out web addresses before you start and try to avoid following interesting but useless leads.

All the research that you undertake should enable you to write a product design specification. Review the criteria included in the AS specification because these basic criteria should be part of the A2 specification too. You should also refer to the further product design specification criteria required at A2 level to ensure that your research will enable you to address these criteria.

Think about this!

Don't forget to reference all secondary sources of information in a bibliography at the end of your coursework project folder.

To be successful you will:
- Carry out a wide range of imaginative research, with a high degree of selectivity of information.

4. Develop a design specification, taking into account designing for manufacture

SIGNPOST
Develop a design specification, taking into account designing for manufacture
Unit 2 page 45

At A2 you may find that the development of your product design specification runs concurrently with writing your design brief, especially if you are working in collaboration with a client or users. Early exploration of the design context and possible discussions with a client may provide enough information for you to develop the design brief and an outline product design specification at the same time. This can be very beneficial because it can help you to target more closely the research you need to do. As your research progresses, you can amend and develop your design specification until you reach your final product design specification. You may find that even this will need to be amended later on, after you have worked on and evaluated ideas for the product, or even after modelling and prototyping your potential product.

Specification criteria

The A2 product design specification should take into account all the criteria that you worked with at AS level. At A2 you are expected to demonstrate a greater understanding of product design. You can do this by developing a product design specification that takes into account more sophisticated specification criteria. The questions below should help you develop your A2 design specification.

- How can you ensure that your product will have market potential? The market potential of mass-produced products relies on developing a **competitive edge** – providing high-quality products at an affordable price, combined with the image consumers want the product to give them.
- What is the life expectancy of your product? Will it require any maintenance? What are the implications for the cost and quality of your product?
- What type of materials will you need to use? How can you ensure availability at the right time and the right cost? Is the use of CAD/CAM specified in manufacture?
- What kind of quality assurance and quality control system can you put in place to ensure your product's quality? Does this system make use of tolerances?
- What legal requirements and/or external standards (British Standards) related to performance, quality or safety do you need to take into account when developing your product?
- What influence will value issues, such as cultural, social, environmental or ethical issues have on the design of your product?

Scale of production

At A2 you should be developing 2D and 3D elements that could be batch or mass produced for users in a target market group and will have to take into account ease of manufacture. Your identified scale of production will, therefore, be either batch or mass production, but you will be making one-off prototypes. These should be as close as possible to the 'real' thing. As a result, you need to make sure that quality is a high priority.

Your product design specification is the connection between research and ideas. It should guide all your design thinking and provide you with a basis for generating, testing and evaluating both your design ideas and your final design proposal. The design specification is therefore a control mechanism that sets up the criteria for the design and development of your product.

To be successful you will:
- Write a clear design specification.

Task
Writing a design specification
Read the list of questions above. Then develop your own product design specification to include the following:

- The product purpose, function and aesthetics.
- Market and user requirements.
- The expected performance requirements of the product, materials and components, including considerations of maintenance and product life span.
- The kind of processes, technology and scale of production you may use.
- Any value issues that may influence your design, such as cultural, social, environmental or moral.
- Any quality assurance and quality control procedures that will constrain your design.
- Time, resource and cost constraints you will have to meet.
- Any legal requirements and external standards you may need to meet, for example British Standards, safety standards, risk assessment.
- Product market potential including considerations for manufacture in high volume.

B. Generating ideas (15 marks)

1. Use a range of design strategies to generate a wide range of imaginative ideas that show evidence of ingenuity and flair

SIGNPOST
Generating ideas
Unit 2 page 46

You are expected to demonstrate a more mature approach to design when generating ideas at A2. This means making closer connections between your research, your product design specification and your ideas.

This is how it works in many industrial situations – the product design specification forms the basis for developing ideas. Although your specification will provide inspiration both for aesthetic and functional considerations for design, it will probably be aesthetic considerations that provide the most freedom when generating exciting ideas.

In your research, you will have investigated:

- user needs relating to preferences about products
- market trends relating to styling, colour, images, or typography
- existing products that are similar to the prototype you intend to design and make.

This kind of research information is vital if you are to generate imaginative ideas. Industrial and commercial designers rarely start from scratch, mainly because they have to meet deadlines and because they have to design products that people will buy. Industrial and commercial designers have to take shortcuts. These shortcuts often involve using other product designs, the work of artists or other designers as inspiration or as starting points for new ideas (see Figure 5.4).

Figure 5.4 *This 1927 poster by A M Cassandre uses powerful geometric forms based on the most famous avante-garde movement of the day – cubism. It very cleverly integrates the company name with the image*

You can work in a similar way, using information about market trends and other aesthetic considerations to inspire ideas. You could:

- produce a '**moodboard**' of visual images to inspire ideas. For example, make a collection of quirky images and products, or a collection of colours, textures, typography and 'swipes' from magazines that suggest a mood or theme – these can be used to give your prototype product an identity or image
- make connections between old technology and new ideas. For example, adapt the style of 1930s or 1960s advertising materials to produce a new look for packaging, point-of-sale or promotional materials
- use the work of a design movement to inspire the imagery for your 2D and 3D elements. For example, reflect the 'Memphis' style
- use an art movement or the work of an architect to inspire ideas. For example, use the art movement 'Futurism' or the work of the Spanish architect Gaudi to develop form or styling
- use the influence of cultural or traditional art or design to inspire ideas. For example, use themes such as 'American Indian' or 'East meets West'
- use themes built around 'value issues' as a starting point for design. For example, 'eco design' or 'recycling'
- use the natural or built environment to inspire ideas. For example, shells, fruit, grids, wrought-iron work.

It is often a good idea to keep a notebook for jotting down ideas as thumbnails or quick sketches to help you work out design ideas. Thumbnails are small rough sketches showing the main parts of designs in the form of simple diagrams.

Sometimes it is helpful to try to put down as many initial ideas as possible in a set time, which is a bit like producing a brainstorm in image form. These initial ideas can be pasted into your coursework project folder, rather than be redrawn – your project folder should show evidence of creative thinking rather than stilted copied-out work. At this stage, the examiner is looking for evidence of your design thinking. Quick sketches need to be produced fast, using pencils, pens, markers or the like. Use arrows and brief notes to explain your thinking and do not include too much detail at this stage.

Think about this!

Many products in an industrial context are modifications, rather than original ideas. Inspiration for this type of designing can come through product analysis and disassembly. In industry it may involve:

- adapting products to compete with 'branded' products
- developing products to appeal to different target market groups
- developing products through following new legal guidelines, i.e. related to environmental or safety issues
- adapting products through the use of new or different materials or processes.

To be successful you will:

- Use a broad range of design strategies to generate and refine a wide range of imaginative ideas.

2. Use knowledge and understanding gained through research to develop and refine alternative designs design detail

As you become more involved with your ideas and start to think about them in greater depth, you may need to produce larger, slightly more detailed sketches. These should still be produced quickly, but may start to show alternative ideas or some parts in more detail. Always add brief notes to explain your design thinking so you provide evidence of how your research influences your ideas. An example of how one student tackled this is shown in Figure 5.5.

SIGNPOST
Use knowledge and understanding gained through research
Unit 2 page 48

At this stage, it may be helpful to use cut paper or simple 3D images to explore ideas. **One-point** or **two-point perspective** or the technique called '**crating**' are extensively used by designers to produce initial ideas. You may wish to add

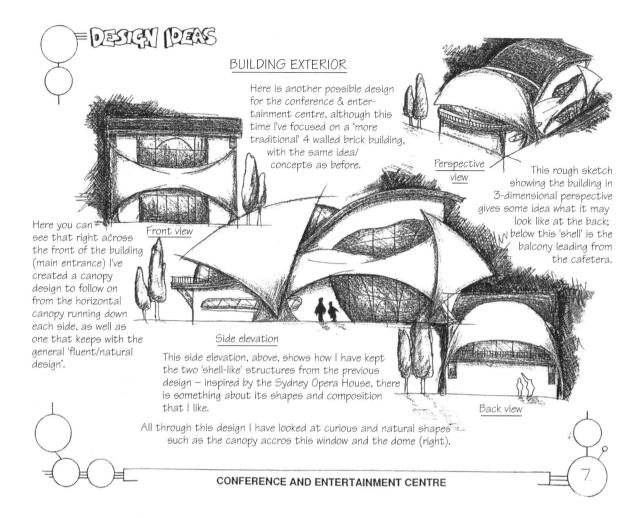

DESIGN IDEAS

BUILDING EXTERIOR

Here is another possible design for the conference & entertainment centre, although this time I've focused on a 'more traditional' 4 walled brick building, with the same idea/concepts as before.

Perspective view

This rough sketch showing the building in 3-dimensional perspective gives some idea what it may look like at the back; below this 'shell' is the balcony leading from the cafetera.

Front view

Here you can see that right across the front of the building (main entrance) I've created a canopy design to follow on from the horizontal canopy running down each side, as well as one that keeps with the general 'fluent/natural design'.

Side elevation

This side elevation, above, shows how I have kept the two 'shell-like' structures from the previous design — inspired by the Sydney Opera House, there is something about its shapes and composition that I like.

Back view

All through this design I have looked at curious and natural shapes such as the canopy accros this window and the dome (right).

CONFERENCE AND ENTERTAINMENT CENTRE

7.

Figure 5.5 *Generating ideas for a conference and entertainment centre*

shading or texture, using pencil crayons or pale-coloured markers to convey information about possible materials, for example to show if they are smooth and polished or matt. Knowledge and understanding about the materials or components you may use and about suitable techniques, processes or finishes can be shown by annotating your drawings.

Sometimes it is helpful to use 2D or 3D modelling in paper or card for developing initial ideas, especially if complicated shapes or nets are involved. Simple modelling can give a real sense of the size and feel of a product, but if you work in this way, be sure to provide evidence in your coursework folder. It is easy to include 2D modelling, but you may have to photograph any 3D modelling work that you do.

Think about this!

Solving design problems is a complex activity because there are many conflicting constraints and possible solutions. You must satisfy the brief, client needs, constraints of manufacture, limitations of materials and equipment and demands of the market place. Designer-makers have to respond to all these constraints and take on the many roles that in industry would be filled by design, marketing, planning and production teams. The simultaneous design of a product and its manufacturing process is called 'concurrent manufacturing'. An individual taking on a range of design and manufacturing roles also works 'concurrently'.

3. Evaluate and test the feasibility of ideas against specification criteria

SIGNPOST
Evaluate and test the feasibility of ideas against the specification criteria
Unit 2 page 49

As your ideas develop you should evaluate them against your design specification. It is often helpful to explain your ideas to others. Always listen to their views and ask them for any criticisms. Their views may be based on 'gut feeling', unless they fully understand the context of your work. However, the views of others are sometimes very helpful as they can provide you with unexpected insights into your work. If you have the opportunity to use the expertise of a visitor from business or are working with people in the local community, they may be able to offer constructive criticism which will help you develop your design work further. Design ideas may also be evaluated by asking potential users of your product about the market potential and feasibility of your ideas.

Testing the feasibility of ideas

In industry, evaluation of ideas is often done by constructing an **evaluation matrix** which compares each idea against the specification criteria. Each idea is assessed using the following:

- + (plus), meaning better than the specification criterion
- – (minus), meaning worse than, more expensive than, more difficult to develop than, more complex than, harder than the specification criterion
- S (same), meaning the same as the specification criterion.

Each idea in turn is evaluated against the specification criteria. Each idea is given a score, either +, – or S. Scores for each idea are added up, to show the strengths and weaknesses of each one. Table 5.2 shows the format of a typical evaluation matrix. Each idea 1–4 is given a score (either +, – or S), against the specification criteria A–H. Total scores for each idea show the strengths and weaknesses of each one.

Table 5.2 An evaluation matrix

Specification criteria	Idea			
	1	2	3	4
A	+	–	+	–
B	+	S	+	S
C	–	+	–	–
D	–	+	+	–
E	+	–	+	–
F	–	–	S	+
G	+	+	–	+
H	+	+	+	–
Total score for + (plus)	5	4	5	2
Total score for – (minus)	3	3	2	5
Total score for S (same)	0	1	1	1

Task
In the evaluation matrix in Table 5.2, explain why Idea 3 is better than Idea 1.

In industry a design team would look at the weaknesses of all ideas to see what could be done to improve them. Very weak ideas are then eliminated, leaving the strong ideas which can be developed individually or combined in some way. This kind of exercise gives the design team a greater understanding of design problems and potential solutions and is a natural stimulus to produce design solutions.

Try using an evaluation matrix to evaluate your initial ideas. This should give you a clearer view about what is worth developing. Use written notes to explain this thinking.

You should also make a note of any further research you need to do. You may need to modify your design specification to take into account your decisions. At A2 level you should be using a more refined approach to focus your ideas so that they meet more closely the requirements of the specification.

C. Developing and communicating design proposals (15 marks)

1. Develop, model and refine design proposals, using feedback to help make decisions

SIGNPOST
Develop, model and refine design proposals, using feedback to help make decisions
Unit 2 page 49

Your aim is to develop and refine your chosen ideas until you find the optimum solution – the best possible solution to your design problem. Be aware that you are developing 2D and 3D elements that could be batch or mass produced so take into account how easy they will be to manufacture. In many industrial situations planning of the production processes is a normal part of the design process. Modelling and prototyping your proposals should play a key role in your production planning. You can use them to trial your ideas, to explore materials or components, to work out construction and assembly processes and for materials planning.

Developing your design proposals will enable you to consider every aspect of your design. Modelling and prototyping will enable you to make judgements about the visual elements of your design, such as shape, form, proportion, styling, images, colour, texture, typography or layout, as well as the size and appropriateness of any components you may use. Many everyday products, such as toothpaste tubes or toys, are modelled to ensure they are the right size and easy for people to use. Figure 5.6 shows an example of 2D modelling.

> *To be successful you will:*
> * Develop, model and refine the design proposal, with effective use of feedback.

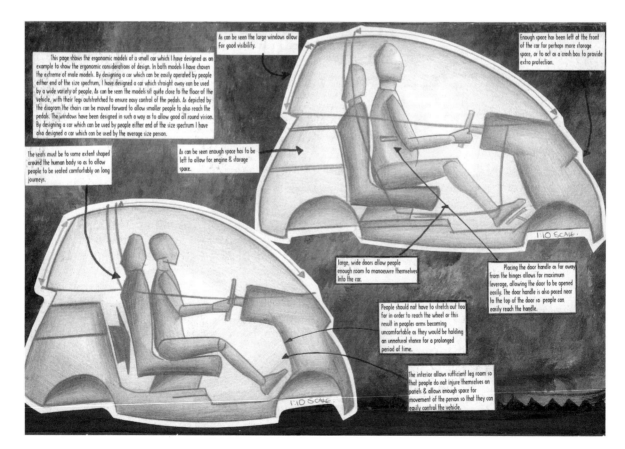

Figure 5.6 *2D modelling is used to ensure that products are the right size for people to use*

2. Demonstrate a wide variety of communication skills, including ICT for designing, modelling and communicating

SIGNPOST
Demonstrate a wide variety of communication skills, including ICT for designing, modelling and communicating
Unit 2 page 49

At A2 you are expected to use a variety of 2D and 3D communication skills to develop, model and refine your design proposals. These should include the use of accurate technical language when you are explaining technological or scientific concepts. You should use ICT if it is available to you, but you will not be penalised for non-use. ICT is useful for recording decisions, data handling, identifying the properties of materials and for modelling ideas in 2D and 3D.

When using 2D modelling you can develop all the techniques you learned during your AS course. This may include modelling your product using:

- **exploded views** to show individual parts and how they fit together or to show hidden detail (see Figure 5.7)
- sections or cutaways to show the inside of the product
- CAD modelling.

Task

Using ICT in industry

The increasing use of ICT by industry has had an enormous impact on design, through the use of CAD systems for computer modelling.

Every aspect of a product's development can be modelled using **electronic product definition (EPD)**, in which all the data required to develop and manufacture a product are stored in a database. This means that even complex products such as cars or aero-engines can be modelled electronically as 'virtual products' and developed directly from the computer screen.

List the advantages to manufacturers of such a system.

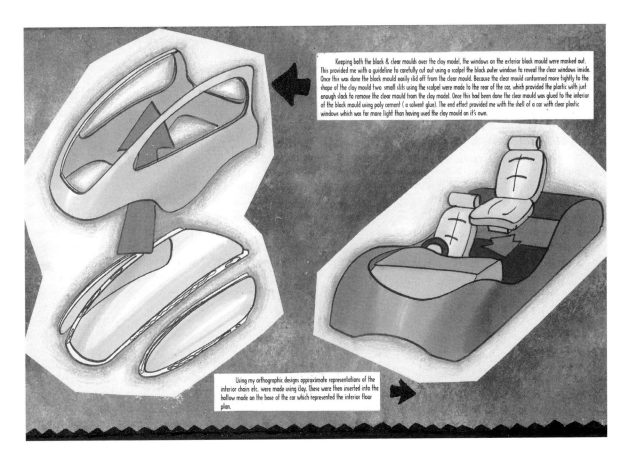

Figure 5.7 Exploded views to show the construction of the car

When using 3D modelling and prototyping, you should also develop the techniques you learned in the AS course. Your modelling should ensure that your product will meet the performance and ergonomic requirements of your design specification.

SIGNPOST
Anthropometrics and ergonomics
Unit 3B1 page 131

Testing ergonomic requirements

Often when testing ergonomic requirements, designers work with 3D models to give a real impression of size and the relationship of the product with the user. Since there is no such thing as an 'average'-sized person, designers often use anthropometric data about measurements of the human body, such as height, weight, reach, leg and arm lengths and strength. If necessary, check out anthropometric data from standards organisations such as the British Standards Institute.

Factfile
Prototyping in industry
As the product develops, it becomes more accurate until it becomes a prototype product. This is a detailed 3D model made to test a product before manufacture. Prototyping is a key industrial process because it enables the product, its construction and the manufacturing process to be designed. It can be used to trial a design proposal to see how materials behave, to test construction processes and to work out costs.

Modelling techniques

You can use a variety of modelling techniques and a range of materials to make 3D working mock-ups or '**lash-ups**'. For example:

- Sheet materials such as paper, card and thin plastics can be used for prototyping ideas as they are quick, easy to use and relatively inexpensive. Look out for useful modelling materials such as recycled packaging, or use flat sheets cut from drinks cans and odd bits of string or wire.
- Simple techniques like cutting, scoring and folding will enable you to explore curved and rectilinear forms. Product designers often prototype products in card and add detail by drawing details like switches or dials on the surface. You could photocopy design details from magazines, cut them out and stick on to give a realistic idea of your product.
- Clay or plasticine may also be used for moulding solid models. Polystyrene foam is useful for making block models.
- Frame models can be built to scale to explore and test structures, using anything that cuts easily, like drinking straws, strips of wood, uncooked spaghetti or wire.

Think about this!
A photocopier is a useful piece of equipment for developing design details. Copy mesh textures for detailing grills on speakers. Copy your outline drawings so you can experiment with different colours and textures to create different surface finishes and materials.

Choosing a colour scheme

Colour is an essential characteristic of many products and most advertising materials. It is a powerful marketing tool because when we look at any product or packaging, it is usually the first thing we notice. Colour can convey strong messages about the product. For example, colour is often associated with specific products such as pastel blue or pink for baby products. Would these colours look 'right' used in the design of breakfast cereals or for point-of-sale advertising?

Some of our responses to colour are through association, for example red for danger, blues are cool, whereas browns and greens may suggest a natural quality. When choosing colours for your product, you need to be sure of the message that you wish to communicate. The creation of moodboards can often be helpful in selecting colours that are appropriate to your product.

It is important to try out your product colour scheme. You could use photocopies or CAD software to try out different colourways. Experiment with different tones and try unusual combinations. Before colouring your final design it may be helpful to test your colours as near to full size as you can. Large areas of colour can look entirely different to a small colour swatch.

Think about this!
In industry 3D prototypes are often made for products such as torches or telephones. Prototypes are made as accurately as possible to represent the appearance and function of the finished product.

When James Dyson was developing his cyclone action vacuum cleaner, he made 5127 prototypes!

To be successful you will:
- Use high-level communication skills with appropriate use of ICT.

3. Demonstrate understanding of a range of materials/components/ systems, equipment, processes and commercial manufacturing requirements

SIGNPOST
Selecting materials Unit 2 page 51
Testing materials Unit 3A page 104
Selection of materials Unit 4A page 174

At A2 you are expected to draw on and use a higher level of understanding about materials, components and systems. You should make relevant and real connections between that understanding and your own design and manufacture. Modelling and prototyping should enable you to explore a range of interesting materials and look for new ways of using them. Testing materials will enable you to select those that are the most suitable for your product. Testing may include researching known properties or the use of **comparative testing**.

Do not forget to annotate your final design proposal, to explain how and why it meets the requirements of your product design specification. You should also identify possible materials and components and show how to manufacture your product.

> *To be successful you will:*
> - Demonstrate a clear understanding of a wide range of resources, equipment, processes and commercial manufacturing requirements.

4. Evaluate design proposals against specification criteria, testing for accuracy, quality, ease of manufacture and market potential

SIGNPOST
Evaluate design proposals
Unit 2 page 53

During the refinement of your final design proposal you will need to check its feasibility against your product design specification (see Figure 5.8). If you are collaborating with others on the development of your product, the views of potential users should provide you with feedback about its viability.

Final idea

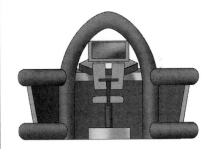

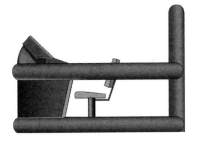

Here you can see some views of my final idea rendered in 3D Studio MAX. This is particularly useful as any pre-production faults that were not apparent in the design stages are now visible.

One such problem that came up while modelling this in 3D was that the circular designs prior to this one did not provide enough working space to do things such as drawing or writing, which would mean that the design fails to meet the brief. This was not an option.

The computer equipment is stored in the main upright beneath the screen and keyboard and is accessed from the rear of the unit to allow easy upgrades. The hardware will be user specifiable and thus always meet the needs of the exact business it is needed for.

The unit will be constructed in two separate parts in the factory and then joined together in the customer's premises. This allows the large units to appear whole but fit through doorways. The customer ticks an option list as on a car and then the unit is built to the exact requirements. There will also be a selection of pre-determined versions that are already constructed for customers who want the units at short notice.

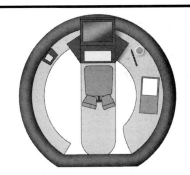

Figure 5.8 The final design proposal rendered using CAD software

Your evaluation should enable you to justify why this is the best solution to the problem. Refer to your AS project. How did you justify that design proposal? You will need to explain:

- how your design proposal's aesthetic and functional characteristics meet the specification
- how its aesthetic and functional characteristics relate to the needs of your intended users (including ergonomics where appropriate)
- how the proposal will meet market requirements and provide a competitive edge
- how the performance characteristics of your proposed materials and the manufacturing

processes will meet the quality, cost and safety expectations of your users
- how easy it will be to manufacture in the time available
- how the processes and scale of production will meet the design specification and market potential.

> **To be successful you will:**
> - Objectively evaluate and test your design proposals against the specification criteria.

D. Planning manufacture (10 marks)

1. Produce a clear production plan that details the manufacturing specification, quality and safety guidelines and realistic deadlines

SIGNPOST
Planning manufacture
Unit 2 page 54

You should produce and use a clear **production plan** that explains how to manufacture your product.

The work that you have already done during the research, design and development stages of your project should enable you to do this. During those developmental stages you needed to take into account how easy your product will be to manufacture. In many industrial situations the planning of the production processes is a normal part of the design process. Your modelling and prototyping will have played a key role in your production planning. You will have used them to trial your ideas, to explore materials, to work out assembly or production processes and for materials planning.

Think about how long each different production process will take. Will you have enough time to make your product? Will you need to simplify anything? Do you need any special materials or tools? How soon do you need to order any materials or components so they are ready when you need them? At this stage, you need to estimate your production costs and a possible **selling price (SP)**.

Producing a production plan

Your production plan should include clear and detailed instructions for making your product. An example of a production plan is shown in Figure 5.9. Planning information should enable the manufacture of identical products, so

planning quality into the product is vitally important. Base your production plan on the one you used at AS level. You should also take into consideration the following:

- Your manufacturing specification should include a dimensioned working drawing (see Figure 5.10). You could use **orthographic** 3rd angle drawings or CAD software.
- You should use quality control methods to ensure that identical products can be made. Plan your quality control using critical control points (CCPs) at key stages of assembly. Use quality indicators to explain how to check for quality. Think about balancing the quality of your product against the cost of making it and the time available. What are the quality requirements of the users?
- Use risk assessment procedures to check the safety of your manufacture. Are there any potential hazards? Does the design specification set out any safety standards that will help you assess risks? Are there any risks attached to the materials, processes and equipment you will use? Do you need to follow any safety regulations? Are there any risks in the use of your product? What about its disposal?

> ## Tasks
> **Risk assessment**
> 1 Draw up a flow chart to show the key stages of manufacture of your product.
> 2 For each stage list the risk assessment procedures required to make your product as safe as possible.

> **To be successful you will:**
> - Produce a clear and detailed production plan with achievable deadlines.

PLANNING THE MAKING

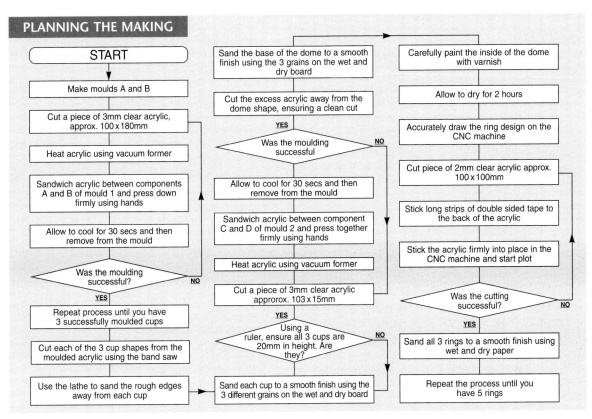

START

Make moulds A and B

Cut a piece of 3mm clear acrylic, approx. 100 x 180mm

Heat acrylic using vacuum former

Sandwich acrylic between components A and B of mould 1 and press down firmly using hands

Allow to cool for 30 secs and then remove from the mould

Was the moulding successful? — **NO**

YES

Repeat process until you have 3 successfully moulded cups

Cut each of the 3 cup shapes from the moulded acrylic using the band saw

Use the lathe to sand the rough edges away from each cup

Sand the base of the dome to a smooth finish using the 3 grains on the wet and dry board

Cut the excess acrylic away from the dome shape, ensuring a clean cut

YES

Was the moulding successful — **NO**

Allow to cool for 30 secs and then remove from the mould

Sandwich acrylic between component C and D of mould 2 and press together firmly using hands

Heat acrylic using vacuum former

Cut a piece of 3mm clear acrylic approrox. 103 x 15mm

YES

Using a ruler, ensure all 3 cups are 20mm in height. Are they? — **NO**

Sand each cup to a smooth finish using the 3 different grains on the wet and dry board

Carefully paint the inside of the dome with varnish

Allow to dry for 2 hours

Accurately draw the ring design on the CNC machine

Cut piece of 2mm clear acrylic approx. 100 x 100mm

Stick long strips of double sided tape to the back of the acrylic

Stick the acrylic firmly into place in the CNC machine and start plot

Was the cutting successful? — **NO**

YES

Sand all 3 rings to a smooth finish using wet and dry paper

Repeat the process until you have 5 rings

Figure 5.9 *A detailed production plan with references to industrial production processes*

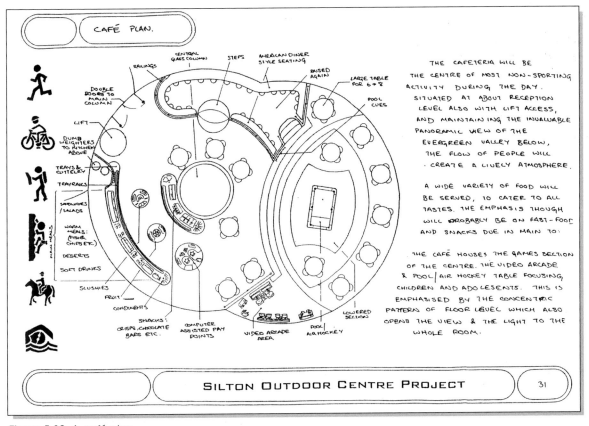

Figure 5.10 *A café plan*

Factfile

Quality assurance (QA) and quality control (QC)

1 The aim of quality assurance is to make identical products with zero faults. A QA system monitors every stage of design and manufacture. QA makes use of written documents such as:

- production plans
- detailed specifications
- work orders
- costing sheets
- quality control and inspection sheets.

2 Quality control is used to test and monitor the production of products.

- QC means checking, at critical control points, for accuracy and safety.
- Quality indicators are used to show how quality is checked. These often use sensory tests such as sight and touch.
- QC makes use of standard sizes, dimensions and tolerances to enable quality to be checked.

and how one activity relates to another. Decisions have to be taken as to which activity needs to be done first; how long each activity will take; when each activity has to be done by; which activities can be done at the same time; which activities depend on the completion of other activities and if any activities are more essential or critical than others.

The critical path is the one that takes the shortest time. It is often helpful to work backwards from the deadline, plotting when an activity has to start in order for you to finish on time. You could use a Gantt chart to plot the critical path of your product. Map each task against the time available, so you can prioritise the critical activities. Using an adjustable Gantt chart, (made from card and reusable adhesive), will allow you to be flexible in response to any changes in the availability of materials, modifications to the design of your product or in manufacturing processes.

> *To be successful you will:*
> - Demonstrate effective management of time and resources, appropriate to the scale of production.

2. Take account of time and resource management and scale of production when planning manufacture

> **SIGNPOST**
> Take account of time and resource management and scale of production when planning manufacture
> Unit 2 page 56

Many of the activities that you undertake during your A2 project will overlap. This is inevitable because designing and making is a complex process. Planning a project is not easy, because it involves estimating how long each activity will take. Use your AS experience of project planning to help you plan your A2 project. Did you have to amend your time plan? Did some activities take longer than you expected? How did you overcome these problems? Did you find it helpful to use a Gantt chart to plan your project? Did you use one to plan your manufacture?

In industry, project managers often use **critical path analysis** to plan the successful outcome of a project. This involves working out all the critical activities that must be undertaken

3. Use ICT appropriately for planning and data handling

> **SIGNPOST**
> Use ICT appropriately for planning and data handling
> Unit 2 page 57

The aim of using information and communications technology (ICT) for planning and data handling is to enhance your design and technology capability. You will not be penalised for non-use of ICT, although you should use it where appropriate and available. If you do have access to ICT, you can use it for a range of activities. You can:

- find out costs of materials using email or the Internet
- use spreadsheets to work out quantities and costs of materials and components
- plan the critical path of your product, using colour-coded Gantt charts.

> *To be successful you will:*
> - Demonstrate good use of ICT.

E. Product manufacture (40 marks)

1. Demonstrate understanding of a range of materials, components and processes appropriate to the specification and scale of production

SIGNPOST

Demonstrate understanding of a range of materials, components and processes appropriate to the specification and scale of production Unit 2 page 58

Classification of materials and components Unit 3A page 66

Working properties of materials and components Unit 3A page 85

Selection of materials Unit 4A page 174

Your AS coursework experience should enable you to select and use materials and components with growing confidence. By this stage of your course, you should have developed a more detailed knowledge of their working properties and characteristics. Your modelling, prototyping and testing of materials, components and processes should also give you confidence in making your product. Most of the problems related to the product assembly should, hopefully, have been ironed out.

Your aim now should be to manufacture a well-made product (see Figure 5.11). This may not necessarily be made to a very much higher quality than the product that you made at AS. However, your A2 product should fulfil the design and manufacturing specifications more closely than your AS product did.

Scale of production

Your 2D and 3D elements will be made as a one-offs, but they will be prototypes that could be manufactured in higher volume. Even though you are making manufacturing prototypes, they will still need to be made and finished to as high a quality as possible.

> **To be successful you will:**
> - Demonstrate clear understanding of a wide range of materials, components and processes.

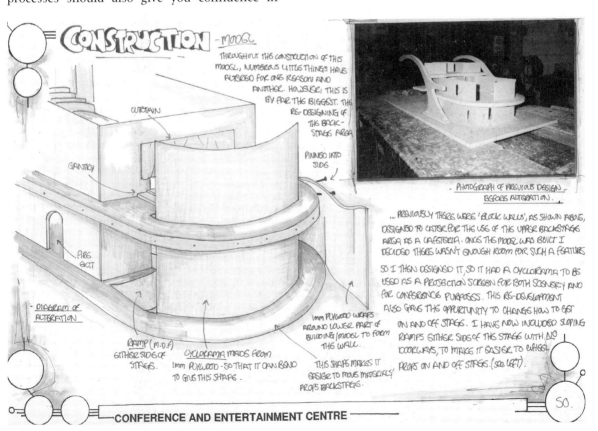

Figure 5.11 Manufacturing a product

2. Demonstrate imagination and flair in the use of materials, components and processes

SIGNPOST
Demonstrate imagination and flair in the use of materials, components, processes and techniques
Unit 2 page 60

You are expected to work creatively, innovatively and imaginatively with materials, components, processes and techniques. You can only do this if you have a good understanding of the materials and processes you use. During the modelling and prototyping stages you may have had the opportunity to explore a range of interesting materials and trial new ways of using them. Your ability to use these materials and processes with flair should result in the production of a quality product that:

- meets the specification and is well finished
- is easy to use with well-designed and innovative detail.

3. Demonstrate high-level making skills, precision and attention to detail in the manufacture of high-quality products

SIGNPOST
Demonstrate high-level making skills
Unit 2 page 60

At A2 you should place a greater emphasis on the prevention of faults through your own use of quality assurance and quality control. This will involve using your production plan to monitor your product manufacture:

- Plan where you will check for quality during manufacture.
- Use tolerances and dimensions to check the accuracy of your product.
- Check the accuracy of machines or sharpness of tools prior to cutting.
- Test any components parts prior to assembly.
- Your making skills should result in the production of a high-quality product (see Figure 5.12). This should be capable of being tested against the specification and used by potential users for the intended job.

Figure 5.12 *The final product*

> *To be successful you will:*
> - Demonstrate demanding and high-level making skills that show precision and attention to detail.

4. Use ICT appropriately for communicating, modelling, control and manufacture

SIGNPOST
The use of ICT in manufacture
Unit 1 page 29
Use ICT appropriately
Unit 2 page 60

You should use ICT to help your product manufacture where it is appropriate and available, but you will not be penalised for its non-use. You are not expected to know how to use specific equipment or programs, but you should understand the benefits of using ICT for manufacture.

> *To be successful you will:*
> - Demonstrate good use of ICT.

5. Demonstrate a high level of safety awareness in the working environment and beyond

SIGNPOST
Demonstrate high level of safety awareness in the working environment and beyond
Unit 2 page 61

Safety should be a high priority in your work at all times. Safe production means identifying all areas of possible risk and documenting safety procedures to manage and monitor this risk. This means that you should:

- identify hazards and use risk assessment procedures to ensure safe use of materials, tools, equipment and processes
- demonstrate awareness of safety in the workshop
- identify safety aspects relevant to the design, manufacture, use and disposal of your product.

Risk assessment

Risk assessment means identifying the risks to the health and safety of people and to the environment. In practice, this means using safe designing and manufacturing processes and making products that are safe to use and safe to dispose of.

Manufacturers use risk assessment procedures to look for possible hazards in their products. They use British Standards to test and monitor production. All possible health hazards to employees have to be eliminated and safety procedures have to be followed to ensure the safety of people at work. Some manufacturers use life cycle assessment (LCA) to assess a product's impact from cradle (raw materials) to grave (disposal).

Task

Risk assessment
Research any appropriate British Standards legislation related to your product type.

To be successful you will:
- Demonstrate a high-level of safety awareness.

F. Testing and evaluating (10 marks)

1. Monitor the effectiveness of the work plan in achieving a quality outcome

SIGNPOST
Monitor the effectiveness of the work plan in achieving a quality outcome
Unit 2 page 62

Your production plan is a key tool in monitoring the manufacture of your product because it enables you to control its quality. Use your production plan to record any modifications to your product or to the processes you use. Even the best laid plans sometimes have to be changed through unexpected problems that arise!

Problems that occur may be due to the availability of materials or equipment or due to time constraints. If, for whatever reason, you find that you are running out of time, you may have to take short cuts or simplify the design detailing on your product. This would obviously have an impact on the quality or 'look' of the product. Record any changes you make, before you forget, so another identical product could be made to the same standard.

To be successful you will:
- Make effective use of your work plan to achieve a high-quality outcome.

2. Devise quality assurance procedures to monitor development and production

SIGNPOST
Devise quality assurance procedures to monitor development and production
Unit 2 page 62

Quality planning is a key process during product development and manufacture and you should use quality assurance and quality control procedures to monitor quality. This includes using quality indicators at critical control points in your product's manufacture.

Refer to the quality requirements that you included in your product design specification. Did you take these into account when designing and developing your product ideas? Refer to the quality control processes that you put into place in your production plan. Use these to monitor the quality of your product.

The meaning of quality

In industry achieving quality means meeting standards. This makes it possible to manufacture identical products that have a zero fault rate. Quality can mean different things to different people:

- Quality for the consumer means a product's **fitness-for-purpose**. This can be evaluated

287

through its performance, price and aesthetic appeal.

- Quality for a manufacturer means meeting the product manufacturing specification and finding a balance between:
  - profitable manufacture of identical products on time and to budget
  - meeting the needs and expectations of the consumer and the environment.

To be successful you will:
- Devise clear quality assurance procedures.

---

3. Use testing to ensure fitness-for-purpose

SIGNPOST
Using testing to ensure fitness-for-purpose
Unit 2 page 62

In industry testing of materials and components before manufacture is part of a quality assurance system. It requires the use of standard tests under controlled conditions. Standard performance tests can be set by the British Standards Institution and by individual manufacturers.

Tests on prototypes are made to test for performance, ease of manufacture, product maintenance and fitness-for-purpose. These tests ensure the production of quality products, avoid costly mistakes and protect the consumer against faulty or unsafe goods

Testing to ensure fitness-for-purpose means testing the performance of the product so it meets the requirements of the specification and the users. You should record the results of any testing that you do during and after manufacture. For example, you could test:

- the performance of the product against the design and manufacturing specifications
- that the quality of the product is suitable for users – this could mean using field trials to test the quality and ease of use of your product
- the performance of the product under working conditions or over an extended period to obtain user feedback
- the quality of the product using external standards, such as British Standards, where appropriate.

To be successful you will:
- Make effective use of testing to ensure fitness-for-purpose.

---

4. Objectively evaluate the outcome against specifications and suggest appropriate improvements

SIGNPOST
Objectively evaluate the outcome against specifications and suggest appropriate improvements
Unit 2 page 63

You should objectively evaluate and justify the success of your product in relation to the design brief and specifications. It can sometimes be hard to be objective about your own work.

If you have been collaborating with others on the design and manufacture of your product, they may be able to give you unbiased opinions about your product's success. Other people can also contribute to your evaluation – talk to potential users of the product, your peers and your teacher or tutor.

You should try to get a range of opinions about your product. This should provide feedback on how successful you have been in meeting the brief and specifications that you set up at the beginning of the project. You should explain and justify your conclusions.

Objective evaluation will help you decide if your product can be improved. As with your AS project, your suggestions for improvement should be based on the product's aesthetic and functional success, its quality of design and quality of manufacture and its fitness-for-purpose.

You should also suggest improvements in relation to the product's market potential. (see Figure 5.13). For example:

- Does it have potential for manufacture beyond prototype production?
- Could it be produced economically?
- Could it be marketed to a range of users?
- Would the product be consistently of a high quality?
- Could it be sold at a price customers can afford?
- Would the product provide customers with an image they want the product to give them?

To be successful you will:
- Objectively evaluate the outcome and suggest appropriate improvements.

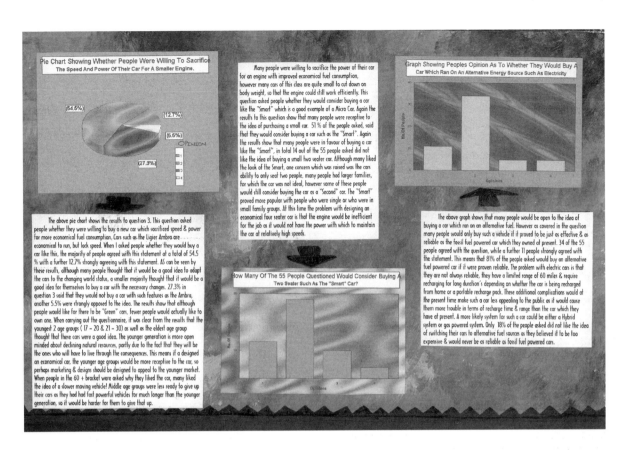

Pie Chart Showing Whether People Were Willing To Sacrifice The Speed And Power Of Their Car For A Smaller Engine.

(54.5%)   (12.7%)   (6.5%)   (27.3%)   OPINION

Graph Showing Peoples Opinion As To Whether They Would Buy A Car Which Ran On An Alternative Energy Source Such As Electricity

No.Of People / Opinions

Many people were willing to sacrifice the power of their car for an engine with improved economical fuel consumption, however many cars of this class are quite small to cut down on body weight, so that the engine could still work efficiently. This question asked people whether they would consider buying a car like the "Smart" which is a good example of a Micro Car. Again the results to this question show that many people were receptive to the idea of purchasing a small car. 51 % of the people asked, said that they would consider buying a car such as the "Smart". Again the results show that many people were in favour of buying a car like the "Smart", in total 14 out of the 55 people asked did not like the idea of buying a small two seater car. Although many liked the look of the Smart, one concern which was raised was the cars ability to only seat two people, many people had larger families, for which the car was not ideal, however some of these people would still consider buying the car as a "Second" car. The "Smart" proved more popular with people who were single or who were in small family groups. At this time the problem with designing an economical four seater car is that the engine would be inefficient for the job as it would not have the power with which to maintain the car at relatively high speeds.

The above pie chart shows the results to question 3. This question asked people whether they were willing to buy a new car which sacrificed speed & power for more economical fuel consumption. Cars such as the Ligier Ambra are economical to run, but lack speed. When I asked people whether they would buy a car like this, the majority of people agreed with this statement at a total of 54.5 % with a further 12.7% strongly agreeing with this statement. AS can be seen by these results, although many people thought that it would be a good idea to adapt the cars to the changing world status, a smaller majority thought that it would be a good idea for themselves to buy a car with the necessary changes. 27.3% in question 3 said that they would not buy a car with such features as the Ambra, another 5.5% were strongly opposed to the idea. The results show that although people would like for there to be more "Green" cars, fewer people would actually like to own one. When carrying out the questionnaire, it was clear from the results that the youngest 2 age groups (17 – 20 & 21 – 30) as well as the eldest age group thought that these cars were a good idea. The younger generation is more open minded about declining natural resources, partly due to the fact that they will be the ones who will have to live through the consequences. This means if a designed an economical car, the younger age groups would be more receptive to the car, so perhaps marketing & designs should be designed to appeal to the younger market. When people in the 60 + bracket were asked why they liked the car, many liked the idea of a slower moving vehicle! Middle age groups were less ready to give up their cars as they had had fast powerful vehicles for much longer than the younger generation, so it would be harder for them to give that up.

How Many Of The 55 People Questioned Would Consider Buying A Two Seater Such As The "Smart" Car?

No. of people / Choice

The above graph shows that many people would be open to the idea of buying a car which ran on an alternative fuel. However as covered in the question many people would only buy such a vehicle if it proved to be just as effective & as reliable as the fossil fuel powered car which they owned at present. 34 of the 55 people agreed with the question, while a further 11 people strongly agreed with the statement. This means that 81% of the people asked would buy an alternative fuel powered car if it were proven reliable. The problem with electric cars is that they are not always reliable, they have a limited range of 60 miles & require recharging for long duration's depending on whether the car is being recharged from home or a portable recharge pack. These additional complications would at the present time make such a car less appealing to the public as it would cause them more trouble in terms of recharge time & range than the car which they have at present. A more likely system for such a car could be either a Hybrid system or gas powered system. Only 18% of the people asked did not like the idea of switching their cars to alternative fuel sources as they believed it to be too expensive & would never be as reliable as fossil fuel powered cars.

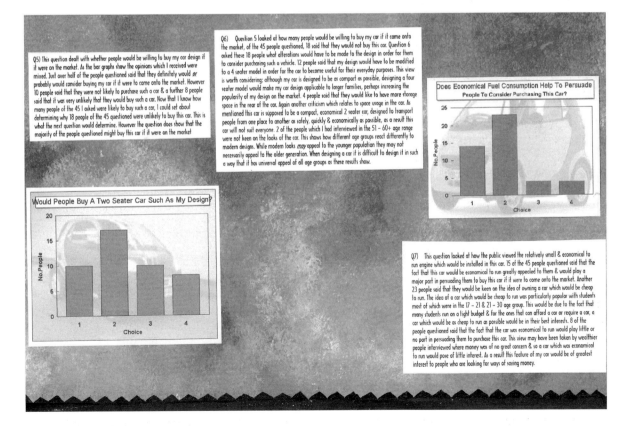

Q5) This question dealt with whether people would be willing to buy my car design if it were on the market. As the bar graphs show the opinions which I received were mixed. Just over half of the people questioned said that they definitely would or probably would consider buying my car if it were to come onto the market. However 10 people said that they were not likely to purchase such a car & a further 8 people said that it was very unlikely that they would buy such a car. Now that I know how many people of the 45 I asked were likely to buy such a car, I could set about determining why 18 people of the 45 questioned were unlikely to buy this car. This is what the next question would determine. However the question does show that the majority of the people questioned might buy this car if it were on the market

Q6)   Question 5 looked at how many people would be willing to buy my car if it came onto the market, of the 45 people questioned, 18 said that they would not buy this car. Question 6 asked these 18 people what alterations would have to be made to the design in order for them to consider purchasing such a vehicle. 12 people said that my design would have to be modified to a 4 seater model in order for the car to become onto the market. This view is worth considering; although my car is designed to be as compact as possible, designing a four seater model would make my car design applicable to larger families, perhaps increasing the popularity of my design on the market. 4 people said that they would like to have more storage space in the rear of the car. Again another criticism which relates to space usage in the car. As mentioned this car is supposed to be a compact, economical 2 seater car, designed to transport people from one place to another as safely, quickly & economically as possible, as a result this car will not suit everyone. 2 of the people which I had interviewed in the 51 – 60+ age range were not keen on the looks of the car. This shows how different age groups react differently to modern designs. While modern looks may appeal to the younger population they may not necessarily appeal to the older generation. When designing a car it is difficult to design it in such a way that it has universal appeal of all age groups as these results show.

Does Economical Fuel Consumption Help To Persuade People To Consider Purchasing This Car?

No.People / Choice

Would People Buy A Two Seater Car Such As My Design?

No.People / Choice

Q7)   This question looked at how the public viewed the relatively small & economical to run engine which would be installed in this car. 15 of the 45 people questioned said that the fact that this car would be economical to run greatly appealed to them & would play a major part in persuading them to buy this car if it were to come onto the market. Another 23 people said that they would be keen on the idea of owning a car which would be cheap to run. The idea of a car which would be cheap to run was particularly popular with students most of which were in the 17 – 21 & 21 – 30 age group. This would be due to the fact that many students run on a light budget & for the ones that can afford a car or require a car, a car which would be as cheap to run as possible would be in their best interests. 8 of the people questioned said that the fact that the car was economical to run would play little or no part in persuading them to purchase this car. This view may have been taken by wealthier people interviewed where money was of no great concern & so a car which was economical to run would pose of little interest. As a result this feature of my car would be of greatest interest to people who are looking for ways of saving money.

Figure 5.13 *The results of a questionnaire 'Would you buy a two-seater car such as this?'*

Student checklist

1. Project management

- Take responsibility for planning, organising, managing and evaluating your own project.
- Ensure that you have photographic evidence of technical details that support your product's quality of manufacture – ideally, you need to show evidence of the processes you used at each stage of manufacture.
- Ensure that your coursework project folder contains only the work related to the assessment requirements.

2. How to make your A2 coursework project a success

A successful A2 coursework project will:

- identify a realistic need and solve a problem for your specified client or target market group
- include research that targets more closely the problem and design brief
- include folder content that represents a conclusion to your research information
- make closer connections between relevant research and the development of ideas
- be a manageable size so you can develop your design and manufacturing skills
- make good use of your coursework project folder to show clearly how your ideas unfold
- use a more refined approach to focus ideas so that they meet more closely the requirements of the specification
- demonstrate a variety of communication skills, including appropriate use of ICT
- show understanding of industrial practices
- make good use of modelling, prototyping and testing
- use a higher level of understanding about materials, components and systems
- demonstrate high-level making skills, using a variety of materials and processes
- manufacture 2D and 3D elements that fulfil specifications more closely than at AS level
- give you time to evaluate your work as it progresses and modify it if necessary
- be well organised so you can meet your deadlines.

3. Evidencing industrial practices in coursework

- Use **industrial terminology** and technical terms.

- Include a range of designing and manufacturing activities that are similar to those used in industry.

4. Using ICT in coursework

- Use ICT where appropriate to enhance your design and technology capability.
- Develop the use of ICT for research, designing, modelling, communicating and testing.
- Develop the use of ICT for planning, data handling and manufacturing.

5. Producing a bibliography

- Reference all secondary sources of information in a bibliography.
- Include references for information found in textbooks, newspapers, magazines, electronic media, CD-ROMs, the Internet, etc. Include references for scanned, photocopied or digitised images.
- Do not use clip art at this level.
- Do not expect to be given credit for any work copied directly from textbooks or other media, including the Internet.

6. Submitting your coursework project folder

- Have your coursework ready for submission by mid-May in the year of your examination.
- Include a title page with the specification name and number, candidate name and number, centre name and number, title of project and date.
- Include a contents page and numbering system to help organise your coursework project folder.
- Ensure that your work is clear and easy to understand, with titles for each section.

7. Using the Coursework Assessment Booklet (CAB)

- Complete the student summary in the CAB. This should include your design brief and a short description of your coursework project.
- Ensure that the CAB contains a minimum of three clear photographs that show the whole product and alternative views.

UNIT 6 Design and technology capability (G6)

Summary of expectations

1. What to expect

Unit 6 is called the 'synoptic' unit because it brings together all the knowledge, understanding and skills that you have learned during the full Advanced GCE course. To meet the assessment requirements, Unit 6 must be assessed in the final year of the Advanced GCE course.

The unit focuses on the knowledge and understanding found within the designing and making process, including knowledge and understanding of product development and manufacture. Since this knowledge and understanding is taught throughout the whole Advanced GCE course, no new learning is expected during Unit 6.

2. How will it be assessed?

You are required to demonstrate your design and technology capability through a three-hour design exam, in which you are asked to produce a design solution to a given design problem and describe how the solution can be manufactured. You should demonstrate your ability to think on your feet, not to recall information.

The assessment criteria cover knowledge and understanding related to designing and making. You will be assessed on your ability to organise and present information, ideas, descriptions and arguments clearly and logically. The style of

assessment will remain the same each year, but there will be a different design problem each time the unit is assessed. Your solution to the assessment criteria should be developed through graphics materials. You should ensure that you are familiar with and understand the assessment criteria shown in Table 6.1.

3. The design exam

The design exam enables you to demonstrate your design and technology capability by producing a design solution to a given design problem in three hours. The Design Paper examines your ability to:

- analyse a design problem and develop a product design specification
- generate, develop and evaluate ideas
- illustrate a final solution
- describe how the solution can be manufactured
- evaluate the strengths and weaknesses of the solution.

Your centre will be sent a Design Research Paper six weeks before the exam. This paper will provide clear and unambiguous guidance about ONE design context to be researched.

In the exam the Design Paper will have ONE compulsory design question that is based on the context you have researched. You may take ALL your research materials into the exam and use them as reference throughout, but this research material is NOT submitted for assessment. The pasting of pre-prepared or photocopied sheets is not permitted. Similarly, ICT will not be assessed in the design paper, so you will NOT be allowed to use ICT facilities during the examination.

During the exam you will be provided with A3 headed sheets, so there should be no need to write headings or fill in borders during the exam. Extra paper will be available.

4. How much is it worth?

The Design Paper is worth 15 per cent of the full Advanced GCE.

Table 6.1 Design Paper assessment criteria

Assessment criteria	Marks
1 Analyse the design problem and develop a product design specification, identifying appropriate constraints	15
2 Generate and evaluate a range of ideas	15
3 Develop the chosen idea into an optimum solution	15
4 Represent and illustrate your final solution	20
5 Draw up a production plan	15
6 Evaluate the final solution and suggest improvements	10
Total marks	90

Unit 6	Weighting
A2 level (full GCE)	15%

Unit 6 requirements in detail

Synoptic assessment

This unit is called is called 'Design and technology capability' because it assesses the knowledge, understanding and skills you have learned throughout the course. This means that no new learning takes place during the unit itself, but you are expected to develop a 'synoptic understanding' of what you have already learned. In other words, you are developing an overview of the knowledge and understanding related to your design and technology course so that you can make connections between individual bits of learning.

This is how industrial designers work, making connections in order to develop their own creativity. For example they need to be aware of past, current and future trends, user and market requirements, keep up to date with events, films, exhibitions, current affairs, environmental and cultural issues. They must also develop, on a regular basis, a creative and technical understanding of materials and processes.

Unit 6 assessment

You are expected to apply your synoptic understanding of the whole course content to a given problem in the design exam. The exam has to take place at the end of the final year of the Advanced GCE course so that it can assess the application of the following types of learning:

- knowledge and understanding of product development and manufacture
- knowledge and understanding of materials, components, processes, systems, technology, scale of production, quality, health and safety, markets, users, cultural, social, moral and environmental issues.

The unit also assesses quality of written communication. You should:

- use clear drawings or sketches
- answer clearly and coherently, using specialist vocabulary where appropriate
- use accurate spelling, grammar and punctuation to make your meaning clear.

Revising Unit 6 learning will help prepare you for assessment. You are advised to practise exam skills to ensure that you are familiar with and understand the assessment requirements. This will enable you to fulfil your potential in the design exam.

Design Research Paper

The Design Research Paper will follow the same format each year of the exam but will be based on different contexts or problems. You will be given a context for design, together with bullet points that give you direction about what to research. This means that you will have a good idea of what to expect and will research appropriate areas. An example is given in the box below.

When a sweets manufacturer develops a new range of confectionery, the presentation and display of the product is a key marketing tool. Investigate:

- design styles and colours used for typography and imagery for confectionery
- point-of-sale displays and promotional materials for sweets and confectionery
- the aesthetic and functional requirements of confectionery packaging materials.

The examiners do not want you to go into the design exam to find that you have researched the wrong information or to find the exam question a complete surprise. As well as targeting your research around the specific areas indicated in the Design Research Paper, you should take into account the assessment criteria. For example, the exam asks you to develop a product design specification, so make sure that you understand how to do this.

Design Paper

In the exam, there will be one compulsory design question that is based on the context you have researched. An example is given in the box below.

You have investigated:

- design styles and colours used for typography and imagery for confectionery
- point-of-sale displays and promotional materials for sweets and confectionery
- the aesthetic and functional requirements of confectionery packaging materials.

Your task is to design suitable graphics for the product 'Candy Chocobars' that could be included on both the packets and promotional material. You also need to design a suitable stand to display 25 packets of the product.

---

1. Analyse the design problem and develop a product design specification, identifying appropriate constraints (15 marks)

Research the context

SIGNPOST
Trends and styles
Unit 1 page 33

Carry out imaginative research and demonstrate a high degree of selectivity of information
Unit 2 page 43, Unit 5 page 271

You should have access to the Design Research Paper six weeks before the exam. Read and analyse the design context so you will have a clear understanding of what you need to do. You should take responsibility for planning, organising, managing and analysing your own research.

The clear guidelines given in the Design Research Paper will enable you to target your research. This should include a variety of primary and secondary research, using a range of sources, such as:

- market research to identify trends and user requirements
- analysis of commercial products and the work of other designers
- research into materials, components, systems, processes and technologies
- research into legal requirements and external standards relating to quality and safety
- research into value issues, such as cultural, social, moral and environmental issues that may have an impact on the problem.

You will find it helpful to think of your research as being part of a design process that you will continue in the exam. Your research information may include notes, images, data, materials information, anthropometric and British Standards data. If, by chance, the question in the Design Paper should bring up something that you have not researched in enough detail, you may refer to any other available material, since this is an 'open book' exam. However, this may have a drawback in that looking up any more information at this stage could take up far too much of your valuable time, so you may be better off working with what you have.

You should analyse your research information before the exam so you are familiar with it. Organise your research information under the same headings as the design specification criteria. Write a conclusion to each section of your research. This process will fully prepare you for the exam.

SIGNPOST
Identify, explore and analyse a wide range of problems and user needs
Unit 2 page 41, Unit 5 page 269

Analyse the problem

In the exam you should read the Design Paper carefully so that you have a clear understanding of the design problem you have to solve.

The design question in the Design Paper will give you a specific task which you must analyse. For example:

'Your task is to design suitable graphics for the product 'Candy Chocobars' that could be included on both the packets and promotional material. You also need to design a suitable stand to display 25 packets of the product.'

An analysis of the above would be specific to the problem of identifying the packaging requirements for a branded chocolate bar. The task gives clear direction about what you have to design, i.e. graphics suitable for use on the packaging and promotional material, as well as a stand to display 25 chocolate bars.

Your analysis should identify the purpose, users and the target market, where appropriate, for the product you are to design. What is the need or problem? Who are the potential users? Where will the product be used? How will it be used? Your analysis should indicate the key points of the problem. This will show that you have a clear understanding about what is required. Analysis may be carried out in a variety of styles, such as writing it in table form or as spider diagrams. Figure 6.1 shows how one student analysed the problem of designing packaging materials and point-of-sale displays for a branded chocolate product.

Making connections

Your analysis of the problem should be a short task. It should enable you make connections between the design problem you have to solve and what you have researched. This will make it easier to pinpoint relevant information that will help you develop a product design specification. In the exam you should NOT copy out extensive research. You must be selective in the use of your research and use only appropriate information.

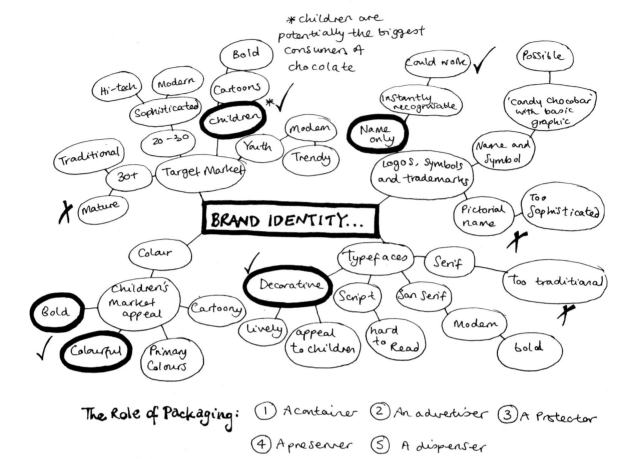

* children are potentially the biggest consumers of chocolate

BRAND IDENTITY...

The Role of Packaging: ① A container ② An advertiser ③ A protector ④ A preserver ⑤ A dispenser

Therefore: Packaging materials have an important Role to play...

Possible Materials.	Possible Advantages	Possible disadvantages	Possible Printing Processes.	Suitability for Packaging...	Suitability for Point-of-sale...	
Plastic...	Tough and durable. Low cost. Recyclable. Easily thermoformed. Lightweight	Initial cost of injection moulding processes	PVC shrink sleeves - flexography/lithography. Direct on to plastic - heat transfer labels. Paper labels - lithography/gravure/flexography.	● ● ● / ● ○	● ● / ○ ○	✓
Paper and board...	Lightweigh. Low cost. Total graphic coverage. Excellent print quality. Enhanced lamination.Corrugation.tough/durable.	?	Lithography/Flexography/Screenprint/gravure/letterpress.	● ● / ● ●	● ● / ● ○	✓
Glass...	Excellent product Visibility. Tough and Durable. Easily formed. High Quality.	Breakages. Relative high cost.	Paper labels - as above. Directly onto glass - ceramic decoration.	○ ○ / ○ ○	○ ○ / ○ ○	
Metal...	Security. Easily Formed. Instant point of Sale. Excellent preserver.	Usually used for liquids.	Paper labels - as above. Directly onto metal - dry offset print.	○ ○ / ○ ○	○ ○ / ○ ○	
Foil laminates...	Easily heat sealed for freshness. Hygeinic. Barrier to moisture.	Cost more than paper and board.	Flexography/gravure.	● ● / ● ●	● ○ / ○ ○	✓

Figure 6.1 *Analysis may be carried out in a variety of styles to identify key points in the design question. This student analysed the problem of designing packaging materials and point-of-sale displays for a branded chocolate product*

Your ability to research the context and to analyse your research material will be assessed through the quality of your product design specification.

Although you can take all your research material into the exam and use it as reference throughout, it is not submitted for assessment. You are, therefore, not allowed to paste, stick or staple any research information, pre-prepared or photocopied work on to your design exam sheets. Similarly, ICT will not be assessed in the Design Paper, so you will not be allowed to use ICT facilities during the examination. However, it may be appropriate to use ICT for research.

Develop a product design specification, identifying appropriate constraints

SIGNPOST
Develop a design specification
Unit 2 page 45, Unit 5 page 272

Your specification is important because it forms the basis for generating and evaluating design ideas and for evaluating the final solution. Look at the example of a design specification in Figure 6.2.

All the work that you have done during the course on specification development should provide you with a firm understanding of how to develop a product design specification in response to research materials. In your specification you should identify the following criteria. These should be specific to the design problem you have to solve.

- Purpose: what the product is for.
- Function: what the product needs to do.
- Aesthetics: how the product should look, its style, form, aesthetic characteristics.
- Market and user requirements: the needs of the target market, user needs, ergonomics, trends, preferences.
- Performance requirements of the product, materials, components, systems: how it needs to perform, its functional characteristics and properties.
- Processes, technology and scale of production: possible processes, possible uses of technology required to make the product as a manufacturing prototype, scale of production required.
- Quality control: how the product will meet the quality requirements of users, use of quality standards.
- Legal requirements and external standards: how the product will meet safety requirements, *referencing* external standards where appropriate (British Standards are NOT to be copied out).
- Cultural, social, moral and environmental issues that may influence your design ideas.

Please note that you are not required to include time, resource and cost constraints, or considerations of maintenance and product life span in your design specification. These criteria are assessed in your coursework.

Figure 6.2 This design specification was developed in response to the problem of designing packaging materials and point-of-sale displays for a branded chocolate product

Design Specification:

The chocolate bar packaging and point of sale display must:

- Fulfil the general packaging requirements by containing, advertising, preserving and protecting the product.
- Have a strong brand identity and incorporate the brand name 'Candy Chocobar'.
- Be aimed at the children/youth target market.
- Be made from quality materials.
- Use full colour printing.
- Be able to be mass and batch produced respectively.
- Be able to contain one 'Candy chocobar' and 25 units respectively.
- Incorporate the necessary legal labelling (1984 food labelling regulations)
- Be priced around 35p and free to retailers respectively.
- Address environmental concerns such as use of materials and disposal.

2. Generate ideas, evaluate each idea and justify decisions made (15 marks)

SIGNPOST
Generating ideas
Unit 2 page 46, Unit 5 page 274

You are expected to generate a range of feasible design ideas, based on the criteria that you have set up in your product design specification. Your ideas should demonstrate a clear understanding of the design problem. Initial ideas should be in the form of quick sketches with written notes. At this stage, the examiner is looking for evidence of your design thinking, in response to the requirements of your product design specification.

SIGNPOST
Use knowledge and understanding gained through research to develop and refine ideas
Unit 2 page 48, Unit 5 page 275

As you gradually refine your first ideas, you may find that you start to think about the possible materials or processes you could use to see if your ideas will work.

You may also start to play around with combinations of ideas or work on the fine detail of some of your ideas. Making use of the information you acquired during your research phase, should enable you to develop and refine alternative ideas to the stage where you can select one (or possibly two) that are the most promising.

SIGNPOST
Evaluate and test the feasibility of ideas against specification criteria
Unit 2 page 49, Unit 5 page 277

All first ideas should be evaluated against the specification, and the design or designs chosen for development should be explained and justified, using written notes. Figure 6.3 shows some initial design ideas.

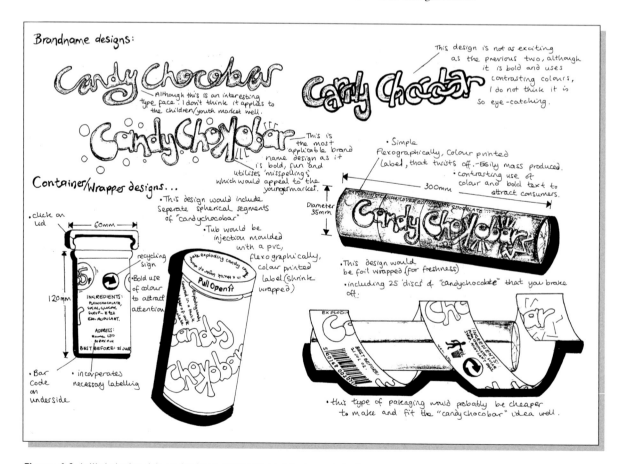

Figure 6.3 *Initial design ideas for brand identity and chocolate bar wrapper*

3. Develop chosen idea into an optimum solution, describe and justify solution in terms of function, appearance, performance, materials, components, systems, processes and technological features to be used (15 marks)

Develop an optimum solution

SIGNPOST
Develop, model and refine design proposals
Unit 2 page 49, Unit 5 page 278

Your aim should be to develop the best possible solution to the problem in the time available. You should clearly show how your ideas unfold by developing, modelling and refining them.

Use a good graphic style to communicate aesthetic and functional aspects of your design. You can use pen or pencil, but should try to resist rubbing out work, so the examiner can see your design thinking. It is often helpful to use a backing sheet of **isometric paper** as a guide for 3D sketches. Only use colour after you have completed a section, or spend the last ten minutes adding colour. If you go wrong completely or make a mistake, state it and attempt to get back on the right track. Avoid throwing away or discarding any work as it is your thought processes that are being assessed. An example of the development of a 3D point-of-sale display is shown in Figure 6.4.

SIGNPOST
Demonstrate a wide variety of communication skills
Unit 2 page 49, Unit 5 page 279

The examiner is looking for evidence of development from your initial ideas and for evidence of your design thinking. This should include how your ideas are influenced by your

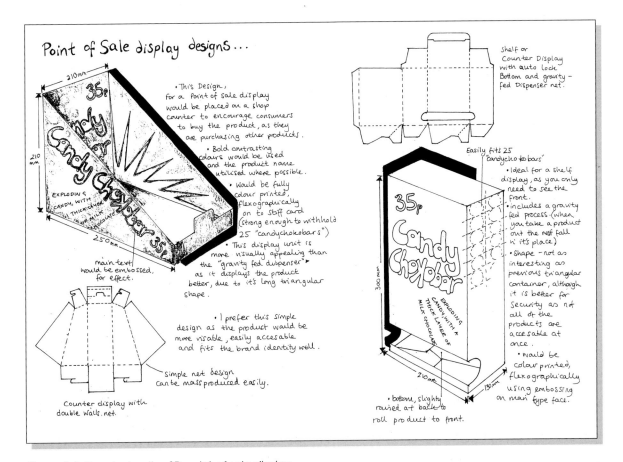

Figure 6.4 *Developing the 3D point-of-sale display*

research. You can show this by annotating your 2D or 3D drawings and sketches to explain hidden details or to give information about how and why materials or components are used. When you annotate your design work, you should use appropriate technical language or **industrial terminology** to demonstrate your understanding of industrial practices.

It is not always necessary to develop totally different ideas, although you should always try to produce variations related to the look of the product. Give alternative ideas of how it will function, especially if there are any moving parts. Think about the materials to be used and how they are cut, shaped and joined.

> **SIGNPOST**
> Materials and components Unit 1 page 21
> Industrial and commercial practice
> Unit 1 page 23
> Demonstrate understanding of a range
> of materials
> Unit 2 page 58, Unit 5 page 281
> Working properties of materials
> Unit 3A page 85
> Selection of materials Unit 4A page 174

Identify the best solution

Make sure that you identify which is your best solution to the design problem and say why you think it is the best. The most effective way to do this is to explain how the solution meets aspects of your product design specification. This should enable you to describe and justify your design solution in relation to:

- what the product is for
- what it needs to do
- how it should look, its style, form and characteristics
- how it meets market and user requirements
- how the performance of the chosen materials, components or systems meet requirements
- the appropriateness of assembly processes
- the appropriateness of construction techniques in relation to the scale of production
- any cultural, social, moral or environmental issues that have influenced your design solution.

> **SIGNPOST**
> Evaluate design proposals against
> specification
> Unit 2 page 53, Unit 5 page 281

4. Represent and illustrate the final solution, using clear and appropriate communication techniques (15 marks)

> **SIGNPOST**
> Materials and components
> Unit 1 page 21
>
> Industrial and commercial practice
> Unit 1 page 23
>
> Demonstrate a wide variety of
> communication skills
> Unit 2 page 49, Unit 5 page 279
>
> Demonstrate understanding of a range
> of materials
> Unit 2 page 58, Unit 5 page 281
>
> Working properties of materials
> Unit 3A page 85
>
> Selection of materials
> Unit 4A page 174

You should produce drawings of your final solution. Your drawings should clearly show what the product will look like and how it will be made. A technical or working drawing is not required.

Appropriate graphic techniques could include line drawings to show back, front or side views. You could also use exploded drawings to show hidden construction details. Do not forget to annotate your final solution. Use appropriate technical language or industrial terminology to justify the use of appropriate materials, components and construction processes. Make sure that your drawings include:

- clear construction and assembly details
- dimensions and sizes
- details and quantities of materials and components.

Figure 6.5 shows a final solution. The examiner is looking for evidence of logical thinking and for knowledge and understanding of appropriate use of materials, components and processes.

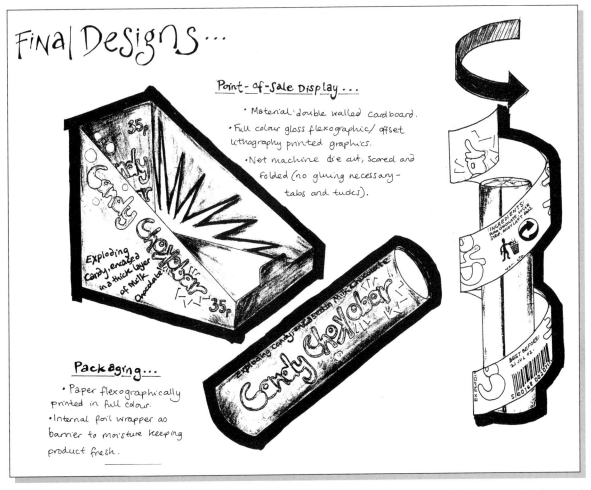

Figure 6.5 *The first page of the final solution – the final solution would also incorporate clear construction and assembly details, dimensions and sizes, as well as details and quantities of materials and components*

5. Draw up a production plan (15 marks)

SIGNPOST
Stages of production
Unit 1 page 24
Planning manufacture
Unit 2 page 54, Unit 5 page 282

This section of the exam asks you to draw up a production plan. In your coursework units you produced detailed plans for the manufacture of your products. In the Design Exam, you should produce a simple production plan in flow diagram or table form to show how the product would be manufactured.

Your production plan should include:

- the major assembly processes
- a **work order** showing the sequence of assembly, with details of tools, equipment and tolerances

- where you would check for quality at critical control points
- how you would check for quality, using quality indicators.

SIGNPOST
Quality
Unit 1 page 31
Devise quality assurance procedures
Unit 2 page 62, Unit 5 page 287

In your product design specification you identified the scale of production for your product. This will have influenced many of your design decisions, although ideas about possible manufacturing processes may have changed as your design developed.

Details of how to make your final solution should be found in your drawings. Here you should identify how your product is constructed, joined and assembled, together with any processes required to manufacture it. For an example of part of a production plan, see Figure 6.6.

The critical control points in your sequence of assembly should show where you would check for quality and accuracy. You should also identify quality indicators that explain how you would check for quality. Quality indicators can include, for example, the use of dimensions and tolerances to check the accuracy of fit and finishes. Other quality indicators include the sensory checks of touch and vision. The quality control processes that you identify in your production plan would ensure the manufacture of a quality product.

In this section of the Design Exam, the examiner is looking for evidence of logical thought and for knowledge and understanding of manufacturing processes and quality control.

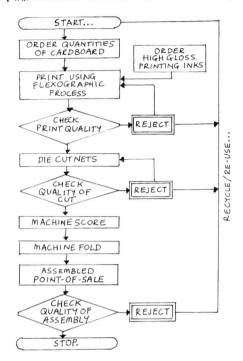

Figure 6.6 *Part of a production plan to describe the requirements of the solution*

6. Evaluate the strengths and weaknesses of the final solution and suggest improvements (15 marks)

> **SIGNPOST**
> Objectively evaluate the outcome against specifications and suggest appropriate improvements
> Unit 2 page 63, Unit 5 page 288

The last section of the Design Exam asks you to evaluate the strengths and weaknesses of your final solution and suggest improvements. Strengths and weakness should relate to your perception of the quality of your design and the **market potential** of your final solution.

Your evaluation may include force field analysis, an evaluation matrix or a star diagram.

> **SIGNPOST**
> Force field analysis Unit 2 page 49
> Evaluation matrix Unit 5 page 277

You can judge the **quality of design** of your solution by evaluating its fitness for purpose in terms of its performance against the product design specification. Explain how your solution meets the specification criteria. Does it meet all of them? What are its best aesthetic features in relation to the design specification? In what way is it a marketable product? How does it meet user and market requirements?

You should suggest improvements to the product's aesthetic or functional characteristics, in response to the market potential you have identified. Figure 6.7 shows how one student evaluated her final solution and suggested improvements.

> **SIGNPOST**
> Products and applications
> Unit 1 page 14
> Developing a commercial approach
> Unit 5 page 269

The examiner is looking for evidence of your ability to evaluate objectively against the product design specification and to suggest imaginative improvements related to quality and market potential.

Evaluation:

Testing the final design against the design specification.

- Fulfil the general packaging requirements by containing, advertising, preserving and protecting the product.

Container- The packaging for my product securely contains the chocolate by wrapping it in foil with an outer paper label.

Advertiser- The paper label offers an excellent printing surface for the full colour graphics.

Preserver - The foil wrapper acts as a barrier to moisture and keeps the product fresh.

Protector - The tight wrappers prevent the chocolate bar inside from sliding around, thus preventing breakages.

- Have a strong identity and incorporate the brand name 'Candy Chocobar'.

The brand identity that I have created clearly shows the name 'Candy Chokobar'. I deliberately mis-spelt the name by using a 'k' so it would look modern and exciting. I have used strong bold colours so it will look effective on the crowded sweet counter. The point-of-sale display also carries the strong brand identity so that there is a clear connection between the two.

- Be aimed at the children/youth target market.

The brand identity is deliberately bold and colourful so that it would appeal to children. To test this further I would survey groups of children to ask their opinions.

- Be made from quality materials

The materials I have used will provide a high quality product. The double walled cardboard point-of-sale display would be strong enough to last an initial promotion but light enough to be transported.

- Use full colour printing

Both the packaging and point-of-sale display use full colour printing using the flexographic or offset lithographic printing processes. The point-of-sale display would use high gloss printing inks for a quality finish. The paper label could be laminated for durability and effect.

- Be able to be mass and batch produced respectively

The materials and processes used for the packaging are ideal for mass production processes. The point-of-sale display is batch produced because it is needed in less quantity than the packaging. It uses machine die cutting, scoring and folding to assemble the net and does not need gluing because it has tabs and tucks to secure it.

- Be able to contain one 'Candy chocobar' and 25 units respectively

The packaging contains one product which is split into 25 discs of chocolate. The point-of-sale display easily contains 25 units and could in fact hold more if the shop-keeper desired.

- Incorporate the necessary legal labelling (1984 Food labelling regulations)

The packaging incorporates all of the necessary labelling, required by law i.e. ingredients, manufacturer address etc.

- Be priced around 35p and free to retailers respectively

The product is priced at 35p although it could be launched at a lower price to attract more customers. The point-of-sale display would be free to retailers i.e. shopkeepers, because it is a form of advertising for the product.

- Address environmental concerns such as use of materials and disposal.

The label would have a recycle logo clearly printed on it and a 'keep Britain Tidy' logo to encourage responsible disposal. The packaging uses two layers; paper label and foil wrapper so there wouldn't be much litter produced. The cardboard used in the point-of-sale display could be recycled cardboard and sent back to the manufacturer for recovery.

Marketing Issues

The product would be aimed at the children/youth market so the consumer will be the child and the customer may be an adult. An effective advertising campaign involving TV commercials on children's TV, billboards and adverts in children's magazines would create a need for the product. The child would then be inclined to pester their parents for the product while in a corner shop or supermarket. One ploy may be to place the product at the supermarket check-out so when the child gets bored it wants the chocolate bar.

Further Improvement/development

Further developments and improvements would obviously call for a good market research campaign. The product needs to be field tested with the target market to see if they like it or not. The views of the target market would then influence the development of the product so that it could be launched nationally and internationally.

Figure 6.7 *Evaluate the strengths and weaknesses of your final solution and suggest improvements*

Exam preparation

Revise what you have learned

Revising and reviewing what you have learned during your Advanced level course will help prepare you for the Unit 6 exam. In Unit 6 you will find 'Signposts' that refer you back to information in the other AS and A2 units.

Practice doing timed design tasks

You should be familiar with all of the design processes that you use during the exam. They are the same or similar processes to those that you use in your coursework.

Read pages 291 and 304 of Unit 6 so you are absolutely clear about the exam requirements. Practise doing timed design tasks that correspond to the design processes you will use in the Design Paper.

Analyse
- Analyse a design problem set by your teacher or tutor.
- Analyse a body of research information related to the design problem – this could be downloaded from the Internet, from CD ROMs or from textbooks.
- Develop a specification based on your research.

Exam tip
Analyse your research before the exam so you are familiar with it. Make a list of the design specification criteria and organise your research information under the same headings. Write a conclusion to each section of your research. This process will fully prepare you for the exam. Practise developing design specifications.

Generate ideas
- Generate a range of design ideas based on your design specification and research.
- Evaluate the feasibility of each idea against the design specification.
- Justify the design you have selected for development.

Exam tip
You should make use of a product design specification. Use it to guide and evaluate your design ideas. Practise producing quick design ideas. Always annotate your drawings and give reasons for your design decisions.

Develop ideas
- Develop a chosen idea into a workable solution.
- Describe and justify the solution in terms of function, appearance, performance, materials, components/systems, processes and technological features to be used.

Exam tip
Show how your ideas develop by annotating your drawings. Explain your reasons for your chosen idea. How does it meet the specification requirements? How might it be made? Will it be easy and practical to make? Are the manufacturing processes appropriate to the scale of production? Are any parts or components the same size or shape? Can their manufacture be easily repeated? Can you use any bought-in parts or components?

Represent and illustrate the solution to a design problem
- Illustrate and annotate the solution using 2D or 3D line drawings.
- Use exploded drawings to show hidden construction details.
- Include dimensions and sizes of component parts.
- Provide details and quantities of materials and components.

Exam tip
You are not required to produce a full working drawing in the exam. Annotate your drawings so that an examiner can understand how you intend the product to be made.

Production plan
Describe how to make the product. Include:
- the assembly processes
- a work order with details of tools, equipment and tolerances
- quality checks at critical control points with quality indicators.

Exam tip
Produce a flow diagram or table to show the major assembly processes for the product. Add relevant details of tools and equipment. Show where and how you would check for quality.

Evaluate
Evaluate the strengths and weaknesses of the final solution and suggest improvements, relating to:
- quality of design
- market potential.

Exam tip
Evaluate your solution against the design specification. How does it meet the requirements? What are its best design features in relation to the design specification? In what way is it a marketable product? How can it be improved?

Practise doing a complete timed exam

Practise a complete exam, working to the same time and under the same conditions as the Unit 6 Design Exam. Choose from the following four Design Research Papers and corresponding Design Papers. The assessment criteria in Table 6.2 should be used with each of the Design Papers.

Table 6.2 Design Paper assessment criteria

Assessment criteria	Marks
1 Analyse the design problem and develop a product design specification, identifying appropriate constraints	15
2 Generate and evaluate a range of ideas	15
3 Develop the chosen idea into an optimum solution	15
4 Represent and illustrate your final solution	20
5 Draw up a production plan	15
6 Evaluate the final solution and suggest improvements	10
Total marks	90

Design Research Paper 1

A kite manufacturer wants to increase sales of a new box kite by basing its design on a cartoon character. The kite is to be advertised on the company website. Investigate:

- existing box kite designs
- styling, colour and images used in the creation of cartoon characters
- website advertising.

Design Paper 1

Your task is to design a new 'sky diver' box kite, together with advertising materials to be used on the company website. The kite needs to be based on a cartoon character and should be suitable for 7–12 year olds.

Design Research Paper 2

Communicating a company's image is essential in today's competitive market, especially for food manufacturers, where branding is an essential marketing tool. Investigate:

- signs, symbols and brand marketing used in graphics for breakfast cereals
- the use of colour, typography, shape, form and layout
- the role of 'promotional gifts' in marketing products for children.

Design Paper 2

Your task is to design a logo that can be used on the packaging for 'Crackle Pops' breakfast cereal. The logo needs to be applied to a flat-pack promotional gift that can be assembled into a 3D toy.

Design Research Paper 3

A torch manufacturer needs to update the styling of a torch and its packaging. The design styling needs to promote an upmarket image so that it maintains a 'competitive edge'. Investigate:

- design styling of existing torches
- typography and images used for torch packaging
- marketing issues related to the promotion of torches.

Design Paper 3

Your task is to design a prototype torch that promotes an upmarket image. The torch packaging should protect and display the torch in a compact form.

Design Research Paper 4

When a sweets manufacturer develops a new range of confectionery, the presentation and display of the product are key marketing tools. Investigate:

- design styles and colours used for typography and imagery for confectionery
- point-of-sale displays and promotional material for sweets and confectionery
- the aesthetic and functional requirements of confectionery packaging materials.

Design Paper 4

Your task is to design suitable graphics for the product 'Candy Chocobars' that could be included on both the packets and promotional material. You also need to design a suitable stand to display 25 packets of the product. The 3D part should include the use of resistant materials.

Student checklist

1. Understanding the research context

- Make sure that you read the Design Research Paper carefully so you have a clear understanding of what you need to research.
- Take responsibility for planning, organising, managing and analysing your own research.
- Make sure that you analyse your research before the exam.

2. Preparing for the design exam

- Review what you have learned about designing and making products.
- Practise doing timed design tasks that correspond to the different sections of the exam paper.
- Practise doing a complete timed exam.
- Make sure that you understand the assessment requirements so you will know how much time to spend on each part of the exam question.
- Prepare a time plan that corresponds with the marks allocated to each section.

3. Showing your capability in the design exam

- Produce a short analysis of the problem you are asked to solve.
- Show the influence of your research in your product design specification.
- Clearly show how your ideas unfold. Annotate your drawings.
- Demonstrate a variety of appropriate communication skills, including graphics, diagrams, flowcharts and written text.
- Allow enough time to evaluate your work as it progresses.
- Use industrial terminology and technical terms to demonstrate your understanding of industrial practices.

4. Exam hints

- Practise any drawing techniques you may need to use.
- Prepare a backing sheet with heavy lines as a guide for written work.
- Prepare a backing sheet of isometric paper as a guide for 3D sketches.
- Collect together all your research work, sketch books, folios, notebooks and technical data compiled as part of the work done during the course. You may also refer to any relevant British Standards. Be sure to analyse your research before the exam.
- During the exam you will be provided with headed sheets. Extra paper will be available. Make sure you write your candidate name and number and centre name and number on each sheet.

5. Exam day

- Arrive half an hour before the exam to get yourself organised.
- Read the exam question carefully and identify key points in the question. Do not copy out the question.
- Use a pencil or pen for writing and sketching. Try not to rub anything out as it is better if the examiner can see your thinking. Take a pencil sharpener with you.
- Only use colour after you have completed a section, or spend the last ten minutes adding it.
- If you make a mistake or go completely wrong, state it and attempt to get back on the right track. Do not throw away or discard any work as it is your thought processes that are being marked.
- You will be assessed on your ability to organise and present information, ideas, descriptions and arguments clearly and logically. You will get marks for quality of communication.

Help! What if the exam goes wrong?

- If your Unit 6 design exam doesn't meet your expectations, don't worry! You can retake the unit and the better result will count towards your final grade.
- If you find yourself in this situation your teacher or tutor will be able to advise you on the best way forward.

Glossary

A

Adjustable sensing inputs – transducers configured in potential dividers to sense changes in the surrounding environment and produce an electrical signal accordingly.

Advertising – any type of paid-for media that is designed to inform and influence existing or potential customers.

Advertising Standards Authority (ASA) – regulates all British advertising in non-broadcast media.

Aesthetics – the sensory responses that people make to external stimuli such as colour, shape, sound, touch, smell and taste; individuals respond differently to these stimuli, finding them pleasing or disagreeable; 'visual aesthetics' refers specifically to those that are concerned with appearance; people respond to 2D and 3D products using a range of senses.

Agitation – the movement of film and photographic solutions in development to ensure an even coating.

American National Standards Institute (ANSI) – administrator and coordinator of the United States voluntary standardisation system founded in 1918.

Analogue signals – type of electronic signal that can exist at any level between the two extreme points of high and low

Annealing – the process that relieves internal stresses within a material as a result of cold working.

Artificial intelligence (AI) – a branch of computer science concerned with developing computers that think/act like humans.

Axonometric – drawings showing an object in 3D including isometric, planometric and oblique.

B

Bandwidth – the amount of HYPERLINK 'data.html' data that can be transmitted or received in a fixed amount of time; expressed in bits per second (bps) for digital devices. The higher the bandwidth the faster the rate of data transfer. See 'ISDN'.

Bar code (data communication tag) – machine-readable pattern of stripes printed on a component part or a finished product to identify it for production processes, stock control, pricing or retail sales.

Binary signal – a digital signal with only two values: high ('on' or '1') or low ('off' or '0').

Bitmaps – images of a surface or solid model produced by an intensity of points called picture elements or pixels.

Biochips – in biochips, semi-conducting molecules are inserted into a protein framework, in which the proteins grow to take up the shape required for an electronic circuit. The manufacture of biochips will lead to the further miniaturisation of electronic products.

Blow moulding – the process of using air, under pressure, to give form to products by 'blowing' plasticised material into moulds.

Brand – a product or service with a marketing identity that belongs to one producer.

Brand loyalty – involves buying a chosen brand of product, because it provides a perceived level of reliability, quality, or lifesyle.

Brand name – (also trade-mark or trade name) protects and promotes the identity of a product or process, so that it can't be copied by a competitor. A branded product usually has additional features or added value over other generic products – making it 'special' in the eyes of consumers.

Break-even point – point at which products sold at a certain price will equal in value the cost of their manufacture.

British Standards Institute (BSI) – an independent, non-profit-making and impartial body that serves both the private and public sectors. BSI works with manufacturing and service industries to develop British, European and International standards.

Buy in – components or sub-assemblies that are 'bought-in' or purchased from another manufacturer for use within a product, e.g. a gearbox may be bought in by a car manufacturer.

Buying behaviour – establishes how people make buying decisions and factors that influence these decisions.

C

CAD modelling – involves a number of representation techniques such as vector or raster graphics. Models are used for variety of purposes ranging from a simple record of a design with critical manufacturing dimensions through to a fully interactive simulated product or system model that can be viewed in operation.

Caliper – used to describe the thickness of card material; measured in microns (mic) which are equal to one thousandth of a millimetre.

Camera ready artwork – has been pasted into position to be photographed for final film ; usually a page containing all the necessary image texts and tints in position needed to produce a printing plate.

CEN – the European Committee for Standardisation. It implements the voluntary technical harmonisation of standards in Europe, in conjunction with world-wide bodies and European partners.

CENELEC – the European Committee for Electro-technical Standardisation.

Cleaner design – aimed at reducing the overall environmental impact of a product from 'cradle to grave'.

Cleaner technology – the use of equipment or techniques that produce less waste or emissions than conventional methods. It reduces the consumption of raw materials, water and energy and lowers costs for waste treatment and disposal.

Closed question – provides a limited number of possible answers to choose from.

Closed-loop control system – a system that has a degree of feedback built into it that enables a degree of checking to see whether actions or processes have been carried out.

Clutch – type of shaft coupling that allows a rotating shaft to be easily connected or disconnected from a second shaft.

Comparative testing – comparing the characteristics and properties of different materials, often in relation to specification requirements; simple comparative testing makes use of standard tests under controlled conditions, so each material is tested in exactly the same way; results can be compared to see which material is the most appropriate.

Compatibility – hardware and software applications capable of being used in combination without any technical problems. If systems are compatible, computers can 'talk' to other computers and to equipment such as CNC machines.

Competitive edge – the reason why a customer might choose a certain product rather than its competitors.

Components – the parts of a product that go to make up the whole.

Composite image – a photo or other graphic image that is made up of a number of images

Compound gear train – a series of gears connected together, some on common shafts, which allow rotational speeds to be increased or decreased.

Computer integrated manufacture (CIM) – a system of manufacturing that uses computers to integrate the processing of production, business and manufacturing information in order to create more efficient production lines.

Computer simulation – method of modelling manufacturing processes, designs or other product characteristics using computer software.

Computer-aided engineering (CAE) – use of computer systems to analyse and simulate engineering designs under a variety of conditions to see if they will work.

Conductivity – the measure of a material's ability to have heat or electricity passed through it.

Consumer demand – the potential for the demand to 'pull' products through the distribution system; often stimulated by marketing and promotion.

Consumer goods – products purchased by consumers for their own consumption.

Consumer society – a social culture in which consumers are encouraged by advertising to buy consumer goods.

Continuous Improvement (CI) – (Kaizen) involves companies systematically looking for opportunities to make manufacturing operations leaner and more cost effective whilst also getting the most from their workforce by involving employees in all aspects of their operation. It aims to develop a culture where all employees are communicating better, working proactively to meet the company's business objectives and where implementing improvements is a matter of course.

Conversion – in timber production, the process in which a felled tree is turned into a useable source of timber.

Co-polymerisation – process in which two or more monomers combine to form a new material.

Corporate Identity Systems – define and describe the way in which an organization presents itself both internally and externally; the system expresses and reflects the values of the company to employees, suppliers and customers; it commonly deals with the permitted variations of use of logotypes and symbols applied in different ways to its products, stationary, delivery vans, uniforms, promotional materials, etc.

Corporate image – the identity or 'personality' of a company or organisation, created through the use of different graphic images.

Crating – way of drawing a product by imagining it being made up of a number of boxes joined together.

Critical control points (CCPs) – points during the production cycle at which a product is monitored, to ensure that it is successfully manufactured to specification; ensures that any faulty components are rejected before they are processed further or built into the final assembly.

Critical path analysis – the breakdown of the whole manufacturing process into an ordered sequence of simple activities.

Critical technologies – technologies that need to be in place for a product to develop, e.g. the availability of well-established sensor technologies is critical in the development of industrial robots.

Cupping – a fault in timber where the board is hollowed along its length.

Current – the flow of an electric charge through a conductor.

Customer profile – profile of potential customers, such as gender, age, family group, income, education, beliefs and attitudes, taste, lifestyle and perception of products.

D

Darlington pair – two smaller transistors connected together, increasing the overall gain and sensitivity.

Data storage devices – such as CD-ROMs that are used for storing digital electronic information known as data.

Demographics – patterns and trends of population and society, such as age, gender or income bracket.

Dendrites – a crystal that has branched during its growth and has a tree-like look.

Design management – the planning of a product to include organisational, economic, legal and marketing considerations, as well as decisions about form and function.

Design specification – sets out the criteria that the product aims to achieve.

Designing for manufacture (DFM) – aims to minimise costs of components, assembly and product development cycles and to enable higher quality products to be made.

Desktop publishing (DTP) – the use of a computer program to design the arrangement of text, images and graphic devices on a printed page; the electronic data included in the resulting files can be passed directly to computerized printing systems.

Desktop videoconferencing (DTVC) – videoconferencing applications that use video cameras mounted on standard desktop computer system such as an Intel-based PC, Apple Macintosh, or Unix workstation.

Die cutting – the process of cutting and creasing sheet materials, so that the can be formed into 3D products such as cartons or packages.

Die cutting tools – required for the cutting and creasing of cardboard or plastic sheet components that make up cartons or boxes; the design for die cutting tools are normally produced on a CAD machine, but the tool is made by hand.

Digital imaging – the process of creating a digital copy of an illustrated or photographic image.

Digital photography – the process of recording images using a digital camera.

Digital printing – involves linking printing presses with computers, bypassing the need for making printing plates.

Digital signal – a signal that can only take fixed values between two points.

Digital system – electronic system that can exist in one of only two states: on or off.

Diode – an electrical component that will only allow a current to pass through it in one direction.

Direct costs – see 'variable costs'.

Downtime – unproductive period when a computer system or machine is not operational usually because of technical problems, maintenance or in the case of machines, setting up new tools or reprogramming tools on a CNC machine.

Download – to get electronic information, software, files and documents from the Internet on to a computer system.

Dynamic images – give a sense of movement to a graphic layout; their shape directs the viewer's eye to different parts of the page and adds impact.

E

Economies of scale – occur when the cost of producing each product falls as the total volume of products produced increases.

Efficiency – a measure of what you get out in relation to what you put in.

Elastic deformation – when a material returns to its original shape and length once a deforming force has been removed, it is said to have been elastically deformed.

Electromotive force (emf) – a source of energy that can cause a current to flow in an electrical circuit or device.

Electronic Data Interchange/Exchange (EDI or EDE) – the electronic transfer of commercial or organisational information from one computer application to another. It is also known as paperless trading.

Electronic Product Definition (EPD) – makes use of CAD/CAM systems in which all of the product and processing data is generated and stored electronically in a database. The whole production team has access to the database, which evolves as the new product is developed. EPD enables the use of computer integrated manufacture (CIM).

Enabling technologies – provide the useful technologies, for example drive motors and power control systems that make critical technologies effective.

Environmentally friendly plastics – plastics capable of bio-degrading naturally.

Enzymes – naturally-occurring proteins, used to create industrial products and processes. These enzymes are the same kind that help us digest food, compost garden rubbish and clean clothes.

ETSI – the European Telecommunications Standards Institute.

European Standards Organisation – joint standards organisation called CEN/CENELEC/ETSI.

Evaluation matrix – used to compare and evaluate a number of ideas against specification criteria. Each idea is given a score showing its strengths and weaknesses. Very weak ideas are eliminated, resulting in the emergence of strong ideas, which can be developed individually or combined in some way.

Expert systems – part of a general category of computer applications known as artificial intelligence (AI). They either perform a task that would normally be done by a human expert or they support the less expert to complete a task.

Exploded – (drawings) shows a product or component pulled apart, laid out in an ordered and linear form.

External failure costs – occur when products fail to reach the designed quality standards and are not detected until after being sold to the customer.

Extranet – an intranet that is partially accessible to authorised outsiders.

F

Fabrication – the joining and fixing together of various materials and components to form a new product.

Feedback – information generated within a system or process to enable modifications to be made to maintain the operation of a system or to ensure a consistent level of production.

Field effect transistors (FET) – electronic voltage amplifier.

Figure – the natural decorative pattern of the timbers grain.

File Server – a computer with data that can be accessed by other computers.

File Transfer Protocol (FTP) – method of transferring information files from Internet libraries directly to a computer.

Final film – the intermediate print production step between artwork and plate making.

Finishing – in commercial printing, this is the way a document is collated, bound, folded or glued.

Finite resources – see 'non-renewable resources'.

Firewall – a hardware or software system designed to prevent unauthorised access to or from a private computer network (see 'intranet').

Fitness-for-purpose – a product's fitness-for-purpose can be evaluated through its performance, price and aesthetic appeal.

Fixed costs – (indirect or overhead costs) fixed costs remain the same for one product or hundreds as they are not directly related to the number of products made. They include design and marketing, administration, maintenance, management, rent and rates, storage, lighting and heating, transport costs.

Force field analysis – maps the forces for and against an idea or concept and the forces for and against changing it.

Form – created when a 'shape' becomes three dimensional, e.g. a circle becomes a sphere or cylinder.

Fourdrinier machine – releases and processes pulp to form a continuous paper roll.

Function – the means by which a product fulfils its purpose.

G

Gantt chart – a simple chart that maps each task against the time available, together with an order of priority.

Geometric modelling – using computer programs for representing or modelling the shapes of three-dimensional components and assemblies. Geometric models are the basis of all CAD/CAM systems.

Glass Reinforced Plastic (GRP) – a matting of glass strands held rigid in a polyester resin; also called fibreglass.

Global manufacturing – the manufacture, by multinational companies, of products that may be designed in one country and manufactured in another.

Global market place – the marketing of products such as washing machines and cars, across the world. To be successful in this global market place, a company has to have a product that appeals to people in different countries and cultures.

Graphic devices – lines or shapes (usually coloured) used in a layout to add visual interest and/or to help identify different sections of information within the design.

Gravity die casting – the process by which molten metal is poured into metal or graphite moulds.

H

Hard sell – a hard sell advertisement has a simple and direct message, which projects a product's Unique Selling Points (USPs).

Hazard – source of or situation with potential harm or damage. Hazard control incorporates the manufacture of a product and its safe use by the consumer.

Heat treatment – the changing of a material's properties and characteristics due to the application of an external heat source.

High-technology production – the production of 'high tech' products, which emphasise technological appearance and modern industrial materials.

HTML (Hyper Text Mark Up Language) – text-based coding system and scripting language used when writing web pages.

HTTP (Hyper Text Transfer Protocol) – a transport protocol used when transmitting hypertext documents across the Internet.

Hyperlink – an electronic connection that allows links between different web pages to be made, usually shown in a different colour and/or underlined.

I

Indirect costs – see 'fixed costs'.

Industrial terminology – includes the use of technical terms, such as critical control point or production plan to demonstrate your understanding of industrial practices.

Injection moulding – highly automated manufacturing process in which a plasticised material is injected into a mould cavity under high pressure

Interface – device that will allow electrical signals into and out of a computer.

Internal failure costs – occur when products fail to reach the designed quality standards and are detected before being sold to the consumer.

Intranet – a network based on the Internet belonging to an organisation that is accessible only by authorised users with user names and passwords.

Inventory – a company's merchandise, raw materials, finished and unfinished products that have not yet been sold.

ISO (International Organisation for Standardisation) – world-wide federation of national standards bodies from some 130 countries, one from each country. IOS is a non-governmental organisation established in 1947.

ISDN (Integrated Systems Digital Network) – a high-speed, wide-bandwidth electronic communications service to carry digital data, digitised voice or video across digital phone lines.

ISO 9000 – a set of management processes and quality standards to ensure that a product meets the customer's requirements.

Isometric paper – has vertical lines, with all other lines drawn at 30° to the horizontal; useful as a backing for sketching an isometric view, in which the product is drawn at an angle with one corner nearest to view. In this type of drawing all vertical lines on a product remain as vertical, while all horizontal lines are drawn at 30° to the horizontal on the paper. No vanishing points are used and the height, width and length are shown as parallel sets of lines.

ISP (Internet Service Provider) – a company providing a connection to the Internet.

J

Jidoka – (autonomation) Japanese term for the automatic control of defects, a machine finds a problem, finds a solution, implements it without outside assistance and then carries on.

K

Kaizen – See 'continuous improvement'.

Kanban – Japanese term for a card signal or visual record, 'Kan' meaning card, 'Ban' meaning signal.

Kerning – the space between characters that can be adjusted, so that parts of the characters overlap; the purpose of this is to make words fit on a line without affecting readability; LY LY and AT AT are called 'kerning pairs' as they can overlap.

Kitemark – a seal of approval by the British Standards Institute, awarded to any product that meets a British Standard, as long as the manufacturer has quality systems in place to ensure that every product is made to the same standard.

L

Laminating – process of sticking sheets of laminate or veneers together in either flat sheets or over curved formers.

Lash-ups – quick, rough models used to work out and test the relationship between different parts of a design.

Lattice structure – the pattern adopted when the atoms of a liquid solidify.

Layout – the arrangement of images and text in relation to each other.

Layout paper – allows images to be traced and is commonly used for sketching.

Level/scale of production – the size of production, e.g. a one-off such as a bridge or thousands such as chocolate bars.

Life-cycle assessment (LCA) – evaluates the materials, energy and waste used in a product through design, manufacture, distribution, use and end-of-life, which could be disposal, re-use or recycling.

Lifestyle marketing – the targeting of potential market groups and matching their needs with products.

Light-dependent resistor – a semiconductor whose resistance changes as the amount of light falling upon it changes.

Linkages – a series of levers connected together to change the direction of motion.

Liquid crystal display (LCD) – numerical and alphanumerical display system used in calculators, digital watches, etc.

Local Area Network (LAN) – collection of computers connected together to share information and other computer resources such as a printer.

Logic – a structured way of thinking or a set of operating principles applied to a manufacturing system or product to allow it to perform a specified task.

Logic gate – series of electronic switches that give a known output when a certain configuration on the input pins exists.

Logistics – the detailed organisation and implementation of a plan or operation such as supplying and moving parts, components and finished products within and from a manufacturing system.

Logotype – a logotype is the use of a distinctive typeface to identify the goods or services of a particular organization, or the brand name of a particular product; often used in conjunction with a symbol.

M

Malleability – the ability of a material to be beaten or pressed into a shape without breaking or fracturing when cold.

Manufacturing specification – clear details of product manufacture, such as accurate drawings, clear construction details, dimensions, sizes, tolerances finishing details, colour tolerances in printing/reproduction processes, quantities and cost of materials and components.

Market driven – the concept that promotional activity and marketing pushes products through the market group to pull products through the distribution system.

Market led – the concept that promotional activity and marketing stimulates demand in customers in a distribution system.

Market potential – the potential for a product to sell into a specific target market group.

Market research – identifies the buying behaviour, taste and lifestyle of potential customers and establishes the amount of money they have to spend, their age group and the types of products they like to buy.

Market segment – a group of people with similar needs who wish to buy a certain type of product or service.

Market segmentation – a marketing technique that targets a group of customers with specific characteristics.

Market timing – attempting to predict future market directions, usually by examining recent price and volume data or economic data, and investing based on those predictions.

Marketing – anticipating and satisfying consumer needs while ensuring a company remains profitable.

Marketing plan – a set of marketing activities developed to match a company's products to selling opportunities; involves developing a competitive edge by providing reliable, high quality products at a price customers can afford, combined with the image they want the product to give them.

Mechanical Advantage (MA) – a mathematical relationship which exists between the load and effort in relationship to levers. The greater the MA, the easier it becomes to move the object.

Media – agencies such as the press, television or posters that carry advertising.

Metal crystals – basic unit cells which make up the lattice structure of a metal.

Metal grains – small crystals that form between dendrites on cooling.

Micro-structure – the structure of material as observed under a microscope.

Milestone planning – project management process involving identification of key points or milestones that needed to be reached in a production process if a product is to be successfully completed on time and to budget.

Miniaturisation – came about through developments in microchip technology, resulting in ever smaller products.

Modelling – visualising design ideas using hand or computer techniques in two dimensions (2D) or three dimensions (3D).

Monomers – a compound whose molecules can join together to form a polymer.

Moodboard – used by a professional designer to explore moods or themes and to give a product an identity; moodboards communicate ideas on design, illustrating themes, trends, form, colours, texture and styling details; these product 'stories' are inspiration for generating design ideas.

Multimeter – measuring device used to measure current, voltage, resistance, and also used as a continuity tester.

Multinational companies – operate in more than one country and used to be mainly associated with mineral exploitation or plantations, such as cotton or food.

N

Niche markets – target market groups for whom products are designed and marketed.

Noise – unwanted electrical signals, e.g. fuzzy TV pictures caused by electrical interference.

Non-renewable resources – finite resources, such as oil or coal, which will eventually be exhausted unless action is taken.

O

Offset printing/lithography – currently the most common commercial printing method; ink is offset from a printing plate to a rubber roller and then to paper.

One-point perspective – the simplest form of perspective drawing in which the front view is drawn as a flat two-dimensional image. All receding lines are then taken back to a single vanishing point, to give a three-dimensional view.

Opacity – how paper is judged in its degree of transparency.

Open-loop system – a control system that incorporates no feedback, being a pure linear progression from the input to the output.

Optical character recognition (OCR) – where an electronic device recognises written letters or numbers.

Orthographic views – see 'working drawing'.

Output transducers – devices such as bulbs, motors and alarms.

Overhead costs – see 'fixed costs'.

P

Parallax error – the mis-alignment of the image when taking a photo; this occurs when using a camera that does not allow the user to view the image through the lens.

Parametric designing – involves establishing the mathematical relationships between the various parts that make up a shape or a product. Once the parameters are determined a designer can model exactly what would happen if particular sizes were redefined because if one measurement is changed all the others are changed in the correct proportion.

Patents – issued by government authority, these documents grant the sole right to make, use or sell a design, making it both unique and protected.

Performance modelling – working prototypes that enable the designer to test the function of a design against the design specification.

Perspective – allows an object to be drawn as it is viewed by the human eye, with parallel lines converging at a vanishing point; only the vertical edge closest to the viewer is in scale.

PEST – part of the basic structure of a marketing plan. It involves analysing values, such as political, economic, social and technological issues related to marketing a product.

Photovoltaic cell – a semiconductor that generates a small voltage when exposed to bright light.

Pictorial – shows the most realistic view of a product, sometimes called an 'artistic impression'.

Piezo-electric actuators/transducers – electronic device capable of generating a small voltage when pressure is applied, or a small movement if voltage is applied to it

Planning horizon – how far to plan forward, determined by how far ahead demand is known and by the times to pass through the manufacturing operation.

Plasticity – the ability of a material to be moulded.

Plates – sheets of treated aluminium alloy on which a print image is chemically etched; one plate is required for each colour; used in the printing process to transfer and ink image onto board or paper.

Poka-yoke – Japanese term meaning a device or procedure to prevent a defect during order-taking or manufacture (also called baka-yoke). The nearest translation is 'foolproofing' or 'mistake-proofing'.

Polluter pays – the concept that those generating, handling and treating wastes should pay large fines if they allow potentially harmful materials to enter the environment.

Polymerisation – chemical reaction that occurs when a polymer is formed.

Post-production – manipulation and addition of elements –usually computer generated – of filmed scenes in studio and on location.

Potential divider – two resistors connected in series, set up to divide the potential in the ratio of resistor one to resistor two.

Potentiometer – three-legged device that can be configured to work either as a potential divider or rheostat.

Presentation drawings – communicate ideas about a product or environment using a variety of suitable drawing techniques; a children's pop-up book could be presented in a simple, colourful style, whereas a high-tech interior may benefit from a more technical style of presentation.

Prevention costs – costs of 'making it right first time'. Prevention costs include those relating to the creation of and conformance to a quality assurance system and the management of quality.

Primary processing – the conversion of raw materials into usable stock for production, e.g. steel making.

Primary research – facts and figures that are collected specifically to provide information and help achieve the research objectives.

Primary sector – concerned with the extraction of natural resources such as mining and quarrying. priority.

Product Data Management (PDM) software – integrates the use of computer systems, including CAD/CAM and computer integrated manufacturing (CIM). PDM software enables the design and development of virtual products on screen. The software organises and communicates accurate, up-to-date information in a database, monitors production and enables fast, efficient and cost-effective manufacturing on a global scale.

Product design cycle – process leading to the design and manufacture of a product involving design, make, redesign and remake, which starts with a perception of need and includes many influences throughout the process such as government policy, manufacturers, advertisers, retailers and consumers.

Product viability – essential to the existence of a manufacturing company and to the employment of its workforce; relates to the cost of manufacture, the product's market potential and the potential profit from manufacturing the product.

Production capacity – the maximum number of products that can be made in a specified time.

Production chain – the sequence of activities required to turn raw materials into finished products for the consumer.

Production plan – shows how to manufacture a product, based on the breakdown of the whole process into an ordered sequence of simple activities; includes all specifications, the stages of production, resource requirements, and the production schedule.

Production schedule – an ordered sequence of processes that are required to manufacture a product.

Production team – flexible, organised, skilled, versatile people, who work collectively, make joint decisions and share the responsibility for the design and manufacture of products.

Productivity – a measurement of the efficiency with which raw materials (production inputs) are turned into products (manufactured outputs). High productivity results in lower labour costs per unit of production and a higher potential profit.

Profit – the amount left of the selling price of a product, after all costs of manufacture have been paid.

Programmable logic controller (PLC) – small but complex systems containing timers, counters and many other special functions capable of almost any type of control application, including motion control, data manipulation and advanced computing functions such as manufacturing plant management.

Protocol – an agreed standard or set of rules. See 'File Transfer Protocol' and 'HTTP'.

Prototype – a detailed 3D model made from inexpensive materials to test a product before manufacture.

Pulley – a circular disc normally with a V-shaped groove cut around its circumference.

Q

Qualitative research – an investigation to find out how people think and feel about issues and why they behave as they do.

Quality – conformity to specifications and ensuring fitness-for-purpose. Making products right first time, every time, with zero faults, to ensure customer satisfaction.

Quality assurance (QA) – a system applied to every stage of design and manufacture; ensures conformance to specifications to make identical products with zero faults.

Quality control (QC) – checking at critical control points against specifications for accuracy and safety, so that a product meets consumer and environmental expectations.

Quality indicators – quality control techniques, such as inspection, testing and sampling, that are applied at critical control points during manufacture to ensure the product meets specifications. Quality indicators may be attributes that can only be right or wrong, such as using the correct type of wood or variables that can vary between specified limits, such as meeting a tolerance of +/–1.0mm.

Quality Management System (QMS) – uses structured procedures to manage the quality of the designing and making process.

Quality of design – a product that is well-designed and attractive to the target market; meets specifications; uses suitable materials; is easy to manufacture and maintain; and is safe for the user and the environment.

Quality of manufacture – refers to a well-made product, that uses suitable materials; meets specifications and performance requirements; is manufactured by a suitable, safe method; is made within budget limits to sell at an attractive selling price; and is manufactured for safe use and disposal.

Quantitative research – an investigation to find out how many people hold similar views or display particular characteristics.

Questionnaire – a standardised set of questions designed to collect data that is relevant to the research objectives.

Quick Response Manufacturing (QRM) – a manufacturing system able to respond quickly at all levels of business or production processes in response to market trends and changing demand patterns.

R

Rapid Prototyping (RPT) – a CNC application that creates 3D objects using laser technology to solidify liquid polymers in a process called stereo-lithography.

Rectification – the process of converting an alternating current into a direct current.

Recycling waste materials – form of waste management in which waste materials from the production process are used in a different manufacturing process.

Relay – device used to interface two separate circuits that operate at two different supply voltages.

Renewable resources – flow naturally in nature or are living things which can be regrown and used again. They include wind, tides, waves, water power, solar energy, geothermal, biomass, ocean thermal energy and forests.

Resistors – electronic component used to control the current flowing in an electrical circuit.

Re-using waste materials – see 'recycling waste materials'.

Right first time – the aim of quality assurance, to make sure the product is right first time, every time. It involves making products that meet the specification, on time and to budget.

Risk assessment – identifying risks to the health and safety of people and the environment.

S

Seasoning – the process of reducing the moisture content in timber.

Secondary processing – the working of a material using engineering processes such as turning or milling.

Secondary research – facts and figures that are already available, having been collected for another purpose by a range of organisations.

Secondary sector – concerned with the processing of primary raw materials and the manufacture of products.

Selling price (SP) – price at which a product can be sold in order to make a profit. It generally includes variable costs, fixed costs and a realistic profit.

Semiconductor transducers – semi-conductors whose resistance changes depending on the surrounding environment.

Server – a host computer that distributes and stores data on a network.

Shape memory alloys – plastics that revert to their original form when heated.

Sketch model – a quick model produced in the early stages of product development, using inexpensive materials such as card, paper, expanded polystyrene, styrofoam and wood.

Sleeve – versatile packaging accessory with applications ranging from the protection of a delicate item to food such as a ready meal; also used as an alternative for a lid; made from card or PVC.

Smart materials – the properties of smart materials can change in response to an input, such as Piezo-electric actuators; provide opportunities for the development of new types of sensors, actuators and structural components, which can reduce the overall size and complexity of a device.

Soft sell – promotes a product's image, with which consumers can identify and is often associated with brand advertising.

Solar panels – these panels normally have water pumped through them that gets heated by the sun's energy.

Specialised components – components that are manufactured specifically for a particular product application.

Standard components – components such as nuts and bolts that are supplied ready to use.

Standards – documented agreements with technical specifications or other precise criteria to be used consistently as rules, guidelines, or definitions of characteristics, to ensure that materials, products, processes and services are fit-for-purpose.

Statutory rights – what consumers should reasonably expect when buying or hiring products and services. Statutory rights are enforced and regulated by a wide range of legislation relating to consumer protection and fair trading.

Strategic technologies – ways of thinking and operating, e.g. artificial intelligence.

Sub-assemblies – component parts of a product that are already made up of smaller components.

Supply chain – companies and organisations that collaborate to produce raw materials, components and end-products for specific end-uses aimed at specific target market groups.

Survey – way of collecting quantitative data, often about behaviour, attitudes and opinions of a sample in a target market group.

Sustainable development – a concept that puts forward the idea that the environment should be seen as an asset, a stock of available wealth. If each generation spends this wealth without investing in the future, then the world will one day run out of resources.

SWOT – part of the basic structure of a marketing plan. It involves analysing a product's strengths, weaknesses, opportunities and the threats from competition.

Symbol – a graphic device used to identify the goods or services of a particular organization, or the brand name of a particular product; often used in conjunction with a specific logotype; as they are easily understandable by most cultures, common symbols are also often used to communicate information such as locations, directions and safety or hazard warnings.

Synoptic assessment – the drawing together of skills, knowledge and understanding acquired in different parts of the whole A level course.

T

Target market group – all the customers of all the companies supplying a specific product.

Target marketing – the process of identifying market groups and developing products for it.

Technical drawing – contains factual information relating to appearance and dimension and is based on British Standard BS7308. See 'working drawing'.

Telematics – a new technology that allows a product to be managed electronically from receipt of the customer order through development, manufacturing, delivery and after-sales support.

Tempering – process of removing the brittleness caused as a result of hardening.

Tertiary sector – concerned with industries that provide a service; employs the most people in developed countries and includes education, retailing, advertising, marketing, banking and finance.

Test marketing – involves introducing a product in a small sector of a target market to test its viability before incurring the expense of a full-scale product launch.

Test models – used to test different parts of a design and are often built from kits to test mechanical, structural or control problems.

Thermistor – a semiconductor whose resistance can change depending on the temperature around it.

3D CAD systems – a computer aided design system that can produce virtual images in three dimensions to present more realistic representations of products and assemblies.

Thumbnail – a small rough sketch showing the main parts of a design in the form of simple diagrams.

Thyristor – three-legged electronic semi-conductive switch that can be used as a latch.

Time bucket – the unit of time on which a production schedule is constructed and is typically daily or weekly.

Time delay circuit – a capacitor/resistor network that is capable of producing an electronic time delay.

Tolerance – the degree by which a component's dimensions may vary from the norm and still be able to fulfil its function.

Total design concept – design using multimedia 'toolkits' to access an integrated on-screen design modelling environment that includes systems linked to production databases to analyse and plan for manufacture.

Total Quality Control (TQC) – the system that Japan has developed to implement Kaizen for the complete life cycle of a product.

Transducer – device for converting physical signals into electrical signals. An input transducer responds to a physical change by producing an electrical signal to represent the change. An output transducer takes an electrical signal from a system and produces a physical change as an output.

Transistor – a semiconductor device that can exist as an insulator or conductor. It can be used as an electronic switch and amplifier.

2D CAD drawings – have length and width but no depth.

Two-point perspective – most common form of perspective drawing in which the vertical lines stay vertical, while all other lines recede to two vanishing points. These points are placed on a horizontal line called the eye level. If a product is drawn below eye level, the top will be visible. If it is above eye level, the underside is visible.

Typography – the design and application of different letter (and number) forms to printed text to ensure appropriate legibility and ease of reading.

U

Unique selling proposition (USP) – a product's unique features and advantages over a competitor's products.

Unix – a computer operating system that was developed by AT&T in the 1960s. The system was used extensively during the establishment of the Internet.

Upload – to send electronic information from your computer to another location via the Internet or other types of network (the opposite of download).

URL (Uniform Resource Locator) – convention used when naming pages on the World Wide Web.

V

Vacuum forming – a plastic processing method in which a softened plastic sheet is pushed down by atmospheric pressure on to a mould, to make products such as baths.

Value analysis – the process of close study of a product in order to reduce manufacturing costs and/or increase the product's perceived value.

Variable costs – (direct costs) variable costs increase with the number of products made. They include depreciation of plant and equipment.

Vector graphics – (object-oriented graphics) are images comprised of a collection of lines rather than dots as in bitmap graphics.

Videoconferencing – a conference conducted between two or more participants at different sites using computer networks to transmit audio and video data.

Virtual reality – combines computer modelling with simulations to enable the development of an artificial 3D product or sensory environment; 3D virtual product can be created and viewed from different angles and perspectives.

Virtual Reality Modelling Language (VRML) – a specification for displaying and interracting with 3D objects on the World Wide Web.

Voltage – a difference in potential between two points in a circuit.

W

Waste minimisation – involves reducing, re-using or recycling materials used in manufacture.

Web browser – software such as Netscape Navigator or Microsoft Internet Explorer that provides the interface between a computer and the Internet; for example, it allows the capture and display of web pages.

Work order – see 'production schedule'.

Working drawing – drawn full size or to scale and contains factual information relating to appearance and dimension. An orthographic drawing is produced to BS7308, which forms the basis of the international standard to which all technical drawings are made. It should include all the necessary information for you or anyone else to make the product.

Work schedule – see 'production schedule'.

World Wide Web (WWW) – part of the Internet consisting of millions of pages of electronically stored information and graphics, complete with hyperlinks. The web has now become a gigantic global marketplace for products, services and self-promotion.

Index

Page numbers in italics refer to illustrations and tables. The AS course is between pages 11 and 169; the A2 course between pages 171 and 304.

Doing A-level Resistant Materials as well?

Having used your copy of *A-level Product Design: Graphics with Materials Technology* you'll have seen just how well it helps you with your course. *A-level Product Design: Resistant Materials Technology* follows the same winning format. It's out now and it's endorsed by Edexcel so you can be sure it covers all the topics you need to know.

There's lots of help for your AS and A2 Resistant Material assessments in the book.

- There is advice on how to plan and organise work, so you can make the best use of your time.

- Each unit explains what is expected and how it will be assessed, providing plenty of advice on what examiners are looking for.

- Revision checklists and practice exam questions give you thorough exam preparation.

To order a copy just contact us direct by:

01865 888068

01865 314029

orders@heinemann.co.uk

www.heinemann.co.uk

Product Design: Resistant Materials Technology
0 435 75770 9   *Available now!*

S 999 ADV 08